T0138043

Code Breaking in the Pacific

Peter Donovan • John Mack

Code Breaking in the Pacific

 Springer

Peter Donovan
School of Mathematics and Statistics
University of New South Wales
Sydney, NSW, Australia

John Mack
School of Mathematics and Statistics
University of Sydney
Sydney, NSW, Australia

Additional material to this book can be downloaded from http://extras.springer.com

ISBN 978-3-319-35982-3   ISBN 978-3-319-08278-3 (eBook)
DOI 10.1007/978-3-319-08278-3
Springer Cham Heidelberg New York Dordrecht London

Printed on acid-free paper

Springer is part of Springer Science+Business Media (www.springer.com)

Margaret Donovan and Vanessa Mack have had to put up with a great deal while the research into the cipher war in the Pacific Ocean was being carried out. And so this book is dedicated to them.

Preface

President Roosevelt, in his January 1943 State of the Union address, said:

The Axis Powers knew that they must win the war in 1942—or eventually lose everything. I do not need to tell you that our enemies did not win the war in 1942.

In the Pacific area, our most important victory in 1942 was the air and naval battle off Midway Island. That action is historically important because it secured for our use communications lines stretching thousands of miles in every direction. In placing this emphasis on the Battle of Midway, I am not unmindful of other successful actions in the Pacific, in the air and afloat—especially those on the Coral Sea and New Guinea and in the Solomon Islands. But these actions were essentially defensive. They were part of the delaying strategy that characterised this phase of the war.

The President could not reveal that Allied interception and decoding of Japanese naval radio messages played a significant role in determining the events of the Pacific War in 1942 and their outcomes. Indeed, the role of Allied Signals Intelligence in WW2 was suppressed for many years afterwards, with the earliest authoritative accounts appearing only in the late 1970s. Even later, publications on the events of WW2 commonly failed to address underlying contributions from this source. Since then Bletchley Park has become a great British icon and is (at long last) receiving appropriate maintenance. European and Atlantic Signals Intelligence has been investigated and explained often enough for its significance to be accepted and appreciated. For example, it is clear that the 1942 Battle of the Atlantic and the 1944 invasion of Normandy involved much use of what is generically called codebreaking and that this was applied to codes based on encryption machines, notably the Enigma and Lorenz SZ42.

The situation with regard to Signals Intelligence in the WW2 Pacific Theatre is different. There is no iconic single site associated with this activity—it involved units around the world and some of these are almost unknown. It is well known for its crucial contributions to the 1942 Battle of Midway and the 1943 shooting down of the aircraft carrying Admiral Yamamoto. Yet it is chastised by some writers for its claimed failure to warn of the 1941 raid on Pearl Harbor, and its overall contribution to the evolution of the Pacific War is much less well appreciated than is the case with its European counterpart.

The methods used in the code breaking behind the successes of Allied Pacific Signals Intelligence are quite different to those used against encryption machine ciphers such as the Enigma. The reason is that the main cipher systems used by both the Imperial Japanese Army (IJA) and the Imperial Japanese Navy (IJN) were based on code books rather than a machine. This book is the first to provide a complete description of those 'additive cipher systems' and the development of the techniques used to break various implementations of them. In doing so, it addresses the last major gap in the literature of WW2 cryptography and most likely the last major gap in the literature of WW2.

The IJA and IJN in fact made little use of codes based on a Latinised alphabet with which most of their thousands of radio operators would have lacked familiarity. Just as Morse code was invented to enable plain language messages written in the 26-letter alphabet to be converted into radio signals, the Japanese devised Kana Morse for the same purpose in relation to the use of Kana syllabary for writing Japanese. Both types of Morse had capacity for transmitting the ten digits 0, 1, 2, 3, 4, 5, 6, 7, 8 and 9. In the code books used by the IJA and the IJN for their major cipher systems, plain language entries in these were matched with code *groups* (or 'words') that were purely numerical—3-digits long for some simple codes, 4-digits long in others and, in the case of the all-important JN-25 naval ciphers, all of length 5 digits.

Having numerical code groups made it possible to employ a second-stage encryption process before transmission. This process concealed each code group in a message by combining it with another group of the same length selected from a second book, called here the additive table, according to prescribed rules. In this book, the code book taken with its *superencryption* process is called an *additive cipher system* and much of the book is directed to an explanation of the procedures used to decrypt intercepts of messages sent in such systems. The code books and encryption systems were normally changed every few months. Numerous minor systems were used as well.

In particular, this book explains why JN-25, the principal series of operational cipher systems used by the IJN, was broken almost immediately after its introduction and continually thereafter. It was the sole source of highest-level Allied Signals Intelligence in the Pacific throughout 1942 and was providing some 60% of this as late as mid-1944. This fact demands explanation, given that all the principal operational cipher systems used by the IJA were of a type similar to JN-25 but defied attack until mid-1943, when one important such system was broken. The others remained unreadable until early 1944 when capture at Sio in New Guinea of a full set of current code documents transformed the situation.

The explanation is both simple and profound. The IJN used only multiples of three for 'code groups' in most JN-25 code books. The IJA did not. This book exposes in considerable detail the disastrous consequences of that IJN practice.

The all-important Signals Intelligence successes obtained from JN-25 and other early Japanese cipher systems depended totally on separate decisions made by both the UK and the USA to develop Signals Intelligence capabilities against Japan soon after the end of WW1. Although these were initially directed at diplomatic messages, concerns regarding Japanese intentions towards China, South-East Asia and the Pacific stimulated development of this capacity in the military and naval sectors. The build-up of Japanese naval power was another source of concern. The 1930s saw both the UK and the USA setting up teams of cryptanalysts with a mathematical and/or scientific background. WW2 was to see the introduction of large-scale professional attacks on Axis communications by high-level profession- als with massive (by the standards of the day) data processing facilities and support staff.

The cryptanalytic developments explained in this book are based on extensive reading of archival documents held in several countries, including oral history interviews, and much published material. Another, totally reliable, source of information used was computer-based experimentation. This is identified wherever it occurs. Both authors have strong backgrounds in classical mathematics coupled with initially quite separate interests in WW2 dating back at least 40 years.

Our research generally supports the conclusion that little useful knowledge came from JN-25 decryption and decoding prior to the raid on Pearl Harbor in December 1941. No support is offered to 'revisionist' (conspiracy theories) about the raid. This is not to deny the existence of other intelligence that suggested that there was a strong possibility that such an attack would happen. Indeed, the extensive diaries kept by Canadian Prime Minister Mackenzie King show that he even tipped the day that war would break out in the Pacific. The US Army Air Force (predecessor of the USAF) could have had its planes fully armed in the air over Pearl Harbor and all major west coast targets every day in that month but did not.

The interception process was complex enough by itself. It involved a number of sites around the globe and required the secure and camouflaged transmission of data to the code-breaking centres. Experience with recording, decrypting and reading messages sent using Morse code and its Japanese equivalent, Kana Morse, was needed to handle messages sent in the Kana system of writing the Japanese language.

While the use of Signals Intelligence to turn an ambush planned by the Japanese at Midway into the Battle of Midway is well known, the critical role played by American submarines in destroying so many Japanese merchant ships is not widely appreciated. But Winston Churchill's *The Second World War* makes it clear that Japan would have been defeated by submarine action alone. These ships were carrying desperately needed raw materials from South East Asia, Malaya and the Netherlands East Indies (now Indonesia) to Japan. The United States Navy (USN) utilised intelligence co-ordinated from analysis of intercepts from several naval codes (JN-25, JN-40 and JN-11) and the important IJA Water Transport Code. This last was often called 2468 after the code group inserted in each signal that showed that this system was in use. The *Central Bureau Technical Records*, the principal report of General MacArthur's Signals Intelligence unit and now available online at the National Archives of Australia website, is an important source on 2468.

The Water Transport Code is unique among WW2 ciphers in that a large quantity of translated decrypted traffic transmitted in it has survived. This material is in the U.S. Archives at College Park, Maryland and includes signal after signal about movement of supplies to various IJA bases. This is in itself an interesting example of progressively building up military intelligence, in this case about the strengths of the IJA garrisons in various places.

This somewhat one-sided 'cipher war' depended on three factors. The first was the experience, skill and (in some cases) genius of the senior Allied personnel involved, exemplified by the outstanding British codebreaker John Tiltman, who first broke JN-25. As already noted, this went back to the long years of preparation. The second was the financial support made available. Perhaps as many as 60,000 people were involved in Allied radio and signals intelligence (including interception) in WW2. The processes involved had to be and were mechanised as much as possible. The third was simply stated by William Friedman, the great American cryptologist, in some lectures to National Security Agency staff in 1959. He noted that while high level US Naval communications security in WW2 was quite adequate for the time, Japanese Naval communications security was quite inadequate and the IJN lacked the 'experience and knowledge' to rectify it.

The contribution of the Bletchley Park cryptologists to the evolution of the war against Germany is by now well known. The successful attacks on the Enigma and the Lorenz encrypted teleprinter emerge as their two greatest achievements. The foundational work on JN-25 is the number three achievement of Bletchley Park. This was followed up by essential collaboration with the Americans in 1941. Alan Turing was involved in this in 1939 and somewhat later developed the statistical techniques that were needed to handle both European and Japanese ciphers.

The work by William Friedman and Laurance Safford in establishing and building up American Army and Navy Signals Intelligence capability, respectively, in the 1920s and 1930s was crucial to the ability of both services to develop it rapidly from early 1942 onwards. The Army units at Arlington Hall Station in Virginia, Central Bureau in Brisbane, Australia and a British unit in Delhi, India led the attack on Japanese Army systems. The Naval systems were attacked by both British and American teams up to late 1942, but the *Holden Agreement* then transferred overall responsibility to the Americans with some British participation.

The high level Japanese diplomatic machine cipher, named *Purple* by the American Army team (the SIS) that broke it in 1940, provided important intelligence, known as *Magic*, to the Allies but this was of much more relevance to the war in Europe than to the Pacific War. Hence it is minimally mentioned in this book.

The following books are valuable in understanding aspects of Signals Intelligence in the Pacific War.

John Prados in *Combined Fleet Decoded* and Ed Drea in *MacArthur's ULTRA* have described in detail the great use made of Signals Intelligence by Allied operational commanders throughout this war. Michael Smith in *The Emperor's Codes* has carefully presented a chronological account in which those involved in codebreaking activities, their locations and achievements are vividly reported, but he does not examine the cryptanalytical techniques behind their successes. Steven Budiansky's *Battle of Wits* gives an account of Allied cryptanalytic successes but mostly in the European theatre of WW2. Those of the European theatre are much better known and understood. This book gives the first reasonably complete account of who did what in Pacific War cryptology and why it could be done at all. Comparisons with European and Atlantic cryptology help put this into perspective.

Sydney, NSW, Australia   Peter Donovan
Sydney, NSW, Australia   John Mack

Acknowledgements

First and foremost, Ralph Erskine of Belfast must be thanked for numerous direct contributions in addition to his valuable published material on WW2 communications intelligence and associated topics. Lee Gladwin, formerly of the U.S. National Archives (NARA), has also earned very special thanks. The following have also contributed:

Bob Anderson   Peter Freeman (dec'd)   Alan Roberts
Deborah Anderson   David Giordano   Donald Robinson
Anna Aquilina   Jay Hannon   Tim Robinson
George Aspden   Peter Hilton (dec'd)   Edward Simpson
Desmond Ball   David Kahn   David Sissons (dec'd)
Tim Blue   Helen Kenny   Michael Smith
Steve Budiansky   Vanessa Mack   Catherine Spencer
Barbara Cathers   Roy MacLeod   John Steele
Hugh Clarke   Robin Michaelson   Dennis Trenerry
Matthew Connell   Judith Pearson   Hilda Treweek (dec'd)
Vic Czernezkyj (dec'd)   Ian Pfennigwerth   Cameron Verrills
Jim Donovan   Doug Pyle   Ian Watson
Margaret Donovan   Kim Rasmussen   Paul Watson
Jim Douglas (dec'd)   Jim Reeds   Leslie Weatherall
Jim Franklin   Joe Richard (dec'd)   James Zobel.

Over time the journal *Cryptologia* has contributed much to the proper understanding of WW2. Its editors and the authors of all cited material are also acknowledged.

Susan Lagerstrom-Fife, Jennifer Malat, Patrick Carr, Rekha Udaiyar and Kulanthaivelsamy Karthick are thanked for their advice and assistance in the carriage of this book from proposal through to publication.

    The technical material in this book has been compiled over 12 years from a variety of sources, of which the long contemporary reports *Central Bureau Technical Records* and *Cryptanalysis of JN-25* are of great importance. The former was strongly influenced by the late Professor T. G. Room, whose family has been of assistance. The latter must have been written by a team, but remains anonymous.

    The National Archives of Australia, New Zealand, the United Kingdom and the United States as well as the MacArthur Museum (Norfolk VA, USA), the US National Cryptologic Museum Library and the Australian War Memorial Research Centre have provided useful assistance and access to various documents of the era. The State Library of New South Wales and the Australian Defence Forces Academy Library were very useful.

Note to the Reader

This book describes a range of issues, some of which are quite subtle. This is inherent in the subject matter: if the methods needed to decrypt so much of the Japanese military and naval communications in WW2 had not been subtle, the blunders that were exploited would not have occurred. In general such matters are consigned to the notes at the end of the chapters.

We have had four types of potential readers in mind when writing this book—(1) the general reader interested in military history, particularly military intelligence, (2) the more specialised reader familiar with the history of WW2, particularly its Pacific aspects, (3) the reader with general interests in pre-electronic cryptography and (4) the reader with special interest in signals intelligence in the years 1919 to 1945.

Some mathematical experience is necessary in order to understand fully the cryptologic aspects of the cipher war against Japan. Parts I, III, IV, V and VI of this book do not require much mathematical expertise. In particular Part VI gives a general account of the Pacific War with some reference to the rest of WW2. The technical material is appreciably harder. The reader without the technical background may well find it easier to defer Part II, read Chaps. 8–13 in the order 8, 11, 13, 9, 10, 12 later and not try to get much out of Chaps. 14 and 15. Some understanding of Chaps. 9 and 10 will at the least give an impression of the seriousness of the error in using only multiples of three as JN-25 code groups.

The more specialised reader will already be familiar with the general background material (Parts I and VI) and will (we hope) find everything except Chaps. 14 and 15 comprehensible with some effort. Chapters 14 and 15 are appreciably more technical, but the former gives a taste of what high-level cryptology of the era really involved while the latter shows that the use of multiples of three in JN-25 codebooks was much more insecure than was considered in 1939–1942.

In view of the size of the Pacific Ocean, the background (Chaps. 21–23) are better read with a globe rather than a paper map to hand.

The enthusiastic follower of WW2 signals intelligence will find much of interest throughout this book.

In view of the widespread misunderstanding of the matter to hand, much documentation is supplied in the form of numbered footnotes throughout the text. Some notes provide further explanation of difficult points.

Acronyms and Abbreviations

1WU (etc)	Number 1 Wireless Unit (RAAF)
A/T	Auxiliary Table (in later JN-25 code books)
ATIS	Allied Translator and Interpreter Service (Brisbane)
AWM	Australian War Memorial
BP	Bletchley Park
BTM	British Tabulating Machine Company
CB, CBB	Central Bureau, Central Bureau Brisbane
CBTR	Central Bureau Technical Records
D/F or DF	Direction Finding
DMI (DNI)	Director of Military (Naval) Intelligence
FECB	Far Eastern Combined Bureau (Hong Kong, later Singapore)
FRUMEL	Fleet Radio Unit Melbourne
GAT(s)	Group(s) As Transmitted
GCCS	Government Code and Cipher School—now called GCHQ
GYP-1	Pacific Cryptology section 1 unit in Op-20-G (see below)
HMS	His Majesty's Ship
HMAS	His Majesty's Australian Ship
IBM	International Business Machines Corporation
IJA, IJN	Imperial Japanese Army, Imperial Japanese Navy
JICPOA	Joint Intelligence Center, Pacific Ocean Area
JN-25B8	Japanese Naval cipher system, series 25, version B, additive 8
NAA	National Archives of Australia
NARA	National Archives and Records Administration (U.S.)
Op-20-G	USN radio intelligence unit
OTP	One-time pad
RAF	Royal Air Force
RAAF	Royal Australian Air Force

RFP (REB)	Radio Fingerprinting
RI	Radio Intelligence
RN (RAN)	Royal (Australian) Navy
SIS (SSA)	Signals Intelligence Service (U.S. Army)
TNA	The (British) National Archives, Kew
US8	U.S. Army unit near WEC, Delhi
USN	U.S. Navy
USAAF	United States Army Air Force (now the USAF)
WEC	Wireless Experimental Centre (near Delhi, India)
W/T	Wireless/Telegraph (British usage)
WW1	World War I
WW2	World War II

Contents

Part I
Build Up

Chapter 1
Communications and Sigint

This chapter gives a brief account of the early development of telegraphic and radio communication systems and of the use of message interception techniques to obtain intelligence of diplomatic or operational value. Thus modern Signals Intelligence was born.

1.1 Electric Telegraph and Radio

The electric telegraph was the genesis of land-based wire and cable-based communication systems. In 1843, the US Congress granted Samuel Morse a licence to operate a telegraph line between Washington and Baltimore. Messages were sent using a system that encoded the 26 letters and the 10 digits in terms of combinations of short and long electrical pulses, a system that became known as Morse code. The importance of this communication system for both military and diplomatic message transmission was quickly realised, and further emphasised when the laying of undersea cables linked the continents and enabled the colonial powers to establish direct contact with their representatives overseas. Such messages were already protected from interception by virtue of their means of transmission—one usually had to have physical access to the telegraph line, or cable, through which they were sent.

The first operational trans-Atlantic cable was in place in 1866. As it was relatively expensive to use, the incentive to develop codes and shorthand notations for commercial communication purposes increased. For example, those monitoring market prices could arrange that messages like 'Cotton price down by small amount' could be abbreviated to one code group[1] (word) such as AWARA. Commercial codes were also created for use by businesses wishing to protect their plans and operations from competitors. Openly published codes used to transmit standard

---

[1] In this book, code groups are groups of 3, 4 or 5 digits only, or of capital letters only.

© Springer International Publishing Switzerland 2014
P. Donovan, J. Mack, *Code Breaking in the Pacific*,
DOI 10.1007/978-3-319-08278-3__1

messages were also available. If nothing else, use of such a code would protect messages from being read casually by cable company staff.

One openly published code was the *Chinese Telegraphic Code* (CTC) which began to be developed soon after a Danish company introduced the telegraph to China in 1871. It evolved steadily over the next fifty years, an indication of its value in commercial transactions between Chinese and Western businesses. It assigned different 4-digit groups (numbers from 0000 to 9499) to each of 9500 commonly used Chinese characters (also used in Japanese writing and known as Kanji). This was done by writing out the characters in slash order (the equivalent of alphabetical order) with their code groups being used in numerical order. The 500 remaining groups not assigned a character in the code book could be allocated special meanings by users. A modern variant of the CTC, the *Standard Telegraph Code*, is still occasionally used when an 'official' western equivalent of a Chinese character is required, as for example with the correct conversion of a proper name.

Cable-based technology was of little use to ships at sea. For them, and for the mercantile and naval fleets of the sea-powers, the invention of the radio around 1895 marked a new era in maritime communications, quickly adopted by all navies. Transmitters and receivers could be tuned to certain carrier frequencies to transmit and receive a message sent on a specific frequency. Initially only Morse code could be sent via radio, but by the early 1920s transmission of plain language communication was becoming available.

The use of radio created a new problem for those wishing to protect their messages. Anyone in a suitable location with the right equipment was able to search the airwaves for messages of possible interest, intercept them and examine their contents. This stimulated development of more sophisticated encryption systems designed to permit only those in possession of a key (an appropriate secret) to decipher messages sent using them.

1.2 Early Military Aspects

In 1883 Auguste Kerckhoffs[2] wrote the seminal account *La Cryptographie Militaire* of military communications security. As well as providing some practical design principles, he gave one absolutely essential requirement: *Do not use new cryptographic systems invented by people without the necessary experience* (that

---

[2] Kerckhoffs' paper is available at www.petitcolas.net/fabien/kerckhoffs in both the French original and in English translation. See the end of Part II. The French original of the free translation given in italics in the main text is 'Ce ne sera que lorsque nos officiers auront étudié les principes de la cryptographie et appris l'art de déchiffrer'. Section 8.19 gives various other maxims for cryptographers operating with WW2 technology.

Kerckhoffs' comment may be compared with the extract from a British report of 12 March 1942 quoted at the end of Chap. 3.

is, without expertise in the breaking of codes). Kerckhoffs knew that such people would inevitably commit blunders.

The first serious military use of intelligence derived from radio interception occurred in the Russo-Japanese War of 1904–05. The Imperial Japanese Navy (IJN) intercepted and used Russian radio traffic to set up a major naval victory.

1.3 Cables in the First World War

Perhaps the most telling early example of the value of having control over or access to the means of communication occurred on 5 August 1914—one day after Britain declared war on Germany. On that day a British cable-laying ship, the *Alert*, was sent to haul Germany's overseas cables from the floor of the North Sea, sever them, and drop them back. From that time, Germany had to use British-owned cables for its overseas traffic. (The last German link, a German-American cable connecting Liberia with Brazil, was cut in 1915.) The only alternatives were to use couriers, postal services, or, where available, radio transmission, all of which were subject to interception. (Likewise on 31 July 1945 the British midget submarine *XE4* cut two vital cables used for Japanese communications. They linked Saigon with Singapore and Hong Kong.)

The British Empire had set up a network of cable communications before 1914 using only Empire territory. This required the use of various minor islands, such as Fanning Island in the Pacific Ocean. The German Navy managed to damage the British cable base on Fanning Island in 1914 but repairs were possible. A German raider was sighted off the Cocos Islands in the Indian Ocean in that same year but a radio appeal for help was answered by the first HMAS *Sydney*. This British network also proved a strategic factor in WW2. Indeed, in order to avoid a repetition of the 1914 incident, on 30 August 1939, just a few days before its declaration of war against Germany, the New Zealand government despatched an army platoon to Fanning Island to guard the cables.

1.4 Tannenberg

The Battle of Tannenberg, 26–30 August 1914, at the very beginning of WW1, overturned Allied expectations that the size of Russia's military forces would always triumph against their German counterparts. With the use of intelligence gathered from daily intercepts of plain language exchanges among the Russian commanders, a German force was able to rout a much larger Russian force, taking some 90,000 prisoners of war and eliminating the threat of successful future Russian action on the eastern front.

Also in the very early stages of WW1, radio intelligence enabled the British and French Armies to avoid being cut off by a German advance along the coast of Belgium and France. This became known as the Race to the Sea and was of considerable importance.

1.5   British Naval Intelligence in WW1: Room 40

Neither Britain nor Germany was prepared for code breaking at the start of WW1. Henry Oliver, the British Admiralty's Director of Intelligence, began to receive intercepts of coded enemy radio messages and realised that he needed to find someone to handle them. His choice of Alfred Ewing, the Director of Naval Education, to take charge of this activity, was a fortunate one. Ewing had an interest in ciphers and had access to the naval colleges, from which he recruited volunteer staff to help set up a unit, known, from its location in the Admiralty, as Room 40. Later it was expanded and renamed ID25. One of the volunteers was Alastair Denniston, who became a principal player in British code breaking in WW1 and was made head of the newly formed Government Code and Cypher School (GCCS) after the end of the war. Initially staffed with veterans from Room 40 and its Army equivalent, he was able to augment it during the 1930s and was thus instrumental in setting up the agency that had such success at Bletchley Park and elsewhere in WW2.

At the beginning of WW1, Room 40s initial lack of both experience and skill in code breaking was quickly offset in the best possible way—by its receiving copies of three major German naval codes! On 13 October 1914, Room 40 had obtained, courtesy of Russian naval action in the Baltic Sea and with the cooperation of the Russian authorities, a copy of the code known as SKM, used primarily in connection with major naval operations. Information derived from it may have helped the Royal Navy (RN) quickly to intercept and sink four old German destroyers in the North Sea, off the Dutch coast. By itself, this would not merit comment, but at the end of November, a British fishing trawler, peacefully plying its trade in the same waters, hauled up in its nets a lead-lined chest, which turned out to contain amongst its treasure-trove a full set of documents related to the second major German naval code, known as VB, used only at high level.

Earlier, on 14 August 1914 near Melbourne, Australian authorities seized the hapless German-Australian steamship *Hobart*, whose crew was unaware that war had been declared. The head of the boarding party then duped the ship's master into revealing the hiding place of his confidential papers and so obtained the code book for a code known as the HVB, used by the German Admiralty and its warships for communication with their merchant craft, and also widely used within its High Seas Fleet. Inexplicably, Room 40 received these documents only at the end of October.

As Patrick Beesly writes in his book *Room 40*:

Now the process was complete, and virtually any wireless signal made by the German Navy which could be intercepted, was to become equally available to Winston Churchill, Admiral 'Jacky' Fisher (now First Sea Lord[3]) and their immediate subordinates. No one, on either side, had anticipated such a swift and overwhelming intelligence defeat for the Germans; it was one from which the Imperial German Navy was never to recover.

Messages in all three of the above codes were subject to what became known as super-encipherment. This means that code words taken from a code book were not transmitted as such, but were subjected to an additional modification which depended upon a 'key' that could be changed more easily than could the underlying code book. Key changes were made later, but these usually presented much less of a challenge to code breakers[4] than did a replacement of the code book.

Room 40 was not initially part of the RN Intelligence Division and its head did not report to the Director of Naval Intelligence. Its reporting channels were constrained and neither the Intelligence Division, nor any of the RN operational units, were permitted direct access to Room 40 decrypts. They were sent, one at a time, only to the Sea Lords, who had expressly ordered this. Only the Sea Lords were authorised to pass on such intelligence to the operational commands. Although a naval officer was appointed later in 1914 to assist in interpreting the decoded messages (the code breakers in Room 40 had no direct experience of navy work), that officer was initially not allowed access to Room 40 or its staff and often was not shown the mass of 'less important' decrypts that help to illuminate the information in the more important ones. The result was that the achievements of Room 40 were not always used effectively.

Improvements, especially in integrating Room 40 with Naval Intelligence, were made when Captain (later Admiral) Reginald 'Blinker' Hall was first made Director of Naval Intelligence and then placed in charge of Room 40. But the problem of not being able to provide the operational commands directly with regular and up-to-date intelligence assessments continued until the end of WW1.

---

[3]In WW1 and WW2 the Minister responsible for the Royal Navy was called the First Lord of the Admiralty while the senior admiral was called the First Sea Lord. A few other senior admirals would also be designated as Sea Lords.

[4]One interesting example of breaking a change of key on a code whose book had been captured was described by Dr. Frederick Wheatley 19 years after his 1914 achievement. The text is preserved by the Australian War Memorial (AWM) as archival item AWM252 A228. He was able to decrypt, decode and translate German naval messages. The difficulty in recovering (working out) a decrypted enemy code book is discussed in Chap. 11.

In the context of additive cipher systems, which were extensively used by the Japanese in WW2, the word 'keys' was sometimes used for what are generally called 'additive tables' in this book.

1.6   Jutland

Winston Churchill was the civilian head of the RN in the initial years of WW1. He noted in volume 3 (1927) of *The World Crisis* the following:

> In the first volume (1923) of this account I recorded the events which secured for the Admiralty the incomparable advantage of reading the plans and orders of the enemy before they were executed. Without the cryptographers' department there would have been no Battle of Jutland.

The 1916 Battle of Jutland was the principal clash between the capital ships of the British and German navies. Although radio intelligence was not used particularly well, the German ships retreated and the British strategy of keeping the major German surface warships bottled up in the Baltic Sea was basically successful. German submarines, with their greater freedom of movement, were the major threat to Allied shipping. Continuing use of radio intelligence to provide information on the locations of German naval ships enabled the RN to become more economical in its use of limited resources by greatly reducing the need for lengthy patrols at sea in search of enemy warships.

1.7   Diplomatic Intelligence: Zimmermann Telegram

Room 40 also contained a very active political/diplomatic section, which in particular examined numerous German diplomatic messages sent in a variety of coding systems. The simplest of these used code books whose entries were arranged in a straightforward way—for example, plain text entries in lexicographic order with corresponding code words being 4-digit numbers arranged in numerical order.

More difficult were those messages whose codes were hatted,[5] that is when the plain text entries were placed in alphabetical order but the code words were in random order. This had the effect that a code group adjacent to a decoded code group would usually have a totally unrelated plain text form or meaning.

Superenciphered codes used a code book together with a systematic and reversible means of additional encryption, usually varying from message to message, to disguise the code groups. The cryptanalyst had to determine the encipherment system and develop an efficient method for removing it before recovery of the hidden code words became possible.

---

[5]Mavis Batey explains in her book *Dilly: The Man who Broke Enigmas* (Dialogue, London, 2009) that this use of 'hatted' originated in the practice of mixing up slips of paper in a hat and then drawing out one to obtain a random choice. Part of the story of the Battle of Midway is that the JN-25B code words for geographical locations were not hatted!

Such codes require two code books in much the same way as, say, a German-English dictionary is complemented by an English-German dictionary.

Recently, evidence has been found[6] which confirms that this section of Room 40 used punched-card machinery (called Hollerith equipment) to speed up both the decrypting of superenciphered codes and the decoding of code groups. Improved models of such equipment greatly assisted the code breakers of WW2 and will be described later.

The section had its greatest triumph early in 1917. On 16 January, Arthur Zimmermann, the German Foreign Minister, sent a secret message, in the most secure German diplomatic code, to the President of Mexico, via the German embassies in Washington and Mexico City. The message disclosed that Germany was about to commence submarine warfare on all merchant shipping (including American ships) in the Atlantic, with a view to cutting off essential supplies to the British and French and forcing a peace settlement in the European war. Anticipating that such action would likely lead to a declaration of war by the USA against Germany, Zimmermann invited the President of Mexico to join the war on Germany's side, promising that, in return, during the peace negotiations, Germany would help Mexico recover territories (Arizona, Texas and New Mexico) that had been lost to the USA. He also sought the President's help in persuading Japan to change sides and attack the Pacific coast of America.

Because of the British action in cutting Germany's cable connection to America, Zimmermann had obtained permission from both Sweden and the USA to use their cable services for messages to the German Embassy in Washington. These services were routed through London where Room 40 was secretly intercepting German messages. Although the code used for Zimmermann's message had been only partially broken, enough of it was decoded for Hall to realise its potential significance and to order a full attack on it. Several weeks of intensive work produced an almost complete message text, which the British Government was anxious to exploit in its quest to bring the USA into WW1.

The message had to be released in a way that protected Room 40s secret access to the cable networks used. Hall reasoned that the German Embassy in Washington would have recoded the message into an older and simpler code before transmitting it to the German Embassy in Mexico City via a commercial cable service. This proved to be the case. A copy of the coded message was obtained and used as a basis for the material eventually shown to President Wilson. A copy of the translated text was released by him to the American public on 1 March. Its impact was enhanced shortly afterwards when German submarines sank, without warning, several American ships. On 6 April 1917 the USA declared war[7] on Germany.

---

[6]The reference is TNA ADM 223/773 *Memo on 'Political' Branch of Room 40*, by George Young. It is cited in Paul Gannon's *Inside Room 40: The Codebreakers of World War 1*, Ian Allen 2010.

[7]Reinforcements from the USA were of great significance in the Western Front in Europe in 1917–1918. This was to happen again in 1942–1945.

1.8   Army Signals Intelligence in WW1

France had been building up its military intelligence capabilities for some decades prior to the outbreak of WW1 and by then had in place effective cryptographic and cryptologic teams, which also worked on diplomatic and naval intelligence. The major focus of these units was in breaking German Army codes and ciphers sufficiently rapidly for the information gained to be of immediate operational use to the Allied field commanders. The outstanding French code breaker, Georges Painvin, was responsible for a number of its major cryptological achievements.

In particular, the final German offensive towards Paris in 1918 was halted with the assistance of operational information obtained from Painvin's work in decrypting some messages sent in the high-level German ADFGVX[8] code, so named because its code words consisted of the 36 pairs of these six letters.

1.9   Morse Code

The standard device for sending radio messages in languages using the Roman alphabet was the Morse key. This instrument was tapped by an operator to produce sequences of short and long pulses, corresponding to the Morse system of dots and dashes used to represent the 26 letters, plus the ten digits and some punctuation symbols. In general the more frequently used letters had shorter Morse code symbols. For example, Q was – – • – while E was just •. The operator would use short gaps to separate successive letters. This eliminated confusions between, for example, S, represented by • • •, and the sequence transmitted for the three letters

---

[8] Having symbolism for 26 letters and 10 digits is useful because military messages commonly include numerical information best communicated by digits. Details on how ADFGVX was broken are available in various places. For example, the invaluable series of books produced by Aegean Park Press includes two accounts of the matter. Of course the Google search engine has no trouble finding useful material about ADFGVX on the web. It was not fully broken in WW1.

The book *The British Army and Signals Intelligence during the First World War* edited by John Ferris (Alan Sutton Publishing for Army Records Society) gives much more information on WW1 Army signals intelligence. In particular it sets out the WW1 use of traffic analysis and describes an interesting technique, based on electric current leakage through the ground, for intercepting front-line telephone communications.

Ferris has recently reviewed in great depth the paper by C. J. Jenner *Turning the Hinge of Fate . . . 1940–1942* in *Diplomatic History* 32(2) (April 2008). The reference is *H-Diplo Article Reviews* number 199. Jenner's paper deals with the situation in North Africa at that time and so is not relevant to this book. However Ferris very usefully explains that the importance of signals intelligence is often exaggerated. The three episodes given pride of place in this book—the Battle of Midway, the USN submarine attacks on Japanese merchant ships and the exploitation of the material captured in Sio, New Guinea, in January 1944 in the 'Leap to Hollandia'—are exceptions to Ferris' general comments. Richard Overy's book *Why the Allies Won*, Cape, 1995 is useful in assessing the overall importance of signals intelligence in WW2.

EEE. Slightly longer gaps were used to separate words and longer gaps used to separate sentences. Once a radio set had been tuned to the transmitting frequency, the receiving operator would hear, via headphones, short and long sounds in spaced clusters.

Although modern technology has little use for Morse code, it was used extensively for radio communications in the first half of the twentieth century. Considerable training and practice was needed to transmit or receive Morse messages, with speeds of fifty words per minute being possible. The critical role played by wartime interception operators warrants explicit recognition. Some understanding of their training, their work and its importance can be gleaned from a reading of Jack Brown's *Katakana Man*,[9] detailing his experience as an interception operator in the Pacific War.

The interception of radio messages was prone to error, particularly when carried out by inexperienced operators. Mistakes by transmitting staff and atmospheric interference were also sources of error. Intercepted messages, or parts of messages, so corrupted, were called *garbles*. Because garbles could affect the application of decryption methods, their identification and possible correction were important but not always essential. This matter is raised again in Sects. 10.5, 10.8 and 10.9.

The enciphered messages considered in this book were usually sent in the form of groups of digits, or 'figures' 0, 1, 2, etc. Standard Morse code for the digits used a combination of five dots and dashes for each one. This was inappropriate for

---

[9] A useful description of training is given by A. Jack Brown in *Katakana Man* (Air Power Development Centre, Tuggeranong, 2006), pages 13–18, while its chapters V & VI give a graphic account of front-line interception work on Leyte in late 1944. Such work was generally carried out under difficulties and deserves explicit recognition. Lt.-Col. Ryan, head of the Australian Army interception unit in WW2, wrote on page 3 of his final report (National Archives of Australia (NAA) Canberra item A10908 2):

'It must be realised that the work of an intercept operator is particularly trying and that efficiency is inevitably governed by such factors as lighting, ventilation, insulation against heat and cold, facilities for proper rest and recreation.'

On page 9, Ryan states that the backs of trucks do not form good workplaces.

Much the same thoughts can be found in NAA Canberra A11093 320/5K5 Part 2: 'Y Intelligence operators are obliged to listen for five consecutive hours to very faint signals which have superimposed on them other signals and a whole lot of interference.' The process imposed great strain. Rotation and leave were needed.

Part J of the *Central Bureau Technical Records* (*CBTR*) (NAA Canberra item B5436 Part J and accessible on line via the Recordsearch facility) is entitled *Field Sections* and deals with the RAAF Wireless Units, the Australian Army sections led by Lt.Col. Ryan and the American Signal Radio Intelligence Companies. It specifies the duties of these units as being:

1. Provision of air-raid warning intelligence.
2. Interception of low-echelon Army traffic and provision of tactical intelligence derived therefrom.
3. Interception of raw material, inaudible in rear areas, for cryptographic solution by CBB.
4. Local exploitation by traffic analysis in Kana for intelligence value.

such messages. Usually, special codes[10] were adopted, in which the ten digits were represented by sequences of at most three dots and dashes.

The usual lengths of such code groups were three, four or five. This last became prevalent in commercial practice after the international telegraph system adopted a fee structure under which enciphered material could be transmitted in groups of at most five letters or digits. These were then charged for as single plain language words.

1.10   Kana Morse, Roma-ji and the Chinese Telegraphic Code

Since the basic building blocks of the Japanese language are syllables rather than letters, standard Morse code is unsuitable for transcribing it. Japanese can be written using the Kana syllabary of about 70 symbols. Based on this, a system known as Kana Morse was developed to enable messages in Japanese to be sent in dot-dash form via telegraph or radio. Kana Morse is more complicated than ordinary Morse

---

The Allied decryption centres Central Bureau and Frumel, to be introduced later, had interception facilities within short distance of their bases in Brisbane and Melbourne respectively.

The only detailed account of equipment used in WW2 interception stations known to the authors is that by Thomas J. Gray entitled *Some Aspects of the 'Y' Service—India 1943–1945*. This appeared on pages 95–102 of issue 18(2) (1988) of the *Journal of the Royal Signals Institution*.

The NAA Canberra file A2617 Section 43/14609 *Harman: Additions to 'Y' Hut* has an interesting diagram of the interception building at the Harman Naval Wireless Station near Canberra.

The *CBTR* fails to mention here the 1st Canadian Special Wireless Group (1CSWG), which operated at Adelaide River south of Darwin in 1945. A short film *The Canucks in Aussie* was shot by members of the First Canadian Special Wireless Group in August 1945 and shows something of their work at Adelaide River. A copy survives in the Canadian War Museum. It is mentioned in NAA Canberra item A6923 16/6/502.

Kenneth Macksey's *The Searchers: Radio Intercept in Two World Wars* provides much information on WW2 interception services, particularly in Britain.

Of course, the other side, transmitting messages, also had problems. Its signal staff would be carrying out repetitive dull work in difficult conditions. Canberra NAA item A705 109/3/817 confirms this. So one cause of the emergence of insecure signalling practices is evident. See also page 6 of the *CBTR*, part A.

[10]See Joe Richard *The Breaking of the Japanese Army's Codes* in *Cryptologia* 28(4), October 2004, 289–308. Jack Brown (see immediately above) comments on this matter on page 16. In fact, the IJN used standard Morse for transmitting its main operational series of codes known as JN-25.

A navy adopting additive cipher systems for the first time could have gained extra security by adopting an unusual patternless allocation of dots and dashes to the ten digits. It would be necessary to prevent trainees from sending practice messages over the air waves. If some such person transmitted 1 2 3 4 5 6 7 8 9 10 11 12 13 14 15 16 17 18 19 20 21 22 23 24 then anyone intercepting the signals would pick up how the digits are represented. Compare the second maxim of Sect. 8.19.

and presented an immediate problem to interception operators familiar only with regular Morse: they needed to learn both the complete underlying plain language set of symbols and their Kana Morse equivalents. These were known to Western intelligence organisations. Special training was needed for those Allied interception staff who handled the Kana Morse code used by the Japanese military in WW2. The CTC was frequently used inside coded Japanese military messages during WW2 when precise reference to a Chinese character was required because of ambiguity in its Kana equivalent. (It sometimes happens that two or more distinct Chinese characters, when converted into the Japanese Kana syllabary, produce the same set of Kana symbols.) Thus cryptologists also had to be able to identify CTC when this occurred.

An alternative to Kana Morse was available in the Roma-ji system in which each Kana symbol was represented by either one Roman letter or a pair or triple of them. This could be transmitted in standard Morse code.

1.11 Baudot Codes, Teleprinters and Teletypes

Following the introduction of the telegraph, a series of inventions led to the development of communications systems based on a code created by Emile Baudot in the 1870s. This code evolved into the International Telegraphy Alphabet No. 2 (ITA2) by the early 1930s.

Each code group is a 5-symbol sequence of the binary digits[11] 0 and 1, readily adaptable to electrical or radio transmission by use of on-off switches. The 32 possible code groups thus available were extended to about 60 by including two groups called, respectively, 'shift to figures' and 'shift to letters'. These told the receiving teleprinter to read the following code groups as figures or letters respectively until the other shift key was received. Specially designed electro-mechanical typewriters and printers were used to convert plain text into code and vice-versa. An important feature of this equipment was its ability to read paper tape in which the input consisted of columns showing patterns of blanks and holes corresponding to the five symbols of the corresponding number or letter as shown overleaf. The machine output could be in punched tape or print form.

A radio version of this system, called radioteletype, was successfully tested by the United States Navy (USN). Both versions of this system were in widespread use by armed services by the end of the 1930s. The teleprinter system was a principal means of communication by the USN throughout WW2.

While Morse code has now almost entirely disappeared, systems based on the Baudot code evolved into the modern computer and its terminals. The Vernam cipher—later given as an early example of a secure additive cipher system—used a

---

[11]The Bletchley Park terminology was 'dot' and 'cross', with the former corresponding to the binary digit 0 and the latter to 1. Thus 'cross' + 'cross' = 'dot', etc.

Baudot code. The following display shows the 32 quintuples with the corresponding letters at the top. The quintuples headed by 'lf', 'sp', 'cr', 'sh' and 'un' were used for 'line feed', 'space', 'carriage return', 'shift' (to figures) and 'unshift' (shift to letters) respectively, while one[12] of the quintuples was not used. Note that 'lf' moves the printing head to the same place on the next line while 'cr' moves it to the first place on the next line.

	E	lf	A	sp	S	I	U	cr	D	R	J	N	F	C	K
0	0	0	0	0	0	0	0	0	0	0	0	0	0	0	0
0	0	0	0	0	0	0	0	1	1	1	1	1	1	1	1
0	0	0	0	1	1	1	1	0	0	0	0	1	1	1	1
0	0	1	1	0	0	1	1	0	0	1	1	0	0	1	1
0	1	0	1	0	1	0	1	0	1	0	1	0	1	0	1

T	Z	L	W	H	Y	P	Q	O	B	G	sh	M	X	V	un
1	1	1	1	1	1	1	1	1	1	1	1	1	1	1	1
0	0	0	0	0	0	0	0	1	1	1	1	1	1	1	1
0	0	0	0	1	1	1	1	0	0	0	0	1	1	1	1
0	0	1	1	0	0	1	1	0	0	1	1	0	0	1	1
0	1	0	1	0	1	0	1	0	1	0	1	0	1	0	1

Another such table is needed for the meanings of the quintuples in shift mode, that is between a shift and an unshift. After the unused, shift, unshift and space characters are taken out, 28 of the 32 are available for the shift mode, and so there is capacity for all ten digits plus brackets, punctuation and the like. The details are not needed here.

1.12 Use of Encryption Systems

Diplomatic and military messages have rarely been sent in plain language by radio. Anyone who could intercept and read the standard system used had access to their contents. Encryption systems have been in use for such written messages for a very long time,[13] and these are generally adaptable for use with cable or wireless

---

[12]In fact the first quintuple, 0-0-0-0-0, was not used in either letter mode or shift mode. This would have been convenient in separating different (unenciphered) messages punched out on the same piece of paper tape. The code set out in the main text is not the original version invented by Baudot but rather a more recent modification. (Some use has been made here of the *Wikipedia* free encyclopaedia and in particular of the entries for *Gilbert Vernam*, *XOR*, *stream cipher* and *one-time pad*.)

[13]A good comprehensive account of 'secret writing' over the centuries is given by David Kahn, *The Codebreakers*, Weidenfeld and Nicholson, London, revised second edition 1996.

transmission. For example, a plain text message in English, encoded by the simple method of 'shifting each letter one place to the right in the alphabet', would yield a new sequence of letters, which is transmitted as before. The recipient, having copied down this sequence, would have to know the encryption rule to decipher it directly, or would have to apply cryptanalytic methods in attempting to decipher it.

The Japanese cipher systems studied in this book required both senders and recipients to have full access to the encryption documentation. Thus coding manuals were needed by field units and were vulnerable to capture. Field unit commanders were told to protect communications security at all times and to destroy all such materials should capture be likely. However, this did not always happen. For example, in August 1942, an American Marine on Guadalcanal managed to rescue IJN cryptological material from a fire. In January 1944 an Australian infantry unit in Sio, New Guinea, found some Japanese Army code books and associated documentation. Both finds were extremely valuable.

While the Japanese Army and Navy also used encryption systems that encoded an original Kana message into Kana Morse, these systems were in general of lesser importance than the superenciphered systems based on the ten digits rather than on letters or syllables. These are called additive cipher systems in this book.

1.13   The Components of Signals Intelligence

The rapid growth in the use of radio during and following WW1 led to immediate interest in ways in which the radio messages (encoded or plain language) of one nation might be listened to by another. The British gave the name *Procedure Y* to this practice when the target messages were of diplomatic or military interest. This activity was soon implemented by the governments of various nations. The British called the fruit of such activity *Y intelligence* or *signals intelligence* or just *Sigint*. Information obtained from code breaking was also called *Special intelligence*. The American usage separated *communications intelligence* (Comint or CI, that obtained with decryption or decoding) from *radio intelligence* (RI, information available from techniques such as direction finding, described below). In 1943, the terminology was standardised by the British and the Americans, with the word Ultra being used to indicate high level Comint. Other intelligence was derived from the practice of secretly obtaining and reading copies of (usually diplomatic) cable messages, using decryption methods if necessary.

The abbreviations Sigint, Comint and RI are used throughout this book.

The principal raw material of Sigint is the content of intercepted plain language and coded radio messages. Therefore its essential infrastructure is the establishment, maintenance and operation of one or more wireless receiving stations, located so that their antennae can pick up often very weak signals from transmitters. This may require some stations to be geographically close to the signal sources. In addition, the way in which radio waves are 'bounced' around the surface of the Earth means that there are, for any given source location,[14] places where signals from it cannot be detected. Finally, since signals can be sent using a wide range of frequencies for the underlying carrier wave, the interception station equipment had to be able to sweep across a range of frequencies quickly in order to detect those frequencies carrying signals.

The interception stations needed secure, rapid and preferably undetectable communication between each other and the message processing unit. Less urgent bulk material was often copied onto microfilm and sent by aeroplane.

As well as the necessary radio equipment and technical maintenance staff at such a station, trained operators were needed to monitor detected signals and to copy down, in real time, a transcript of what they heard. From the 1920s onwards, it was possible to attach equipment to a receiver to record messages that could then be played back for transcription and possible decryption. But in most operational situations and especially when the interception operator had to make an instant decision on the likely importance of one signal over others, and then its immediate transcription, a high level of training in this skilled activity was essential.

Radio messages usually contained information identifying the sender, the intended recipient(s), time of dispatch, and the length of the message. All of these items had to be identified in a message if information was to be extracted from it. For example, each station in a radio network identified[15] itself via one or more call signs. An important task for the cryptanalysts trying to extract intelligence from these messages was that of building up a list of stations in a network and their call signs. Information relating to the numbers and lengths of messages between two or more given stations and variations in the patterns of this provided the extremely useful source of Sigint named *traffic analysis*[16] (TA) by the British. Changes in the volume of message traffic between two particular stations often suggested that some new operational activity was being planned with their involvement. For example, a sudden increase in traffic between a naval headquarters and a fleet unit at sea was likely to indicate a pending operation involving that unit. This would then suggest

---

[14]For example, the British interception stations at Hong Kong and Singapore in the 1930s had great difficulty picking up signals emanating from the Japanese Mandated Islands in the Pacific.

[15]Encryption and regular changes of call signs could only enhance security: this was done to some extent by the Germans in the European war.

[16]The importance of traffic analysis as a source of signals intelligence was recognised in WW1. The very distinguished American cryptanalyst William Friedman (see Chap. 4) served as an officer in the US Army intelligence unit on the Western front late in WW1. He subsequently wrote 'Its utility in deriving intelligence about enemy intentions from a mere study of the ebb and flow of enemy traffic, without having to solve the traffic, was of unquestionable value'. (See *The Friedman Legacy: A Tribute to William and Elizebeth Friedman*, NSA Sources in Cryptologic History, Number 3, page 109.)

As an example of its use in WW2, Hut 6 at Bletchley Park had a TA unit called *Hut Sixta*. James Thirsk's book *Bletchley Park: An Inmate's Story* (Galago, 2008) explains the development of the *Fusion Room* which merged information from Hut Six and Hut Sixta. So-called 'operator chat' was also useful. Another such book is Joss Pearson's *Cribs for Victory* (2011), which is an edited and augmented version of a text written earlier by her father, Neil Webster.

that more attention be paid to traffic between them for other intelligence purposes. Of course, the opposing side might well send out dummy traffic (that is, encrypted meaningless messages intended to create a false impression of activity) in slack periods to confuse the traffic analysts.

TA was of much greater value when it was combined with another type of Sigint called *direction finding* (DF). The pattern of aerials at a receiving station can be designed so that it can identify the direction at which a given signal is most strongly received. This direction is then drawn on a globe of the Earth, or on a suitable map projection of it. (Gnomonic[17] projection works best, because, in it, given directions appear as straight lines. This is not the case, for example, with the more commonly used Mercator projection.) If DF can be applied to the same message at two or more geographically separated stations it is possible to pin down the location of the sending station. The various direction lines, drawn on a gnomonic map, should converge either on or close to a single point that then identifies the sender's location. In its early days, DF was notoriously unreliable, with ships sometimes being located well inland. By the late 1930s it was a reasonably accurate technique.

Most DF was carried out at the high frequency end of the radio wave spectrum. It was often referred to as HF/DF or huffduff. The equipment used, including the antennae, was cumbersome. So navy commanders confidently assumed at first that a ship-board version of HF/DF could not be produced. But the RN had introduced an effective ship-board version by late 1942 and the USN had its own in action by mid-1943. Ship-board HF/DF played an important role, alongside radar, in the Battle

---

[17]In the paper mentioned in Note 9, Gray describes the large map used in a dedicated room at Bangalore to plot 'fixes' from DF stations.

Some account of the various projections used for constructing on flat paper maps of portions of the surface of the planet may be found in Wikipedia and other appropriate websites. Of these the most familiar is the standard Mercator projection, which has the undoubted merit of representing meridians (great circles through the poles) by vertical lines and parallels (small circles parallel to the equator) by horizontal lines. It has the much more subtle property of being conformal, that is representing angles correctly. But it distorts the regions near either pole, misrepresenting the areas of Canada and other such countries. And, for the purpose of direction finding, it represents great circles by excessively complicated curves. Hence it is quite inappropriate for locating the source of a radio transmission from the directions observed at two or more stations.

The gnomonic (also called gnomic) projection is constructed by projecting from the centre of the planet on to the tangent plane at a fixed point on its surface. This ties in better with conventional latitude and longitude if the fixed point is taken to be on the equator. For the Pacific one may as well take the point 0°N, 180°W or perhaps 0°N, 150°E. This projection, which can handle only one half of the planet at once—scarcely a problem—represents the meridians by vertical lines and the parallels by arcs of hyperbolas. It has the practical merit of representing great circles by straight lines.

In practice a gnomic chart would have just selected meridians and parallels (perhaps every second degree of arc) marked, together with dots for the direction finding stations. The coastlines of islands and continents would be harder to draw and not particularly useful. The approximate location of a radio source would be identified on the gnomic chart, its latitude and longitude read off, and the work would then be transferred to a conventional Mercator map.

of the Atlantic, especially during those periods when information about German U-boat movements could not be provided in time by the Bletchley Park and USN cryptanalysts.

Additional refinements were developed to provide further information from an intercepted signal. One such, known as radio-fingerprinting (RFP or, more confusingly, REB), assisted in identifying the particular radio transmitting set being used to send a message by its idiosyncrasies in sending out its radio signals. This was used to check if a given call sign was coming from its usual transmitter or was suddenly coming from a different one in a different location. For example, an aircraft carrier dispatched on a mission might try to disguise its movements by having its usual call sign transferred to a stationary shore base location. The above technique, if available, would spot the change immediately. It also helped to link a new call sign to a previous one, by identifying use of a common transmitter.

A second technique developed a method (called TINA from Serpentina) for analysing the way in which a Morse key operator actually tapped out each keystroke, by displaying the transient electrical signals on a screen similar to a TV screen. This pattern was a stable identifying characteristic of an operator. This could enable an interception station to detect that, although a call sign had changed, the operator had not. Since it was prudent for a military or naval organisation to change call signs regularly, this technique also assisted in relating a new call sign to its predecessor.

Combining TA and DF enables the current location of specific senders to be determined. For example, the senders corresponding to various bases and headquarters of a Navy could be identified, as could the senders on various naval ships. Movements of these ships could be plotted if they sent regular radio messages that could be intercepted and tracked using DF. The listing of the warships in an enemy's navy, or in a stand-alone section of it, such as a fleet, was known as an Order of Battle. Updating this whenever possible was a critical task of a Naval Intelligence Department, as was the corresponding task of listing enemy army units for Military Intelligence.

A radio interception operator could, theoretically, simply transcribe onto paper the dot-and-dash beeps of intercepted messages as received and hand the results over to someone else for conversion into their underlying texts, which might well appear to be meaningless if the original messages had been encrypted. Generally, operators often did the first stage transcription from the Morse or Kana Morse into the underlying linguistic equivalent, and also, with experience, were able to spot some characteristic features of current interest. For example, during a heavy traffic situation, with simultaneous transmission of multiple messages on different carrier frequencies, attention might be directed to those on a particular frequency, or with specific call signs, or in a particular code.

1.14 Hollerith Equipment

By the late nineteenth century, the storing and processing of increasingly large amounts of data stimulated the development of technologies enabling these tasks to be executed efficiently and effectively. The most successful of these processed punched cards with a suite of specialised electro-mechanical machines invented in the following decades. There were several different types, such as card punches, sorters, tabulators and collators.

The first tabulator was designed by Herman Hollerith (1860–1929) to compile the American census of 1890. This needed specially designed punchable cards. Essential census data collected for each person could be coded and then entered in a systematic way via the punching of holes into given spots on each card, storing individual data on individual cards.

Punched cards could not conduct electricity but the holes would enable electric contacts to be made. Cards could be 'read' at high speed by wires brushing against them and relay switches would produce desired effects.

Once the cards were punched in this way, determining, say, the number of 6-year-old girls resident in Rhode Island would be achieved by simply running all the cards through a machine wired to count those with the appropriate combination of holes punched. Thus the tabulators had plugboards (also called jackboards) which could be wired to achieve the desired count. It became obvious that the one tabulator would be more efficiently used if a number of plugboards were available as accessories. Any standard use would require an appropriate pre-wired plugboard for insertion into the tabulator. The pre-wired plugboard was the predecessor of modern computer software. The basic aim was to count cards with a fixed arrangement of punched holes. Later, semi-automated accountancy became possible. The final model of such a tabulator was introduced in 1949.

By 1915 the political/diplomatic section of Room 40 confronted the problems of correctly guessing the meanings of a large number of code words obtained from many German messages sent in a few hatted codes. This task required both intelligent staff and tiresome work. Gannon[18] quotes from the British National Archives (TNA) file ADM 223/773:

> It was not realised that this form [hatted] code required special treatment until May 1916, when leave was granted to set up a special staff of educated women to work machinery by which the guessing process could be accelerated.

The key phrase in this quote is 'educated women'. As later use of this 'special treatment' confirmed, routine tasks in code breaking usually required intelligent, educated personnel, whether or not machines were employed in such tasks. In both world wars such women were available whereas most suitable men were already called up by the armed services.

---

[18]Gannon, *Inside Room 40*, page 168.

Hollerith's company later merged with others into a company subsequently named the International Business Machines Corporation[19] (IBM) which has survived into modern times. In Britain the British Tabulating Machine company (BTM) developed its own products and also negotiated with IBM an exclusive right to market its Hollerith equipment in the UK. This continued until 1948. BTM subsequently became part of International Computers Ltd (ICL). 'IBM machines' became a generic name for the types of equipment described above.

1.15 IBM Cards

The cards that were punched to provide input for the tabulators had to be of a standard size. From 1928 the IBM card measured 187×86 mm (7.25×3.375 in). It had 80 columns, each made up of 12 rectangular spaces which could be punched out by a card punch machine. The spaces were called X, Y, 0, 1, 2, 3, 4, 5, 6, 7, 8, 9. A column could be used to represent a digit 0, 1, 2, 3, 4, 5, 6, 7, 8, 9 by punching out only the rectangle in the corresponding space. The 26 capital letters could be represented by punching out two rectangles. Thus A was represented by punching out X&1, B by X&2, C by X&3, ..., J by Y&1, etc. Specific purposes could be allocated to the columns. Thus the words and phrases for a code book could be typed in columns 1–60 and code numbers in columns 66–70. Indeed, a primitive form of word processing was available.

Tabulators were phased out in favour of electronic machinery around 1963–1965. Cards continued in use as a method of inserting programs and data into the early computers until the modern terminal made them obsolete.

Other commercially available punched card systems were in use before WW2, but eventually it was decided[20] that Allied intelligence units would standardise their operations to use IBM cards only.

The cards had thin margins on the left, lower and right sides and a thick margin on the top. This last was used for the interpretation, that is the digit, letter or other symbol corresponding to the holes punched in the column immediately below. The top left corner was cut off.

Schematic diagrams (with 66 columns instead of 80) of (1) an unused IBM card and (2) a used card are given on the next page.

The interpretation of the used card is:

0123456789 THE QUICK BROWN FOX JUMPS OVER THE LAZY DOG

---

[19]One suitable reference is Chap. 3 of the book by Frederick Brooks and Kenneth Iverson: *Automatic Data Processing*, John Wiley and Sons, New York, 1963.

[20]An interesting letter on this matter may be found in General Blamey's papers at the Australian War Memorial. Copies of some of Blamey's Sigint material are to be found in RG457 in NARA. See also Chap. 19, notes 9 and 20.

which is the classic sentence that uses all 26 letters. The facility to distinguish between upper case and lower case versions of a letter was not available until much later.

```
x x x x x x x x x x x x x x x x x x x x x x x x x x x x x x x x x x x x x x x x x x x x x x x x x x x x x x x x x x x x x x x x x x
y y y y y y y y y y y y y y y y y y y y y y y y y y y y y y y y y y y y y y y y y y y y y y y y y y y y y y y y y y y y y y y y y y
0 0 0 0 0 0 0 0 0 0 0 0 0 0 0 0 0 0 0 0 0 0 0 0 0 0 0 0 0 0 0 0 0 0 0 0 0 0 0 0 0 0 0 0 0 0 0 0 0 0 0 0 0 0 0 0 0 0 0 0 0 0 0 0
1 1 1 1 1 1 1 1 1 1 1 1 1 1 1 1 1 1 1 1 1 1 1 1 1 1 1 1 1 1 1 1 1 1 1 1 1 1 1 1 1 1 1 1 1 1 1 1 1 1 1 1 1 1 1 1 1 1 1 1 1 1 1 1
2 2 2 2 2 2 2 2 2 2 2 2 2 2 2 2 2 2 2 2 2 2 2 2 2 2 2 2 2 2 2 2 2 2 2 2 2 2 2 2 2 2 2 2 2 2 2 2 2 2 2 2 2 2 2 2 2 2 2 2 2 2 2 2
3 3 3 3 3 3 3 3 3 3 3 3 3 3 3 3 3 3 3 3 3 3 3 3 3 3 3 3 3 3 3 3 3 3 3 3 3 3 3 3 3 3 3 3 3 3 3 3 3 3 3 3 3 3 3 3 3 3 3 3 3 3 3 3
4 4 4 4 4 4 4 4 4 4 4 4 4 4 4 4 4 4 4 4 4 4 4 4 4 4 4 4 4 4 4 4 4 4 4 4 4 4 4 4 4 4 4 4 4 4 4 4 4 4 4 4 4 4 4 4 4 4 4 4 4 4 4 4
5 5 5 5 5 5 5 5 5 5 5 5 5 5 5 5 5 5 5 5 5 5 5 5 5 5 5 5 5 5 5 5 5 5 5 5 5 5 5 5 5 5 5 5 5 5 5 5 5 5 5 5 5 5 5 5 5 5 5 5 5 5 5 5
6 6 6 6 6 6 6 6 6 6 6 6 6 6 6 6 6 6 6 6 6 6 6 6 6 6 6 6 6 6 6 6 6 6 6 6 6 6 6 6 6 6 6 6 6 6 6 6 6 6 6 6 6 6 6 6 6 6 6 6 6 6 6 6
7 7 7 7 7 7 7 7 7 7 7 7 7 7 7 7 7 7 7 7 7 7 7 7 7 7 7 7 7 7 7 7 7 7 7 7 7 7 7 7 7 7 7 7 7 7 7 7 7 7 7 7 7 7 7 7 7 7 7 7 7 7 7 7
8 8 8 8 8 8 8 8 8 8 8 8 8 8 8 8 8 8 8 8 8 8 8 8 8 8 8 8 8 8 8 8 8 8 8 8 8 8 8 8 8 8 8 8 8 8 8 8 8 8 8 8 8 8 8 8 8 8 8 8 8 8 8 8
9 9 9 9 9 9 9 9 9 9 9 9 9 9 9 9 9 9 9 9 9 9 9 9 9 9 9 9 9 9 9 9 9 9 9 9 9 9 9 9 9 9 9 9 9 9 9 9 9 9 9 9 9 9 9 9 9 9 9 9 9 9 9 9
```

and

```
x x x x x x x x x x x x ▮ x x x ▮▮ x x ▮ x x x x x ▮ x x x x x x x x x x x ▮ x x x ▮▮ x x ▮ x x x ▮ x ▮ x x x x x x x x x x x x
y y y y y y y y y y y y y y ▮ y y y ▮ y y ▮▮ y ▮ y y ▮ y y ▮ y ▮▮ y y ▮ y y ▮ y y y y y ▮ y y y y y ▮ y y y y y y y y y y y y
▮ 0 0 0 0 0 0 0 0 0 0 ▮ 0 0 0 0 ▮ 0 0 0 0 0 0 ▮ 0 0 0 0 ▮ 0 0 ▮ 0 0 ▮ 0 0 ▮ 0 0 0 ▮ 0 0 0 0 0 ▮▮ 0 0 0 0 0 0 0 0 0 0 0 0 0 0 0
1 ▮ 1 1 1 1 1 1 1 1 1 1 1 1 1 1 1 1 1 1 1 1 1 1 1 1 1 1 1 ▮ 1 1 1 ▮ 1 1 1 1 1 1 1 1 1 1 ▮ 1 1 1 1 1 1 1 1 1 1 1 1 1 1 1 1 1 1
2 2 ▮ 2 2 2 2 2 2 2 2 ▮ 2 2 2 2 2 2 2 ▮ 2 ▮ 2 2 2 2 2 2 2 2 2 2 2 2 2 2 2 2 ▮ 2 2 2 2 2 2 2 2 2 2 2 2 2 2 2 2 2 2 2 2 2 2 2 2
3 3 3 ▮ 3 3 3 3 3 3 3 3 3 3 ▮ 3 3 3 3 3 3 3 3 3 3 3 3 ▮ 3 3 3 3 3 3 3 3 3 3 3 3 ▮ 3 3 3 3 3 3 3 3 3 3 3 3 3 3 3 3 3 3 3 3 3 3
4 4 4 4 ▮ 4 4 4 4 4 4 4 4 4 4 4 4 4 4 4 4 4 4 4 4 4 4 4 4 ▮ 4 4 4 ▮ 4 4 4 4 4 4 4 4 4 4 ▮ 4 4 4 4 4 4 4 4 4 4 4 4 4 4 4 4 4 4
5 5 5 5 5 ▮ 5 5 5 5 5 5 ▮ 5 5 5 5 5 5 5 5 5 5 ▮▮ 5 5 5 5 5 5 5 5 5 5 5 5 ▮ 5 5 5 5 ▮ 5 5 5 5 5 5 5 5 5 5 5 5 5 5 5 5 5 5 5 5
6 6 6 6 6 6 ▮ 6 6 6 6 6 6 6 6 6 6 6 6 6 6 ▮ 6 6 6 ▮▮▮ 6 6 6 6 6 6 6 ▮ 6 6 6 6 6 6 6 6 6 6 6 6 6 ▮ 6 6 6 6 6 6 6 6 6 6 6 6 6
7 7 7 7 7 7 7 ▮ 7 7 7 7 7 7 7 7 7 7 7 7 7 7 7 7 7 7 7 7 7 ▮ 7 7 7 7 7 7 7 7 7 7 7 7 7 ▮ 7 7 7 ▮ 7 7 7 7 7 7 7 7 7 7
8 8 8 8 8 8 8 8 ▮ 8 8 8 ▮ 8 8 ▮ 8 8 8 8 8 8 8 8 8 8 8 8 8 8 8 8 8 8 8 8 ▮ 8 8 8 8 ▮ 8 8 8 8 8 8 8 8 8 8 8 8 8 8 8 8 8
9 9 9 9 9 9 9 9 9 ▮ 9 9 9 9 9 9 9 ▮ 9 9 9 9 9 9 9 9 9 9 9 9 9 9 9 9 9 9 9 ▮ 9 9 9 9 9 9 9 9 9 9 9 9 9 9 9 9 9 9 9 9 9 9
```

It was essential to be able to sort cards. Cards prepared for a code book as above would need to be sorted into alphabetical order by word and also numerical order by code group (code number). In practice free-standing sorters were used. These could handle only one column at a time and had to be used twice when the column had been used with two holes to represent a letter.

The tabulators of 1940 were basically capable of various uses, mostly in accountancy. Cards could be used as lines on a spreadsheet. Cards would be prepared manually by card punch machines and checked by a verifier or mechanically prepared by some other process using cards already punched. The tabulators would be used as a final step after cards had been prepared through various preliminary processes on the other machines. In practice the staff would have written instructions for the task to hand, and would be manually transferring substantial sets of cards from one machine to another.

The collator was the key unit in IBM data processing at this time. It had been introduced only in the 1930s. It could, for example, check the sorting of a large deck of cards. The Type 077 collator, introduced by IBM in 1937, proved to be immensely useful to cryptanalytic units. Described as an 'automatic filing clerk', its principal function in ordinary use was simultaneously to feed and compare two sets of punched cards so as to match or merge them. It could sort matching and non-matching cards into two different groups. It could input a single deck of cards already sorted on specified fields and identify those that coincided on those fields. This might indicate a common code word occurring in different messages.

The jargon was to call the program cards (those instructing the machine) master cards and the data cards were called detail cards. Use of the collator might well mix up detail and master cards. The simple device of having a 1 in the first column of each master card and a 2 in that of detail cards made it possible to separate them again by a single run of the sorter.

As already noted, the cryptology of WW1 used some automated data processing. In the early 1930s the USN Sigint unit Op-20-G developed cryptological techniques using IBM cards and machinery. The US Army Sigint unit started working with such techniques a year or two later. The British GCCS was also using such machinery in the mid 1930s. By 1940 the IBM[21] company had put much effort over many years into developing high-performance card handling machinery suitable for data management and accountancy work. As an interesting and possibly profitable sideline it financed a facility carrying out scientific calculations in astronomy at Columbia University. A book entitled *Punched Card Methods in Scientific Computation* by Wallace Eckert of that unit was published in 1940. This clarified the extra capacities of IBM cards and machinery. Although these were not ideal for scientific or cryptanalytical usages, they were technically very polished, readily available and were to play an essential role in the developing cipher war.

By the end of 1942 it had become clear that the American and British[22] Sigint units had become major users of such equipment. Certain modified[23] versions of standard machines were produced in secret with appropriate new capacities. IBM, at the request of the USN, modified a number of its standard machines to meet specific cryptographic requirements. The resulting machines were known as Naval

---

[21] Stephen Budiansky has written a useful paper *Codebreaking with IBM Machines in World War II*, which is published in *Cryptologia*, 25(4), October 2001, 241–255.

[22] The GCCS had to find secure methods of disposing of 2 million used cards per week from Bletchley Park! Secure acquisition was also a problem. In a report on a visit to Bletchley Park, Friedman praised the productivity of the machine unit.

[23] Although electricity moved at the speed of light, the cards did not. They were transported by rollers and other devices that were continually stopping and starting. The IBM company had developed methods of doing this at commendable speed. Rewiring an existing device was relatively easy but getting totally new card processing technology under way would have taken more time than was available.

Note 46 of Chap. 9 is not easy reading but it gives a rare contemporary description of one of these special purpose electro-mechanical machines.

Change (NC) machines. For example the NC1 machine was designed to insert serial numbers in unused columns in a deck of IBM cards.

The most complete report[24] on the use of IBM machinery in WW2 Sigint work is that written, at the request of GCHQ,[25] by Ronald Whelan, formerly deputy head of the Bletchley Park data processing unit. It survives in TNA as HW 25/22. William Friedman, head of the US Army Sigint unit, wrote[26] on 4 October 1943 a *Report on IBM Operations at Bletchley Park*. His report included a paper, written at his request by Frederick Freeborn, head of the card processing unit there.

1.16 Encrypted Code Book Systems

Attacking a message sent in an encrypted code book system is a two-stage process. First, the cryptanalysts must have determined that the intercepted messages indeed used such a system. (In practice, various systems might well be used simultaneously, leaving the enemy to sort the messages by system.) Then they had to work out exactly how the superencipherment had been done. When this was well under way the intercepts had to be 'stripped' back to the original code groups (commonly called book groups.) As the decryption process here attacks an encipherment and not an encoding this book refers to the systems under study as ciphers rather than codes.

Once the stripping had been achieved or perhaps only partly achieved, the cryptanalysts handed the resulting encoded text over to linguists and intelligence experts whose all-important task was to work out the correct meanings for each of the identified book groups. In the Pacific War this task was made even more difficult by the shortage of expertise in Japanese signal jargon.

Hence the decoding of messages sent in a partly broken encrypted code book system could be a much slower process than decoding messages sent in a broken machine system such as Enigma. For such machine traffic, once the design of the machine is known and methods for determining the initial machine settings have been found, an intercepted signal can be turned quickly into its original plain text. But the production of a full plain text version of messages sent using a broken

---

[24]For example, section 20 of Whelan's report is headed *Positional Double Repeat Search*. It describes the processes used 'to find messages in common with others by virtue of each having a pair of cypher groups (GATs) repeated in another and where the left to right separation of one of these groups from the other was the same in a given pair of messages.' The American jargon was 'double hits' rather than 'double repeats'. As these 'cypher groups' were of five digits, this is almost certainly about processing JN-25 intercepts. In the next section Whelan deals with the mechanical production of difference tables, of which more later.

Edward Simpson's *Bayes at Bletchley Park* (*Significance*, 7(2) June 2010 76–80) confirms that tabulating equipment was used at Bletchley Park in processing JN-25 intercepts.

[25]The Government Code and Cipher School (GCCS) was renamed the Government Communications Headquarters (GCHQ) soon after the end of WW2 and is still called GCHQ.

[26]NARA RG457 Box 1126.

encrypted code book system may never happen: some less common code groups may never be understood (decoded).

Understanding this distinction is essential to any appreciation of why the work done on the JN-25B system throughout 1941 need not have produced Sigint relevant to the IJN raid on Pearl Harbor in December of that year but was crucial in the provision of invaluable Sigint between March and May 1942.

1.17   Restrictions on the Use of Comint

It was clear from pre-war experience that Comint had to be kept secret. An enemy would presume that radio intelligence, such as traffic analysis and direction finding, was being used as much as possible. However, any information that codes and ciphers were being read was likely to provoke their users into replacing them. A warning on this matter is to be found in the National Archives of Australia (NAA) Canberra file A6923 SI/10:

> The utmost secrecy is to be used in dealing with Ultra information. Attention is called to the fact if from any document that might fall into the hands of the enemy or any message that the enemy might intercept, from any word that might fall revealed by a prisoner of war or from any ill-considered action based on it, the enemy were to suspect the existence of the Ultra source, that source would probably be lost for ever to our cause. This loss would vitally affect operations on all fronts, not only on the particular front on which the sources had been compromised.

Detailed regulations follow. These include:

> In general, if any action is to be taken based upon Ultra information, the local Commander is to ensure that such action cannot be traced back by the enemy to the reception of Ultra Intelligence alone. ... No action may be taken against specific land or sea targets revealed by Ultra unless appropriate air or land reconnaissance has also been taken.

Problems existed at various stages with the distribution lists for Comint. In theory a 'need to know' principle was in force, but in practice it was not always easy to work out in advance who needed to know a particular piece of information. Also, there were operational situations where the use of such information must have been guessed. For example, the crews of US submarines, ordered time and time again to station themselves at points in the ocean where convoys of Japanese merchant ships miraculously appeared, must have been able to work out that secret information from Comint was being used.

The distribution of Comint as background information was strictly monitored. Yet the senior politicians and administrators had to be assured that extremely expensive interception and decryption facilities were giving value for money. For example, Canadian Prime Minister Mackenzie King[27] had to be kept well informed.

---

[27] Mackenzie King kept a very extensive typed diary over many years. This is now on the website of Library and Archives Canada. The entry of 26 May 1942 includes the following text: 'I opened one

High-level intelligence units had to operate on full information. It was not easy then (and remains so today) to achieve the correct balance between security and utility. Note 10 of Chap. 21 discusses the likely but not certain decision made in May 1942 not to pass information about Japanese submarines near Sydney Harbour to those responsible for its defence. Some damage was done by the submarine raid on 31 May 1942 but the secrecy surrounding the ambush of the IJN aircraft carriers at Midway a week later was maintained.

The sheer size of units like Bletchley Park or the Naval Communications Annex in Washington must have revealed to anyone who worked there or just regularly walked past that something significant was going on. It seems however that neither the Germans nor the Japanese became aware of the activities at these locations or of their scale. In particular, Bletchley Park was never the target of a major bombing raid.

In retrospect, the aspects of Sigint that were accorded the highest levels of security during WW2 appear to be its technical aspects, as is the case today. These included precise descriptions of a particular enemy code or cipher system, the identification of any weaknesses in a system that might render it vulnerable to cryptanalysis, and, of course, the techniques developed to enable messages sent in a given system to be decoded and then converted into accurate operational information. Only long after the publication of some accounts referring to the Allied use of certain code breaking methods to deliver operationally useful intelligence in WW2 did public accounts of the decryption processes used become available.

In particular, this book is the first to investigate fully why the use of only multiples of three as book groups (code words) for the critically important JN-25 series of IJN operational encryption systems let Allied cryptanalysts obtain from JN-25 messages much of the operationally useful Comint that decisively influenced the evolution of the war against Japan.

1.18   Other Sources of Intelligence

In the Pacific War distance made RI and CI particularly significant. However, military intelligence teams used a variety of sources, including aerial reconnaissance and photography,[28] submarine-based examination of possible landing sites, and captured

---

marked "For immediate delivery: most secret" from the Naval department. The message was from the US liaison officer of the Pacific Command. It indicates that the US Intelligence believes there will be an attack against Midway, Hawaii and the Aleutians probably during the period 20 May to 20 June.' Mackenzie King was very concerned about any possible Japanese attack on Alaska. He discussed the matter with his War Cabinet.

[28] A memorable device used in the European war was to use Spitfires without guns to fly at great heights taking photographs of the ground below. As otherwise the camera lenses would be covered in condensation from the cold air around the plane, photographs were taken through the exhaust stream of the engine. The aircraft used were said to be 'cottonised' in view of the role of the

documents. General MacArthur's South West Pacific Area (SWPA) command set up the Allied Translator and Intelligence Service (ATIS), to translate selected captured documents. It interrogated Japanese prisoners of war also. Other information was gathered from spies or old travel books and maps.

Intelligence from all sources needed to be compiled, analysed and presented to the relevant commander promptly. The commander then had to decide whether to proceed with a proposed operation. Even then, there was always risk that things could go badly wrong.

As already noted, information obtained from breaking enemy codes and ciphers is particularly sensitive. The technology of radio finger-printing was also closely guarded. On the other hand, direction finding was scarcely a secret, in the same way that nowadays the use of satellite photography is widely known.

1.19   The Personnel

During WW1, the British brought into the code breaking arm of signals intelligence a small number of academics talented in other languages. When war seemed inevitable in the late 1930s, plans were again made to bring immediately into intelligence work a group of academics, this time including mathematicians, to augment an already highly skilled team of linguists and cryptanalysts. When the UK declared war on Germany in 1939, a number of previously selected academics from Cambridge and Oxford immediately left for Bletchley Park, and were steadily augmented there throughout the war by selected younger staff and students from across the universities. Alan Turing and Gordon Welchman, usually mentioned in the context of their work at Bletchley Park on German ciphers, are two notable such recruits. Peter Hilton, who was selected for GCCS work while a student, commented to the authors in 2007:

> In WW2 Britain certainly employed its best thinkers far more effectively than any other nation.

This would appear to underestimate[29] the talent recruited by the American armed forces, although it is quite clear that Germany and Japan did not employ such appropriate personnel in their communications security units.

---

redoubtable Australian aviator Sidney Cotton in getting photographic intelligence going. In the Pacific War the distances to enemy sites were often too great for photographic intelligence to be feasible until planes from carriers were put to this use.

[29] Attention has to be drawn to the obituary of Andrew Gleason published in the *Notices of the American Mathematical Society*, 56(10) November 2009 1236–1267. Gleason worked in WW2 for Op-20-G and achieved much in decrypting Seahorse Enigma, used for communications between Berlin and the German Naval Attaché in Tokyo. This involved the exploitation of redundant encryption, which is discussed in Chap. 14.

Chris Christensen's *U.S. Naval Cryptologic Mathematicians during World War II*, in *Cryptologia* 35(3) (2011) 267–276 is a useful general reference here.

In the USA, between the wars, Laurance Safford and William Friedman, for the US Navy and US Army respectively, developed their own methods for identifying potential cryptographers and cryptanalysts. Suitable people were recruited and trained. Central Bureau (CB or CBB, see Chap. 17), the Comint unit attached to General MacArthur's SWPA command, had a staff of over 4,000 by mid 1945. Although only a small proportion of these would have been engaged directly in cryptographic work, files held in the National Archives of Australia (NAA) reveal that most of these were selected on the basis of above average intelligence and education.

Staff employed at Bletchley Park to handle the high-level clerical tasks, including the operating of IBM and other machines, again needed to be competent for the task. Freeborn[30] wrote in mid-1942 to Commander Travis, the Director of GCCS, about this. The following is an extract:

> I must confess that I find it incredible at this late stage that every reference, verbal or written, which is made to the staff here still speaks of 'Punch Operators'. Until this can be changed, as I have repeatedly asked, so we will be sent unsuitable, uneducated workers who would be much better serving the country in the munition plants. I made it quite clear at the meeting with Miss Sharp and Miss Moore and various Ministry of Labour officials that I must have intelligent young women and that the work can best be described as 'clerical', pure and simple. Can steps be taken finally to ensure that the requirement of this section is recorded as approximately ninety Clerical Staff, and not Punch Operators?

This may be compared with an extract from Friedman's previously noted report on the IBM unit at Bletchley Park:

> The personnel at the IBM unit consists very largely of women. A four-weeks' training course is given them during which they spend two weeks on the machines learning how to sort, list, reproduce, etc. Another two weeks are devoted to taking them through a standard job. Mr. Freeborn stated that they have learned that if a girl cannot learn all the operations in four week's time she will never learn it at all. The type of personnel is, in general, not one with very much educational background, there being many women who were formerly shop girls, hairdressers, etc. They have, however, turned out to be very satisfactory.

A relatively small number of successful code breakers and linguists could maintain a very large code breaking operation, if they had the talent needed to break into the encryption systems and if the extensive infrastructure of supporting personnel and machinery was there to operate with them. Extensive support was really needed once an important operational system had been broken. The need to obtain immediately useful information was paramount. Very extensive support work was needed at all stages for the digital cipher systems described in this book.

---

[30]The original source is TNA HW 14/35 and it is cited in Paul Gannon's *Colossus*, page 365.

There is no doubt, as Hilton has stated, that the sheer quality of the principal code breakers employed by the Allies during WW2 was a significant factor in the progress of this war in all its theatres. These key people were subject to high levels of stress.[31] But the exploitation of Sigint relied on the use or neglect of intelligence supplied to the military leaders, and on the actual outcomes of the battle plans they developed.

---

[31] William Friedman (see Note 16 above and Chap. 4), is known to have had at least one breakdown caused by stress and overwork in WW2. Athanasius Treweek of Frumel, the naval Sigint base in Melbourne, gave a detailed interview on his WW2 experiences to his son-in-law Alan Roberts on 2 September 1985, with the following being recorded:

> Treweek: 'Well, they tried to get me to stay on, but I said the mental strain of this was so great—it affected my digestion for the rest of my life. The mental strain of doing this work was so great that I couldn't have faced it.'
> Roberts: 'What was the strain?'
> Treweek: 'Trying to solve a problem against time—feeling if you could go on for another half hour you'd have the answer, and you're getting more and more tired. That's what it is. It took a great effort to say "I have to stop working on this now." People may be dying or something as a result.'

Essentially the same point was made by Captain Arthur McCollum, USN, in a hearing of the Congressional Investigation into the Pearl Harbor Attack. The reference is volume 8, page 3400 of the transcript:

> 'This type of work is one of the most trying mental exercises you can have. We have had a number of our officers and a number of our civil people break down rather badly under continual punching on this sort of thing and it is a continual concern of officers who handle these people to keep them from coming to a mental breakdown on this type of work.'

Chapter 2
Japanese Expansion 1895–1941

The first Sino-Japanese war of 1894–1895 had demonstrated Japan's growing military and naval power. This chapter briefly describes the steady growth of Japanese strength and territorial ambitions up to the outbreak of the Pacific War in 1941.

2.1 From 1895 to 1914

The 1894–1895 war resulted in China yielding territory to the Japanese, including the island of Formosa (now Taiwan) in 1895. China had also agreed to allow Korea to become an independent state.

Japanese concerns were then directed against the activities of the Russians in the Far East. In 1902, Japan formed an alliance with Britain aimed at stopping European governments from intervening in any war involving Japan in the Far East (that is, with Russia). Relations between Japan and Russia continued to deteriorate and resulted in the Russo-Japanese war of 1904–1905. The Imperial Japanese Army (IJA) suffered horrendous losses but forced the Russians back. The IJN routed the Russian Baltic fleet, which had been sent to reinforce the blockaded Eastern Fleet, at the battle of Tshushima in May 1905. As a result, Japan gained half of Sakhalin Island from Russia. Russia also surrendered its access to Port Arthur in China and recognised Korea as under Japan's 'sphere of influence'.

Although not interfering in the war, European and British governments did influence the peace settlement. This heightened Japan's antagonism towards the

© Springer International Publishing Switzerland 2014
P. Donovan, J. Mack, *Code Breaking in the Pacific*,
DOI 10.1007/978-3-319-08278-3_2

Western powers, as it felt that it should have gained more territory on mainland Asia. Japan annexed Korea in 1910.[1]

The IJN used its links to the RN to continue to model its naval organisation and operations on what it saw as RN best practice. It is not surprising that the RN decided to maintain an interest in the development of the IJN from that period onwards. More generally, from the 1890s, the USA, noting the rise of Japanese militarism and its territorial ambitions, instituted strategic consideration of how the USA might respond to future aggressive actions taken by Japan in the Pacific region. The USA was also concerned that Japan might foment unrest among the Japanese settled on American Pacific territories, the Territory of Hawaii and even on the mainland west coast. War Plan Orange, as this strategy came to be called, was subject to regular review as Japan's empire and power grew.

2.2   Japan During WW1 and Its Aftermath

Japan declared war on Germany on 23 August 1914. By late September, British, Australian and Japanese troops were moving to take over the German ports and islands in the Far East. Japan had been guaranteed eventual control over the Caroline, Marianas and Marshall Island chains, which lie strategically just north of the Equator in the Western Pacific Ocean. IJN warships also served with Allied forces in the Mediterranean. Following the defeat of Germany, Japan became one of the permanent council members of the proposed League of Nations. So Japan participated in the 1919 Treaty of Versailles, emerging with a Mandate over the above-mentioned island chains and also with possession of former German territories on the Shandong peninsula in China. This latter gain came despite serious objections from China and with opposition from the USA, and was one of the grounds for the USA's refusal to ratify the Treaty or to join the League of Nations.

However the USA was very concerned with limiting the post-WW1 size of the major navies, in particular those of Britain and Japan, on terms favourable to itself. President Warren Harding convened a conference in Washington late in 1921, resulting in a naval treaty signed in February 1922, at which it was agreed that all capital warship building be halted for 10 years and that existing numbers of capital ships in the US, British and Japanese navies be held in the proportions of 5 to 5 to 3. The success of the American negotiators in holding Japan to this ratio stemmed from the ability of its cryptanalytical team to break the Japanese diplomatic cipher and thus to keep them apprised of the Japanese Government's bottom-line position. A further outcome of the negotiations was that Japan agreed to return to China the former German territories in Shandong.

---

[1] A more thorough account of some of the material in this chapter is to be found in chapters 1 and 2 of Peter Elphick's *Far Eastern File: The Intelligence War in the Far East 1930–1945*, Hodder and Stoughton, London, 1997.

2.3 Japanese Actions During the 1930s

Japan was facing a population explosion and a shortage of resources. The solution favoured in Japan was for it to annex Manchuria as a puppet state to which Japanese colonists could be sent and from which, following economic development, greater exports could be sent home to Japan. From 1928 to 1933 Japan implemented this solution, naming its puppet state Manchukuo and seceding from the League of Nations when it belatedly and ineffectually protested against this occupation.

In 1930, the London Naval Conference resulted in Britain, the USA and Japan retaining the 5:5:3 proportions for numbers of capital ships but relaxed it to 10:10:7 for light cruisers and lesser categories. This time, it was the British cryptographers who broke the relevant Japanese diplomatic codes and assisted in the final outcome. By the mid-1930s, however, Japan had warned that it would no longer be constrained by any of the previous naval treaty obligations. In December 1936, Japan, having alienated itself from the Western democracies, and fearful of the threat of Communism developing from Mao Tse-tung's military campaign in China and the Soviet Union's growing strength, signed an Anti-Comintern pact with Germany.

By this time, Op-20-G and other sources had provided the USN Chief of Naval Operations with the clear message that the IJN had complete naval supremacy in the seas around Japan and in the Western Pacific.

The British reacted to these developments in 1935 by setting up a branch of the GCCS, the Far East Combined Bureau (FECB), in Hong Kong.

Japan launched its second major military offensive against China in mid-1937, attacking from Manchukuo and along the China coast. China signed a non-aggression pact with the Soviet Union in August of that year, and Mao declared that his forces would fight alongside those of Nationalist General Chiang Kai-shek against the Japanese. By late 1938, the Japanese had overrun much of northern and eastern China, forcing Chiang Kai-shek to move his government and headquarters to Chungking in the west. The British colony of Hong Kong was now effectively an isolated outpost on the coast of Japanese-held China. The USA, deeply concerned about the progress of Japanese forces in China, and horrified by their actions following the fall of Nanking in December 1937—the Rape of Nanking—was anxious to see effective opposition to them continue. It granted $25 million in aid to Chiang Kai-shek, and this began a program of US support that continued throughout the Pacific War (1941–1945).

There were sporadic clashes between Russian and Japanese forces in the northern Manchukuo-Outer Mongolian regions from 1937 onwards. In August 1939 General Zhukov routed the Japanese Kwantung Army and effectively terminated Japanese expansion into Mongolia.

Once Japanese forces had sealed off the Chinese coast to deny supply for Chiang Kai-shek and his Nationalist Army, he had to use two communication lines for receipt of military aid. These were a rail link from the port of Haiphong in French Indo-China (Vietnam), and a more difficult route in Burma, the so-called Burma

Road, via Rangoon and Lashio, to Kunming in China. After the surrender of France to Germany in mid-1940, Japan prevailed upon the Vichy government to allow its warships and forces to enter parts of northern French Indo-China, immediately cutting off the rail supply route to Chiang Kai-shek. Japan also briefly prevailed upon the British to place a temporary closure on the Burma Road route, so for some 3 months the Nationalist Chinese were isolated. Also, at this time, Japan sought to establish the Greater East Asian Co-Prosperity Sphere—a plan to incorporate Burma, Thailand, Malaya, the Netherlands East Indies, the Philippines and some smaller territories into the Japanese Empire and so make it self-reliant in raw materials.

In September 1940, Japan signed the Tripartite Pact with Germany and Italy. The sentence 'Germany and Italy recognise and respect the leadership of Japan in the establishment of a new order in East Asia' hardened American attitudes towards Japan and increased US aid to China. In April 1941, Japan signed a non-aggression pact with the Soviet Union.[2]

In July, following further extreme pressure upon the Vichy government from Japan, its virtual occupation of northern Indo-China was expanded southwards, permitting new military bases and airfields to be established there. The USA, using diplomatic signals intelligence to conclude that this was an aggressive act, froze all Japanese assets.[3] Prior to this action the USA had been a major supplier of oil to Japan. When Britain and the Netherlands joined the trade embargo, Japan was deprived of almost all its oil supplies as well as losing access to the mineral and food resources of Malaya and the Netherlands East Indies (NEI, now Indonesia).

British intelligence had estimated that Japan's oil reserves would suffice for less than a year if it went to war. So the oil embargo alone would have forced Japan's hand not just to attack South-East Asia but to achieve a quick conquest of the NEI. Finally, after further negotiations failed, the USA in late November 1941 demanded that Japan withdraw from China. There would be no turning back for Japan now, and all parties realised that outright war, at least against the Dutch and British, was now just a matter of time.

Japan now had additional incentives to expand its aggression. The German Naval Attaché in Tokyo, Admiral Wenneker, handed over to the Japanese in December 1940 documents captured in the previous month in the Indian Ocean by the raider *Atlantis* from the British merchant ship *Automedon*. These documents, destined for the British High Commission in Singapore, included a copy of the minutes of recent British War Cabinet discussions on the Far East, concluding that Britain would not be able effectively to meet any major Japanese action in that area: it could

---

[2]Richard Sorge, a spy for the Soviets based in Japan, had warned Stalin of the coming German invasion. But this warning had been disregarded. However later advice from Sorge was a factor in the Soviet decision to move troops out of Siberia and back to the defence of western Russia and the Ukraine.

[3]The book *Bankrupting the Enemy* by Edward S. Miller (Naval Institute Press, 2007) is highly relevant here.

not reinforce Singapore and both Hong Kong and Borneo were indefensible. Soon after that, Hitler's Foreign Minister Ribbentrop (who must have been aware of the British War Cabinet document) urged the Japanese to launch an immediate attack upon Singapore. The pressures upon Japan to take military action thus grew steadily throughout 1941.

So, by November 1941, Japanese plans to attack the American Pacific territories and to invade South-East Asia and the NEI were complete, as was an intention to invade Burma and so cut off Chiang Kai-shek's supply line to his armies fighting the Japanese. Once Japan initiated hostilities, the Allies had a paramount reason to keep that supply line open: a high proportion of the IJA troops were tied up in the China campaigns, thus keeping them out of the Pacific War!

The extent to which these plans were known to or could have been inferred by the future Allies is discussed in Chap. 7.

2.4 Timperley and Kennedy

It could be argued that WW2 began[4] in China in 1937 rather than in Poland in 1939. The 1937 Japanese offensive into China received massive attention in the United States and the United Kingdom. Harold Timperley, an Australian journalist working for *The Manchester Guardian*[5] in Nanking, described some of the terrible events in messages to his newspaper. The Japanese Army used its control of the Shanghai telegraph office to censor such reports. Interestingly, a somewhat garbled account of a report from Timperley ended up being sent in code to the Japanese Embassy in Washington, and was read by the US Army cryptographic team. Timperley responded to the censorship of his reports by writing the book *What War Means: The Japanese Terror in China* (Gollancz, London, 1938). This claimed (page 72) that events like those in Nanking had occurred all across northern China.

Malcolm Kennedy, employed by the British GCCS as an expert on Japan, kept a diary[6] of those quite momentous years. The entry for 23 January 1937 shows the despair of a great admirer of Japan:

---

[4]This chapter has emphasised the provocation that led to the economic embargoes being placed upon Japan and the sheer size of the conflict in China.

The Rape of Nanking led to the GCCS (renamed the Government Communications Headquarters or just GCHQ in 1946) taking more interest in attacking Japanese Army ciphers. See Sect. 3.2. This work was to have a critical impact on its subsequent attack on the IJN series of ciphers, now known as JN-25, introduced by the IJN in June 1939. The saga of JN-25 is discussed in Chap. 9 and elsewhere in this book.

[5]The *Manchester Guardian* is now published as *The Guardian*.

[6]A paper by John Ferris published in *Intelligence and National Security*, 4(3), 1987, 421–450, describes Kennedy's background. The Kennedy diary may be consulted via the website of the University of Sheffield Library. See also Note 7 of Chap. 3.

Seems that the Navy in Japan, fearing that the budget with the estimates for the new naval program would have to go by the board if the Diet (Parliament) was dissolved, has done its utmost to effect a compromise between the Army and the politicians. Although dissolution has been avoided for the time being, the Government has resigned, so heaven knows what the final outcome will be. Meanwhile Moscow is once again staging a monster show trial ...

Here Kennedy only hints at the troubles characteristic of the 'Government by Assassination' era in Japan. The Japanese Government did not have full control of its Army. The 'monster show trials' in Moscow had the unintended effect of encouraging the Nazis to believe that the Soviet Union would be a soft target for invasion.

The diary entry for 6 October 1937 shows the effects in the West of the aggression of the coming Axis countries:

Roosevelt, speaking at Chicago yesterday, came out with a strong denunciation of Japan, Italy and Germany—by implication. Last night too, Japan received further reproof and condemnation at the Albert Hall meeting and, although the Archbishop was more moderate than might have been expected, he indicated his belief that 'economic pressure' should be applied if protest fails. A MOST DANGEROUS SUGGESTION.

But economic pressure was to be applied and the situation did become most dangerous.[7]

---

[7] One may speculate whether the envisaged danger or dangers included sudden attacks upon British, Dutch and perhaps American interests in South-East Asia and the Pacific.

Chapter 3
The GCCS 1919–1941

Admiral Hugh Sinclair, the Director of RN Intelligence (DNI), appointed Alastair Denniston as the head of the newly established Government Code and Cypher School (GCCS) in 1919. The GCCS was transferred to the Foreign Office in 1922, but Sinclair regained control after being appointed head of the Security Intelligence service in 1923. Denniston remained the effective head of GCCS until early 1942 and in 1944 wrote a memoir[1] on the GCCS between the wars. This remained secret for some 50 years. The account below rests largely on information provided by his memoir.

3.1 Early Activities

The overt function of the GCCS was to advise on the security of codes and ciphers used by all government departments and to assist in their provision. This role (sometimes called construction) was maintained and managed by Edward Travis, deputy to Denniston and later his successor.

However the GCCS had a secret purpose: to study the methods of cipher communications used by foreign powers. Some of the British WW1 radio interception stations were retained, and new legislation created a requirement that all cable companies operating in the UK allow inspection of all cable traffic passing through the UK.

---

[1] The Denniston account of the GCCS is dated 2 December 1944 and reprinted in *Codebreaking and Signals Intelligence*, edited by Christopher Andrew, Frank Cass, London, 1986 and also in the first issue of the journal *Intelligence and National Security*. It is also reprinted in Robin Denniston's *Thirty Secret Years: A. G. Denniston's Work in Signals Intelligence 1914–1944*, 2007.

This chapter concentrates on aspects of GCCS that became relevant to the Pacific War (1941–1945). Attention should be drawn to Mavis Batey's *Dilly: The Man Who Broke Enigmas*, which deals with Dilly Knox (1884–1943) and his career in cryptanalysis. It covers numerous other aspects more relevant to the European Theatre (1939–1945).

© Springer International Publishing Switzerland 2014
P. Donovan, J. Mack, *Code Breaking in the Pacific*,
DOI 10.1007/978-3-319-08278-3_3

The British Army, which had operated a number of wireless interception units during WW1, consolidated its Middle Eastern operations at Sarafand in Palestine, at Baghdad in Iraq, and at Simla in northern India during the 1920s. These sites monitored foreign military, diplomatic and naval traffic, especially that related to the Soviet Union. In a sense, the 'Great Game', played between England and Russia in the nineteenth century via spies and secret agents, continued into the twentieth century via radio interception! Simla was also an important centre for radio communications around the British Empire. Information from this Army interception network was sent back to London for examination.

A classic early example of the unanticipated destruction of a valuable source of intelligence occurred in London on 12 May 1927, when uniformed police raided the offices of the Anglo-Russian Co-operative Society Limited, generally known as Arcos. This large organisation handled trade between the United Kingdom and the Soviet Union and was also a Soviet front for espionage. Quantities of files were removed. The following Tuesday the Home Secretary and other ministers justified the Arcos raid at length in the House of Commons. Decrypts[2] of enciphered Soviet diplomatic messages were read out.

This should not have surprised the Soviet authorities, because as early as 1921 Trotsky had written[3] to Lenin:

> England has organised a network of intercept stations designed particularly for listening to our radio. This accounts for the deciphering of more than 100 of our codes.

The Arcos raid caused a major upgrade of Soviet diplomatic cryptography. One-time pads[4] were introduced, resulting in complete loss of access by the GCCS to any further information from that source.

The Arcos raid, coupled with the publication in 1931 of Yardley's book[5] on the *American Black Chamber* and that of Hugh Hoy on the naval communications intelligence unit in London in WW1, made clear to anyone interested that secure radio and telegraph communications were vital.

---

[2]An interesting contrast is provided by the 1954 Petrov saga in Australia. This did not result in any official disclosure by the Australian Government of breaking of Soviet diplomatic messages. Indeed the Venona project, in which the predecessors of the American National Security Agency were able to exploit the recycling of one-time pads (see Note 4 below), remained secret for many years afterwards.

[3]This is taken from the introduction of Robin Denniston's book *Churchill's Secret War*. On page 9, he gives more information on the use of one-time pads by the Soviets.

[4]One-time pads were an almost totally secure albeit clumsy method of communication. They are explained in Sect. 8.17.

[5]The year after the Yardley book was published Hugh C. Hoy came up with *40 O.B., or How The War Was Won*. After this, British naval Sigint in WW1 became generally well known. Apparently 'Room 40' was called '40 O.B.' (Old Building) at the time.

In fact the quotation in Chap. 1 about the Battle of Jutland taken from Winston Churchill's *The World Crisis* (1923–31) reveals that he was responsible for making public the use of naval Comint in WW1.

3.2   Building Up Sigint Strength Against Japan

By 1922 the Admiralty had acquired some special interest in Japanese[6] diplomatic and naval attaché traffic and had provided GCCS with an officer to assist as a Japanese interpreter. In order to increase its access to this traffic, the Admiralty established an interception facility in the Far East using ships of its China Squadron, based at HMS *Tamar* in Hong Kong.

The introduction of a personnel rotation system by the GCCS within the new facility built up a core of staff trained to work on Japanese signals. Denniston writes of this period as follows:

> From 1922 onwards we always had one naval officer working in the Japanese Section, reading the diplomatic and naval attache telegrams. By 1925 we even had officers still on the active list and a scheme was arranged whereby such came to us for two years and then joined the China Squadron in a ship where there were facilities for local interception. Thus a first start was made on Japanese naval traffic.
>
> From then onwards there was a flow of traffic by bag to London where the various codes were segregated and broken as far as possible, and a return flow of officers with skeleton books to carry on the work locally. I believe that by 1930 they were able to be of definite use to the C-in-C, China. Finally the Admiralty sent out Captain Campbell Tait (then DDNI) to study the Far East question, and, in consequence of his report, set up a small bureau for interception and cryptography at Hong Kong (moved to Singapore in September 1939) where Captain Shaw, now retired, headed the first party to exploit Japanese naval signals, to which was added the beginning of Japanese military. The Diplomatic Section had followed the diplomatic and attaché developments, including the introduction of mechanical devices, successfully, and thus it can be maintained that in early 1939, GCCS had full control of diplomatic and attaché traffic, were reasonably fluent in their reading of all the main naval ciphers and knew quite a lot about Japanese army ciphers as used in China.

The 'small bureau' established by GCCS in Hong Kong in 1935 and known as the Far East Combined Bureau (FECB), strengthened its interception facilities by setting up stations at Stonecutters Island in Hong Kong Harbour and at Kranji in Singapore, with further input from the Canadian station at Esquimalt. These were also used for DF and TA intelligence, which had greater prominence from the mid-1930s onwards.

3.3   Eric Nave

Nave's career, including his extensive contributions to Sigint against the Japanese, is very clearly described in Ian Pfennigwerth's 2006 book *A Man of Intelligence*. Born in South Australia in 1899, Nave entered the Royal Australian Navy (RAN) in 1917 for training in administrative duties. In 1919 he seized an opportunity to

---

[6]Section 6.2 discusses the 1919 reports of Admiral Jellicoe on the Australian and Canadian navies. These reports may well have influenced the development of the GCCS.

study Japanese, and in so doing exhibited his exceptional linguistic talent. This led the RAN to send him to Japan in 1921 for two years of further language study under a program maintained by the British Embassy in Tokyo. His excellent final results were noted by the senior personnel, including Ragnar Colvin, head of the RAN from 1937 to 1941.

On his return to the RAN in 1923 Nave held a range of positions of value, including that of escort officer to the Commander of a visiting Japanese Naval Squadron. He also became involved in early attempts by the RAN to intercept Japanese radio messages, including some sent in Kana Morse.

In 1925 the GCCS through the RN sought his release to enhance its China Squadron program of serious interception and study of IJN wireless traffic from shipboard bases in the China Sea. The RN obtained an extension of his secondment in 1927 and moved him to the GCCS in London, where he worked on both diplomatic naval attaché traffic and on building up the then current IJN code book (known to the USN as the Red Book). Nave was persuaded to transfer to the RN in 1930, and during the 1930s worked in London and later with the FECB in Hong Kong, examining a number of Japanese codes (including the new IJN code introduced in December 1930 and named the Blue Book by the USN) and ciphers. He was a key figure in the GCCS attack on the Japanese diplomatic traffic relevant to the 1930 London naval conference. His conclusion from this episode was that, henceforth, Japan would consider the USA as its major naval opponent and would focus its naval strategy on neutralising any threat from that source.

Nave's health deteriorated in the late 1930s and he left FECB in Singapore in February 1940 for sick leave in Adelaide. He was advised never to return to the tropics and spent the rest of his life in Australia, despite attempts by the GCCS to recall him to London. This was fortunate for Australian Sigint development in 1940–1941 and for the whole intelligence enterprise in the Pacific during WW2.

His work in setting up a Special Intelligence Bureau in Melbourne in 1941–1942 is described in Sect. 6.4. Nave gave training courses in cryptography in 1942. He also worked on Japanese air-ground communications. Coding and encryption done in aircraft had to be relatively simple.

The American naval Comint unit Cast was evacuated from the Philippines to Melbourne in three tranches. It was in operation there by March 1942. Its commander, Rudolph Fabian, had an early and serious falling-out with Nave, apparently because Nave was unfairly blamed for a serious security breach (see Sect. 19.6). Full control of this unit was transferred to the USN in November 1942 and at this time Nave was excluded from it. As his value was well appreciated by the Australian and American army authorities, he was transferred to Central Bureau (Chap. 17) and continued his essential work on air-ground codes there. Nave was the only Australian continuously involved in Sigint work against the Japanese from the 1920s until the end of the Pacific War. He was awarded an OBE for his various achievements. After the war he was transferred to the Australian Security and Intelligence Organisation.

3.4 Japanese Naval Attaché Ciphers

Upon arrival at GCCS in January 1928, Nave learnt that his main task was to decipher and translate those telegrams to and from Admiralty Tokyo to Japanese naval attachés in Europe that had 'passed through British hands'. This phrase referred to copies of such cables that had been sent via a cable station in London or elsewhere in the British Empire, as in WW1. A good reason for this would have been the coming 1930 London Naval Conference. Similarly acquired intelligence had been of value to the future Allies at the 1922 Washington Conference.

Nave noted that these messages were encoded, and that he had to build up the code book. Evidently this presented little challenge to him because he was quickly able to decipher and translate these messages on a daily basis and forward them to the Intelligence section. He must have found his years of work on this task well rewarded at the time of the conference. He recorded in reminiscences preserved by the Australian War Memorial:

> However I came upon one prize, a long telegram in many parts explaining to the Naval Attaché and the Conference Delegates why Japan was satisfied with the ratio of 3 against 5 for the United States, but must insist on maintaining that balance. Their plan was that when the United States Fleet set off across the Pacific on the warlike operations, the ships would be harassed by Japanese submarines en route suffering some losses. On reaching the Western Pacific, it would be short of fuel and water and other supplies, their crews war-weary and thus the Japanese would be satisfied with their ration of three fifths. This message created quite a sensation as well it might and I received many congratulations. I might add that it took days of sustained research and help from my colleagues Hobart-Hampden and Parlett, and there were many new groups added to my new Japanese Naval Attachés code book as a result.
>
> However, for Japan to sign a treaty accepting a ratio 5:5:3 of capital ships vis-a-vis America and Britain meant this was not only a naval but official government decision based on the strategy set out in this telegram. This was valuable intelligence indeed as it gave a picture of naval strategy for war in the Pacific.

Prados, in Chapter IV of *Combined Fleet Decoded*, describes this decision as key to the evolution and implementation of the IJN's strategy of the 'Decisive Battle' that was to influence strongly its actions in the coming war. The text also gives hints of the illusion that the main strategic role of submarines was attacking naval vessels. Much more on this is given in Chap. 16.

Nave continued to record various changes to the JNA code, noting in particular that Hugh Foss had solved a system change in 1933. His next comment is of considerable interest:

> We had another change of coding system; this time it took long messages to get repeats (which were what cryptographers looked for to solve the system). The basic work was done by Strachey and Hugh Foss. The latter, a delightful character, took to growing a beard during the war and, when loaned to the Americans, his tall slim figure became known as 'Lend Lease Jesus'. This was our first experience of machines: there were two wheels, one containing 20 consonants and the other six vowels, Y being a vowel. The first trial was made in the office using a brown foolscap folder cover with a collar-stud retrieved from a returning laundry parcel, a piece of string and slots cut in the cover for the letters. This worked, so we asked the Signal School at Portsmouth to help and received some expertly

finished models in bakelite. To find the starting point or key the Japanese language proved a good start. Japanese is not constructed like most European languages on a consonant-vowel basis, but is monosyllabic. (Yo ko ha ma is four syllables.) The word *oyobi* was a great help, giving a sequence of three vowels and at times four on the smaller wheel. Then when one of these vowels represented itself we could have a zero position necessary to place the enciphered text under each other in sections of sixty symbols to establish the order of letters.

Foss and Oliver Strachey applied knowledge gained from solving this kana-based machine in the 1934 attack on a similarly designed new Romaji-based diplomatic code called the Type A machine by the Japanese. By August 1935 GCCS was using a locally constructed device, the J-machine, which decoded intercepts sent via a Type A machine. The naval attaché machine was solved by the USN in 1935 and named the Orange or M-1 machine, while the US Army solved the Type A machine early in 1936.

The later work on JNA codes mentioned in this extract refers to joint British-American work on CORAL, a machine cipher related to the Purple diplomatic machine broken by the Americans (see Sect. 4.2).

3.5   John Tiltman

An excellent biography of Brigadier John Tiltman by Ralph Erskine and Peter Freeman appeared in *Cryptologia* 27(4), October 2003, 289–318. This book focuses on his remarkable contributions to the breaking of Japanese codes.

John Tiltman was born in Scotland in 1894 and joined the British Army in 1914. He was wounded in France. From 1921 to 1929 he served in the Army of (British) India as a cryptanalyst. He then transferred to GCCS where he was the leading expert on breaking into non-machine encryption systems. He became Deputy Director in 1944 and continued on at GCCS and its successor, GCHQ, until 1964. He then retired to live in the Washington area to be closer to his daughter and took the opportunity to be a consultant with the National Security Agency (NSA) from 1964 to 1980. He died in 1982.

His WW2 achievements include setting up the Bedford[7] organisation that ran 6-month Japanese language courses for talented young students of the classics. Some, such as Hugh Melinsky, were transferred to General MacArthur's Central Bureau (CBB) from 1944 onwards.

Tiltman was a genius at cryptography and totally dedicated to his task. Peter Filby, one of his assistants, recalled that Tiltman was only interested in the work to hand and sometimes persisted into the evening standing at a specially built

---

[7]The entry for 21 February 1942 in the diary of Malcolm Kennedy gives an interesting comment on the special Japanese language course given at Bedford for top students of classical languages. One graduate of the Bedford course, Hugh Melinsky, has published a book *A Code-breakers Tale* (1998) on his war-time experiences. Sue Jarvis has a useful account of the matter in *Captain Oswald Tuck RN and the Bedford Japanese School* (Bletchley Park Reports no. 19, June 2003).

high desk. Leo Marks, on page 194 of his somewhat irreverent memoir *Between Silk and Cyanide*, writes that 'his speed was the equivalent of the one-minute mile'. Marks also recalled that, 12 months before D-day, Tiltman had told an astonished Head of the Special Operations Executive (SOE) to continue to ensure that agents gave absolute priority to cutting telephone wires. This forced the Germans to communicate by radio and gave Bletchley Park a chance to read their messages.

From 1933 Tiltman worked on Japanese military systems. In 1938 he worked on the additive cipher system introduced by the IJA in December 1937. With the help of Patrick Marr-Johnson, Tiltman discovered (see TNA HW 67/3) that this relatively simple cipher system used as book groups only four-digit numbers which were multiples of three. Later he discovered that the IJN cipher introduced in June 1939 and eventually to become known as JN-25A used as book groups only five-digit numbers which were multiples of three.

Although the GCCS had begun to use IBM-type machinery in the early 1930s, it seems that Tiltman, after inspecting it in operation, felt that his own work was better done without it. However he was to make effective use of it by the late 1930s. The GCCS dramatically increased its use during WW2.

Tiltman had trained Marr-Johnson, later head cryptanalyst at the Delhi Wireless Experimental Centre (WEC), and Geoffrey Stevens. The latter was a significant figure at FECB who later returned to Bletchley Park.

Tiltman received various honours, including the British CMG and the American Legion of Merit. He is currently the only non-American to have his name listed in the Hall of Honor set up by the NSA.

3.6 Movement in 1939

The range of activities and volume of traffic studied at GCCS increased through the 1930s, and its staff complement had also grown from an initial 50–125 at the time of the German invasion of Poland on 1 September 1939. The GCCS had just moved most of its operations to Bletchley Park, a site procured by Admiral Sinclair in 1938, reputedly at his own expense. The academics previously recruited by him, including the mathematicians Alan Turing, Peter Twinn and Gordon Welchman, arrived soon after.

Concern over the location of the FECB in Hong Kong being too close to Japanese military incursions into China led to its transfer to Singapore in 1939. There, together with other intelligence and security activities, it continued to work on diplomatic, naval and military traffic. GCCS encouraged FECB to focus on the opportunities provided by the special structure of the JN-25A cipher system.

This was a most significant action, because the weaknesses in the construction and use of the JN-25 systems and their operational importance after the Pearl Harbor raid made these systems the single most valuable source of Sigint in the Pacific War, particularly in its early years. The work done by GCCS and FECB on these systems in 1939–1941 was crucial to the ultimate Allied intelligence victory. Even though a full explanation of the jargon and certain technical parts of the following account by Nave is not given until later, it seems best inserted here:

> [At FECB Singapore from about October 1939.] The new 5-figure cypher the Japanese Navy had introduced became known as JN-25. The messages were reciphered with random numerals in pads of 100 pages. With these systems the indicators as to page and line used are in the first two groups with at times checks at the last two groups. The breaking was a well reasoned matter. We plotted the page indicators and thus obtained true page and line, and then arranged the coded messages accordingly to give us depth. We knew sufficient about the messages to work out the subtractor groups applicable to make sense in each message, and we were constantly building up the pad of fresh indicators. These subtractor numerals applied to the transmitted message gave a clear text of groups that valued at three: every true group was a multiple of three. The numbers, dates and so on were in sequence and with two solved you could write the whole set. 00303, 00306, 00309 and so on. In the early stages it was easy and we built the book at a terrific rate, which was good. At the first change [to JN-25B] they brought in a hatted book which prolonged solution as it was the main naval cipher [sic]. However the very important signals such as new landing operations continued to come in the C-in-C's transposed 5-letter book, and I read these personally.

Nave's text correctly suggests that the various JN-25 systems could be and mostly were successfully attacked and that the intelligence supplied was a factor in determining the outcome of WW2.

3.7   Alan Turing

Alan Turing (1912–1954) is by far the best-known of the WW2 cryptanalysts and arguably the key initiator of computer science. He showed remarkable mathematical and scientific aptitude in his primary and secondary education and graduated in mathematics from Cambridge University, gaining a teaching position at King's College in 1935. He then worked on the problem of limitations to what can be done with mechanical calculation processes, now generically called algorithms. For this he needed to work out a rigorous concept of what an algorithm was. This led to his paper *On Computable Numbers, with an Application to the Entscheidungsproblem* (1936). An (apparently) quite different approach to this problem was published in 1936 by Alonzo Church of Princeton University. Turing's work is now considered more natural but both papers have become classics. Turing went on to obtain a PhD degree at Princeton. He appears to have established some contact with the GCCS before September 1939. The overwhelming probability is that Professor F. E. Adcock of King's College arranged this contact. Adcock had been involved in WW1 cryptanalysis and is known to have been a *de facto* recruiter for GCCS in WW2. As Denniston wrote later:

> At certain universities, however, there were men now in senior positions who had worked in our ranks during 1914–18. These men knew the type required. Thus it fell out that our most successful recruiting was from these universities.

Turing's mathematical work concerned the general theory of computation, and may have inspired some later work (1938–1939) on developing a special-purpose calculating device. But Turing was much more than a theoretician. He was deeply interested in matters such as the practical applications of punched cards, what special purpose machines could do to assist cryptology and how to improve the performance of the 'bombe' machines that attacked naval Enigma. At Bletchley he wrote the remarkable *Treatise on the Enigma* (TNA item HW 25/3 and accessible on the web). It incorporates work of 'Dilly' Knox.

Turing also contributed some extremely useful statistical techniques to the work of Bletchley Park. He may have learnt of the utility of what are now called Bayesian methods while still at Cambridge University. Indeed Cambridge Professor Harold Jeffreys (1891–1989) was a strong proponent of them, despite the then ongoing controversy about their validity. Jeffreys' book *The Theory of Probability* promoted such methods. I. Jack Good (1916–2009), his assistant on statistical matters for much of WW2, published a paper *A. M. Turing's Statistical Work in World War II* (*Biometrika* 66(2) (1979) 393–396) describing Turing's input and explained terminology such as decibans, Bayes factor and sequential analysis. This work was quite ahead of its time. Turing's 1941 account was released in 2013 as TNA items HW 25/37 and HW 25/38 and is explained in Appendices 2 and 3 of Chap. 10. It is the basis of the 'Hall weights' alignment method of Sect. 15.3.

Another quite remarkable document, written by Turing in December 1942 or shortly thereafter, is headed *Report on a Visit to the NCR Factory at Dayton, Ohio.* Most of this report deals with submarine Enigma, and shows that Turing had a keen interest in the 'bombes' built to break into it. However the following extract[8] is of enormous relevance to this book:

> SUBTRACTOR MACHINE. At Dayton we also saw a machine for aiding one in the recovery of subtractor groups when messages have been set in depth. It enables one to set up all the cipher groups in a column of the material, and to add subtractor groups to them all simultaneously. By having the digits coloured white, red or blue according to the remainders they leave on division by 3 it is possible to check quickly whether the resulting book groups have digits adding up to a multiple of 3 as they should with the cipher to which they will apply it most. A rather similar machine was made by Letchworth for us in early 1940, and, although not nearly so convenient as this model, has been used quite a lot I believe.

---

[8]The full text is printed in the journal *Cryptologia*, 25(1), January 2001, 1–17 together with an article by former NARA researcher Lee Gladwin on the discovery. It may be found via Google by asking for Turing + Dayton. DeBrosse and Burke on page 95 of *The Secret in Building 26* mention a group photograph circa July 1943 showing Turing back in the Naval Computing Laboratory in Dayton. There may have been other visits. The word 'depth' is explained in Sect. 9.3.

Andrew Hodges has written a biography *Alan Turing: The Enigma* (1983). His 2003 paper *The Military Uses of Alan Turing* is very relevant to the material in this book. The November 2006 issue of the *Notices of the American Mathematical Society* gives much space to a tribute to the non-military work of Turing. There is a good 'novelisation' of his life by Janna Levi: *A Madman Dreams of Turing Machines*, Orion, 2008.

It will be noted that Turing did not follow modern usage[9] in distinguishing between a 'code' and a 'cipher'. Letchworth was the site of the BTM factory which did much work for Bletchley Park. The quoted paragraph is the clearest (and possibly only) evidence[10] that Tiltman asked the mathematicians how best to exploit the multiples of three phenomenon he had discovered in JN-25A.

3.8   Growth of GCCS

After the outbreak of war with Germany, staff numbers at Bletchley Park grew rapidly and continued to do so, being over 3,000 by the end of 1942 and nearly 9,000 at the start of 1945. The number of true code breakers remained quite small in all the major WW2 Sigint centres. The expeditious and effective production, analysis and transmission of the total information derived from message intercepts and other intelligence work required most of the extra staff. Further increases in equipment and personnel also arose from the need to expand the number of interception and direction-finding wireless stations to improve both the quantity of useful message data available and the speed and accuracy of message source location. Finally, there was a steady expansion in the use of data processing machinery, supplied by BTM. The innovative and specialist devices designed for specific tasks (such as the bombes and other machines made for the work on the Enigma ciphers), required recruitment and training of ever larger numbers of operating and maintenance personnel. The FECB had some card processing equipment supplied to it late in 1941 but appears not to have made it operational before the unit was evacuated from Singapore in 1942.

---

[9]The word 'cipher' was written 'cypher' up to the end of WW2. This book uses the modern spelling except where original text is being quoted.

[10]Tiltman wrote some *Reminiscences* apparently for a privately circulated NSA journal. Fortunately a copy has survived in Box 1417, file 4632 of RG457 in NARA College Park. He mentions there that although he was not a mathematician he had never been afraid to seek professional advice from these little understood people. This may allude to approaching Turing and his colleagues about JN-25.

Martin Campbell-Kelly on page 111 of *ICL: Technical History*, OUP 1989, states that in early 1940 the BTM works at Letchworth had very considerable machine tool capacity.

Mavis Batey's book *Dilly: The Man Who Broke Enigmas* (page 96) gives some information on what Turing was doing between September 1939 and February 1940. He was working in the section headed by 'Dilly' Knox on various Enigma systems. Turing was investigating possible mechanisation of the process, and so in contact with BTM.

3.9   Combined Bureau Middle East

Another external station for GCCS was established at Heliopolis, near Cairo in Egypt, late in 1940. It was called the Combined Bureau Middle East (CBME) and had an adjacent interception facility. A small number of Australian service personnel received training in radio interception work there during 1941 and were extremely useful later when their units were recalled to Australia after the outbreak of the Pacific War. One highly perceptive Army officer, 'Mic' Sandford,[11] was able to observe and learn quite a lot about signals intelligence. John Rogers, a WW1 veteran and later in Australia the Director of Military Intelligence (DMI), was also in a good position to observe and learn.

3.10   Kamer 14

Prior to WW2, Indonesia was a Dutch colony generally called the Netherlands East Indies (NEI). The Dutch Army headquarters there was located in Bandoeng (now Bandung) in central Java. A Sigint unit called Kamer 14 was set up there around 1933, with a staff of about three. Extra people were recruited as the Pacific War approached.

Rudolph Fabian of the USN cryptological team Cast in Corregidor was evacuated to Bandoeng briefly in February 1942. In a report of June 1942 to Commander Safford in Washington, he described the Kamer 14 team as being very good at DF and fairly good at radio intelligence, that is TA. Its success in breaking military codes and ciphers was limited.

Kamer 14 had been reasonably successful in breaking some diplomatic codes. Around December 1940 there was some agreement[12] on co-operation with the FECB and the Australian Naval Intelligence Department (NID). Thus to some extent the work of Kamer 14 contributed to the diplomatic code group that was led by Nave in Melbourne until November 1942. Its later re-organisation is discussed in Sect. 18.8.

---

[11]Geoffrey Ballard's excellent book *On Ultra Active Service* about Central Bureau does not explain the depth of experience the British Army had in Sigint in 1940–1941. However it explains the key role played by Sandford in gaining expertise that was to prove so useful in Central Bureau. Sandford's talents had been recognised quite early.

[12]The British Archives (TNA) items HW 14/16 and HW 67/2 give some information on the dealings between Kamer 14 and FECB. In particular, the latter reveals that in December 1940 Kamer 14 handed over all the information it had on diplomatic codes. NAA Melbourne item MP1185/8 1937/2/415 records a co-operation agreement between Kamer 14 and the RAN on signals intelligence. As noted in Sect. 6.1, Nave arranged for Professor Room and Lieutenant Jamieson of the Melbourne Special Intelligence Bureau (SIB) to visit Kamer 14 in September 1941.

The RAAF liaison officer in Bandoeng wrote a report on 13 December 1941, just five days after the raid on Pearl Harbor, describing the place as 'absolute inefficiency and chaos' (Canberra NAA A1196 37/501/52 part 2). Kamer 14 was better than this.

At least some of the Kamer 14 people appear to have reached Australia ahead of the Japanese capture of Java. Their contribution after February 1942 is rather obscure. A few Dutch submarines operated out of Australia.

3.11   Japanese Army Ciphers

A report[13] from FECB of 29 January 1941 in the British Archives states:

> Marr-Johnson is very despondent about the Military codes. Stevens has tried everything with the main cypher, without a glimmer of hope, and is having a rest from it for a little while.

An adjacent file explains that up to March 1937 Japanese Army material received at FECB was sent back to London. In March 1937, John Tiltman was sent to Hong Kong to get FECB started on this task. He remained until September 1937 and returned to comment that he had 'found no-one at FECB capable of handling Japanese Army ciphers of increasing complexity'. In the (northern) winter of 1937–1938, the GCCS Military section initiated a Japanese sub-section led by Patrick Marr-Johnson and Geoffrey Stevens—both of whom went on to FECB early in 1939. Starting on 1 December 1937, a new Japanese Army high-grade cipher was introduced which by the following (northern) Spring had replaced all the earlier types. This had a 4-digit code book and additive tables. The first break came at GCCS in September 1938. In January 1939 Tiltman flew to Hong Kong, and handed the material on this system to Marr-Johnson and Stevens.

At the FECB in 1941 Marr-Johnson would have seen some of his colleagues working on JN-25B and making real progress courtesy of major blunders by the IJN. He would have been aware that one of these blunders (the use of only multiples of three as book groups) had been observed in the Army system broken by Tiltman in 1938. When FECB staff left Singapore in January 1942, Marr-Johnson went to the British Wireless Experimental Centre (WEC) in Delhi, which became the focus for British work on Japanese Army codes. In a report dated 12 March 1942 he commented that the Allies should maintain touch with Japanese Army mainline ciphers. He made the prescient comment:

> There is always the possibility that the present system may be changed to one that is easier from our point of view, for I am by no means convinced that the Japanese (Army) ciphers are constructed by cryptographers. If they were to have a scare as to their cipher security they might well change to a less secure system.

---

[13]This is TNA HW 67/2, 29 January 1941, Shaw (FECB) to Denniston (GCCS). The 'adjacent file' is HW 67/5, circa 1941: *Notes on Military Anti-Japanese 'Y' Policy*. The Marr-Johnson report is in TNA WO 208/5074. Attacks on Japanese Army 'mainline' systems other than the Water Transport Code produced little Comint until much material was captured at Sio, New Guinea, in January 1944.

This came to pass when there was a switch to the Water Transport Code in December 1942.

The following text is inscribed now on a plaque in the mansion at Bletchley Park. It commemorates the achievement of Knox, Tiltman, Turing, Tutte and their colleagues:

Bletchley Park was occupied by the HQ of
Britain's cryptanalytic and signals intelligence
organisation, the Government Code and Cypher
School, between August 1939 and March 1946.
Here some of the best brains of Britain were
pitted against the enemy's enciphered
communications during the Second World War.
Their success forged for Britain a decisively
powerful intelligence weapon which saved
countless lives and helped significantly
to shorten the war.
The King hath note of all that they intend
by interception which they dream not of.
Henry V Act II Scene II.

Chapter 4
William Friedman and the US Army

William Friedman (1891–1969) was introduced to ciphers in an unexpected way. As a young man he developed an interest in genetics and studied this field at Cornell University from 1911 to 1915. He was then put in contact with George Fabyan, a wealthy man with many interests. Friedman commenced work in the genetics section of Fabyan's Riverbank Laboratories on his farm in Illinois, set up with the aim of improving some of the farm's products. Fabyan had also brought Elizabeth Gallup to these Laboratories, with financial support to assist her attempts to prove that the real author of the body of works attributed to William Shakespeare was in fact Francis Bacon. Fabyan's interest in this was to provide Friedman with a new career.

4.1   From Riverbank to Washington 1917–1930

Friedman[1] became involved in the work of Fabyan's 'cipher' section in 1916, perhaps because Elizebeth Smith, a young English literature graduate, had just then been recruited to it by Fabyan to work with Mrs Gallup. Smith and Friedman married in 1917 and he was appointed section head in that year.

The work of this section became known and various US Government agencies began to seek its assistance in decoding encrypted messages soon after America entered WW1. Friedman's achievements in this led in particular to successful prosecution of a dissident Indian Hindu group which was operating, with some

---

[1]Ronald W. Clark, *The Man who Broke Purple: The Life of the World's Greatest Cryptologist, Colonel William F. Friedman*. It is hard to dispute Clark's thesis that Friedman was a true genius. Indeed, more on this is to be found in Chap. 20. The website www.nsa.gov gives access to the publications of the history unit of the modern National Security Agency: these include material on the Friedmans.

© Springer International Publishing Switzerland 2014
P. Donovan, J. Mack, *Code Breaking in the Pacific*,
DOI 10.1007/978-3-319-08278-3_4

German support, within the USA. As his reputation as a code breaker grew so did demand for his services, especially as a failsafe tester of proposed new encryption systems.[2] To much chagrin, Friedman regularly showed these systems to be insecure.

Additionally, Friedman was asked in 1917 to teach some Army officers the rudiments of cryptology. In order to do this, and probably also to assist himself in setting down his understanding of various aspects of the subject, Friedman began to write a series of technical monographs. Initially these were published anonymously under the aegis of Riverbank. In 1918, he was commissioned as an officer in the US Army and served in France for 5 months in the codes section of the headquarters of the American Expeditionary Force there. He returned to Riverbank in 1919.

A *Final Report of the Radio Intelligence Section, General Headquarters, American Expeditionary Force, 1918–1919*, prepared by Colonel Frank Moorman, summarised the achievements of the WW1 group and then argued for the establishment of a permanent Army code and cipher cadre. In fact, throughout most of the 1920s, this cadre had a professional staff of one, William Friedman. He did have some secretarial support. The Friedmans left Riverbank and moved to Washington in January 1921.

Friedman's most important early work was *The Index of Coincidence and its Applications in Cryptography* (1922) which made explicit systematic use, for the first time, of statistical methods in cryptanalysis and forever changed the relationship between this field and statistical/mathematical knowledge.[3] Chapter 15 of the present book, entitled *The Scanning Distribution*, describes an intellectual

---

[2]Pages 40 and 51–52 of Clark's book describe Friedman's encounter with the Vernam enciphered teleprinter mentioned in Sect. 8.5.

[3]An early indication of the sophistication of the statistics underlying WW2 cryptology is to be found in Jack Good's paper *The Population Frequencies of Species and the Estimation of Population Parameters* published in *Biometrika* 40(3) (1953) 237–264 and repeatedly cited ever since. Good had worked with Alan Turing at Bletchley Park: 'The formula (2) was first suggested to me, together with an intuitive demonstration, by Dr. A. M. Turing several years ago.'

The GCHQ released in April 2013 two reports by Turing. They are in the British National Archives:

• HW 25/37 *Report on the Applications of Probability to Cryptography*, and
• HW 25/38 *Paper on the Statistics of Repetitions*.

Much of HW 25/37 is at least implicit in TNA files HW 25/4 and HW 25/5, which are available on line and contain the 1945 *General Report on Tunny*.

The index of coincidence now seems to be known as the Simpson index. This is an outcome of the paper published in *Nature* in 1949 by the British statistician and later civil servant Edward Simpson. Simpson was employed at Bletchley Park from 1943 to 1945, initially on Italian communications and later on the JN-25 series of IJN ciphers.

descendant of this work. One piece of the complex WW2 cryptological jargon[4] is 'i.c.', meaning 'index of coincidence'. It is now usually called the Simpson index.

From 1921 onwards Friedman served the US military as a major figure in cryptanalysis and generally in Sigint. (He invented the word 'cryptanalysis', meaning the art of decrypting enciphered material.) Because his focus was initially on the preparation of coding systems for the US Army, he had to investigate all proposals for such systems put to the War Department. He was responsible also for inducting officers[5] into the area of military codes and ciphers. His 1924 Army textbook[6] *Elements of Cryptanalysis* provided the first systematic analysis of the types of codes and ciphers then known, and of their methods of production and decryption. Documentation of various technical and pedagogical material in cryptology was compiled by Friedman and his students over the following 35 years. These were made available to the USN. He also wrote a useful survey of the various commercial codes that had flourished since the invention of the electric telegraph.

In 1929, a decision taken elsewhere in Washington was to have an impact on Friedman's future career. This concerned the cryptanalytic activities of a contemporary, Herbert Yardley.

---

[4]Researching Pacific War Sigint requires the collection of a list of jargon. Some of the more significant items were: cages, click, conversion square, Copek, decheesing, dendai, depth, Dexter, discriminant, dragon, E.5., eclectic, false addition, females, Fish, flag, flash, forcing, garble table, GYP-1, Hypo, indicator, kiss, Kodak, krack, NC4, Negat, Nocke, null, REB, recorders 'W', scanned, stripping, Susan, Thumb, Ultra and Yoke. The function of the 'Y' Hut at Harman radio reception station near Canberra should be clear. The 'GT' room in the building used for naval Sigint in Melbourne in 1944 is harder, but eventually one works out that it was named after the Op-20-GT section of the USN unit Op-20-G and so it must have been used for traffic analysis. A photograph of it is in the Canberra NAA as part of digitalised item A10909 3 as well as in the NARA RG457 (Historic Cryptographic Collection) material at the end of SRH-275.

In section 8 of Part A of the *CBTR*, quoted in Sect. 17.8, the use of 'BJ' or 'UBJ' for 'report' may confuse. Presumably this practice is derived from the earlier British practice of encasing diplomatic Sigint in blue jackets.

[5]In general, Army officers needed to know something about traffic analysis and also the basics of elementary encryption systems, such as the Playfair method. Dorothy Sayers' novel *Have his Carcase* gives an account of a Playfair code being broken. See NARA RG457 Box 936 Item 2699 for the more secure double Playfair cipher.

[6]This is now available as eight (C–11, C–40, C–42, C–43, C–44, C–45, C–60 and C–61) of the extremely useful Aegean Park Press series of booklets on cryptology. Another (C–65) in this series contains Safford's SRH-149 report which states that the USN received and used Friedman's 'instructional literature' (pages 7 and 17). Those with experience in teaching mathematics know the value of a bit of written material. Friedman published several notes in the *American Mathematical Monthly* between the wars. His student Abraham Sinkov contributed a chapter to the 1939 revised edition of the classic Rouse Ball book *Mathematical Recreations and Essays* and much later wrote a book *Elementary Cryptanalysis*. This was updated and republished in 2009.

4.2 American Black Chamber and Diplomatic Codes

Herbert Yardley[7] (1889–1958) had started his quite remarkable career in codes and ciphers as a coding clerk for the American State Department before WW1. He had advanced his standing by breaking some of the primitive codes of the era. These had become much more significant with the onset of the war.

Soon after the USA entered WW1 in 1917, Yardley persuaded the War Department to set up a cryptanalytic unit in Washington. One of its tasks was to train cryptanalysts for the American Expeditionary Force in Europe. Late in 1918 Yardley went to Europe to seek information from Allied cryptanalytical units. On his return, he gained support from both State and War Departments to set up in 1919 a new code breaking unit, which became known as the American Black Chamber.

A major task for the unit was the development of a good knowledge of the Japanese language in order to attack the then current Japanese codes and ciphers. This paid off handsomely when the Japanese diplomatic codes in use at the time of the 1921 Washington Naval Conference were able to be read. The derived intelligence greatly helped the American negotiators.

The unit's funding and importance dwindled during the 1920s. In 1929, Secretary of State Henry Stimson (1867–1950), appointed by the Republican President Hoover, received translated intercepts from Yardley, who must have been fishing for more funding. Stimson then made the most celebrated utterance concerning twentieth century cryptology:

> Gentlemen do not read each other's mail.

So Yardley's unit was closed down in 1929 and, unknown to him, its files were transferred to the care of William Friedman, the cryptanalyst for the US Army.

This story had three remarkable sequels.

Yardley's 1931 book, *The American Black Chamber*, told the story of the interception and reading of messages between Tokyo and the Japanese delegation to the 1921 Washington Naval Conference. Its translation was popular in Japan. Moreover, it undoubtedly influenced the Japanese into developing a more secure,[8] cipher-based system for their diplomatic correspondence. The book upset Friedman for two reasons. He felt that it had unfairly indicted the work of the American Army Signals Intelligence group in WW1. He also felt strongly that the revelations in the book would be prejudicial to national security.

The second sequel was that the Democrat President Roosevelt, who took office in 1933, recognised Stimson's great experience and steadfast public opposition to Japanese expansion. Roosevelt appointed Stimson Secretary for War in 1940, 29 years after starting a first term in that office. Stimson then oversaw a major growth in reading the mail of other gentlemen.

---

[7]This book can spare little space for Yardley. However David Kahn has written a biography, *The Reader of Gentlemen's Mail*, Yale University Press, 2004.

[8]But not secure enough! See Chap. 8.

The third sequel was Yardley's brief tenure in 1941 as head of the new Canadian cryptologic organisation, the Examination Unit. His book made him unacceptable to Alastair Denniston, head of the GCCS, because of the links between GCCS and other cryptologic units in the British Empire and the USA. Yardley's 6-month contract was not renewed.

(The use of either of the phrases 'British Empire' and 'British Commonwealth' in reference to the 1930s tends to be anachronistic. An evolution from the former to the latter was under way.)

4.3   The Signals Intelligence Service

After the State Department disbanded Yardley's cryptanalytic unit, its files were transferred to the Signal Corps. Friedman was appointed director of a new entity, the Signals Intelligence Service[9] (SIS). Finance was provided to expand the unit. Its broad mandate covered both code construction and code decryption, in operational and research senses. No doubt much of the credit for this development was due to J. O. Mauborgne, a veteran of Sigint in France in 1917–1918 and by then head of the US Army Signals Corps.

By 1931, Japan was becoming a threat to peace in the Pacific, Germany looked unstable and Italy under the Fascists posed another threat. Friedman was able to recruit and train three appropriate people with substantial mathematical and statistical skills. These were Frank Rowlett, Abraham Sinkov and Solomon Kullback. As attempts to find someone with both mathematical aptitude and a knowledge of the Japanese language failed, John Hurt was chosen for his basically self-taught capability in that language. The expertise in all aspects of cryptography and cryptanalysis that would be so important in WW2 was being built up.

Friedman's creativity was also directed towards making a reasonably secure cipher machine. Eventually in 1933 US Patent 6,097,812 was issued for a machine design which, for security reasons, was kept secret by the NSA until 2000:

> This invention relates to cryptographic systems and has for its object the provision of means for automatically and continuously changing the cipher equivalents representing plain-text characters so as to prevent any periodicity in this relationship.
>
> Another object of the invention is the improvement of existing cryptographs employing a series of juxtaposed, rotatable, connection-changing mechanisms which provide an enormous number of alternative paths for the passage of an electric current corresponding

---

[9]The SIS later became the Signals Security Agency (SSA) based at AHS. The first name survives in the somewhat rare book: *Special Intelligence Service in the Far East 1942–1946: An Historical and Pictorial Record*, published by the S.I.S. Record Association, New York, 1946. This gives a pleasing account of Central Bureau, Brisbane (CBB) (see Chap. 17) without revealing what its functions, if any, actually were. However the reader is told that 'Actually most sections combined the aspects of such a laboratory, a grammar school and the Ford assembly line. The work often provided the sort of thrill that is known only to scholars and labyrinth makers. Independent and individualist characters of several nationalities were always around to provide local color.'

to a message character, from the transmitting contacts of the keyboard to the indicating elements of the recording mechanism.

A further object is to provide means for the irregular and permutative displacements of the members of a set of circuit changing mechanisms so as to eliminate any predictable factors in the movements of the circuit changing mechanisms with the result that unauthorized persons, even though they may possess identical cryptographs, will be unable to decipher messages so enciphered.

A working machine[10] (known in the Army as Sigaba and the USN as ECM) was developed a few years later, with both Rowlett and the USN becoming involved. Because this machine made American high and middle level signals secure, it was as important as breaking enemy communications.

Although the US Army provided appreciably less financial resources[11] to SIS than the USN provided to its corresponding unit, Op-20-G, SIS staff numbers gradually increased. Some interception facilities were developed in Panama, Hawaii and the Philippines. The use of IBM tabulating equipment[12] was introduced. However the SIS budget in the 1930s was always much smaller than that of the corresponding USN unit.

At this stage the SIS had no contact with the British GCCS. It did good work on the early Japanese diplomatic ciphers, including the high level 'type A' machine called Red by the Americans. In 1940 the Friedman team reconstructed the more complex 'type B' machine which remained in use up to 1945. Frank Rowlett played a leading role in this remarkable achievement. The American version of the machine was called Purple. The matter is discussed further in Sect. 8.2.

Initially the processing of Magic—as the important information obtained from Purple was called, was shared between the SIS and Op-20-G. In the second half of 1942 the SIS was given sole responsibility for Magic.

In 1942, Abe Sinkov was seconded to the Sigint unit Central Bureau being set up under General MacArthur in Melbourne. On 30 August 1945, Lt-Col Sandford of CBB reported on the great achievers of this unit. The text has survived in the Canberra NAA as file A6923 16/6/289 *Administration of Central Bureau*, which fortunately may be read via the NAA website.

---

[10]Chapter 9 gives more information on the principal WW2 cipher machines. The full text of the patent may be found by asking Google for Friedman 6097812.

[11]One may speculate that if the USA had instead spent more money on Friedman's unit and less on the naval Op-20-G it would have had a much more difficult time in the Pacific in 1942.

The *CBTR* refers to the call signs on IJN messages being less secure than those on IJA messages.

The relative weakness of IJN communications security is of the greatest importance in the understanding of Allied Comint in the Pacific War.

[12]NAA Melbourne file MP150/1 544/201/217 gives a commercial proposition made to the RAN in 1942 to provide a card-based payroll system at Garden Island naval base in Sydney. Once the cards—not standard IBM cards—were punched and verified the tabulator could process 4,800–9,000 cards per hour. Clearly the significance of civilian use of card-based accountancy systems was not widely understood in Australia in 1942.

First and foremost is Colonel Sinkov ... whose phenomenal technical capacity and his untiring co-operation with the Australian component in all aspects of the work demands special recognition.

The 'phenomenal technical capacity' resulted from great natural ability and ten years working in the environment created by William Friedman.

Solomon Kullback became head of an enlarged Japanese Army cipher unit, which had been moved to Arlington Hall Station (AHS) in Virginia, a former girls school. Its staff grew to many thousands by the end of the War. It used a large machine facility in a basement to go as far as was possible with mechanising the processing of intercepts, through to the stage of providing a plain-language message text based on meanings assigned to identified book groups in the original code. Frank Rowlett headed a separate unit responsible for a wide range of codes and ciphers.

4.4 Elizebeth Friedman

Elizebeth Friedman[13] carried out various cryptological assignments for various government agencies. Although best known for breaking into codes used by some of the 'bootleggers' smuggling alcohol from Canada into the United States of the Prohibition era, she worked in cryptography until well after the end of WW2. In 1957 William and Elizebeth Friedman published the book *The Shakespearean Ciphers Examined*, which showed that the theories of Mrs Gallup and others were not based on any substantial scientific method. The following extract (page 286) provides a fitting conclusion to this chapter:

The Cryptologist must discipline himself to follow certain procedures and to submit to certain checks. Like the experimental scientist he is observing phenomena or occurrences to determine whether they are random or systematic, and if systematic how they work— what principle can be detected in them. ... He is trying to formulate an exact statement about the phenomenon before him. ... Cryptology is an application of scientific method.

---

[13]William and Elizebeth Friedman were each inductees in the initial 1999 batch of eight in the NSA Hall of Honor. The others were Herbert Yardley, Laurance Safford, Frank Rowlett, Abraham Sinkov, Solomon Kullback and Ralph Canine.

Chapter 5
Early American Naval Sigint

Before WW1, the USN had little or no expertise in breaking codes or ciphers. From its alliance with the British Royal Navy in 1917–1918 it learned of the strategic importance of Sigint. It allocated some resources to this activity in the 1920s and greatly expanded it during the 1930s. Plans for a prospective war in the Pacific had existed for many years and considerable preparation had been undertaken. The value of the Sigint component of this preparatory period is shown by the successful collaboration with the British code breakers during 1941 in developing expertise against JN-25 and the remarkable successes of 1942.

5.1 The 1920s and the 1930s

After the Battle of Midway, Laurance Safford,[1] the principal initiator of American naval Comint from about 1922, received a letter[2] from Rudolph Fabian, who was by then commander of the USN group in Melbourne:

---

[1]The primary source on early USN Comint work is the booklet *US Naval Communications Intelligence Activities* by Laurance Safford, J. N. Wenger and at least one unidentifiable author. It contains SRH-149, SRH-150, SRH-151, SRH-152 and SRH-197, all of which can be found in NARA College Park in the RG-457 series. The booklet is number 65 in the Aegean Park Press cryptographic series.

SRH-149 by Safford does not mention the replacement on 1 December 1940 of the original JN-25 code book—called JN-25A—with the second one, JN-25B. Thus, Allied cryptographers had a bare 12 months before the Pearl Harbor raid to determine the entries in the new code book, and their meanings, rather than the two and a half years since the introduction of JN-25A. Also, much less JN-25 traffic was available before the commencement of hostilities. This oversight may have contributed to various theories that at least some senior American naval or political personnel had prior knowledge of the raid on Pearl Harbor 12 months later.

© Springer International Publishing Switzerland 2014
P. Donovan, J. Mack, *Code Breaking in the Pacific*,
DOI 10.1007/978-3-319-08278-3_5

My congratulations for the most remarkable success of the organisation you alone are responsible for. Justifying it through the lean years must have been a job—but I bet they almost eat out of your hand now.

Fabian was not alone[3] in crediting Safford with the later achievements of his organisation, later called Op-20-G. Its evolution is described briefly below.

American naval code breaking had a rather non-intellectual start. The safe of the Japanese Consulate-General in New York was opened by lock-picking in 1922. It was found to contain an IJN code book, which was photographed and returned to the safe. Its contents were translated over a 5-year period by a missionary couple and the final document became known as the Red Book. This had limited use in translating intercepted messages because of a lack of Japanese linguists but it proved a valuable resource by providing information on the items likely to be included in an IJN code book and how these might be arranged in it.

The Red Book was replaced by a new code in December 1930 and this time Op-20-G had only cryptanalysis available to determine code book entries, the encrypting second-stage cipher and its keys. (These terms are discussed fully in Sects. 8.9–8.13.) Within 2 years the system was broken by a team led by Agnes Driscoll with help in data processing from a new unit operated by Thomas Dyer.

Prior to 1924, the USN had relied on Yardley's Cipher Bureau for Sigint information about foreign navies. Dissatisfaction with this arrangement by 1923 had led it to encourage, in an *ad hoc* fashion, some of its shipboard radio officers to intercept IJN messages. In 1924 it established a Research Desk within its Code and Signals Section in the Navy Department in Washington. The name was changed in

---

In general the technical details on how and why ciphers of the JN-25 series were relatively easy to break must have been considered one of the deepest secrets of the war. Analogous details for other enemy cipher systems are also difficult to find. Thus it is not particularly surprising that Safford leaves them out.

Malcolm Burnett's report of 19 January 1943, to be found in TNA file ADM 223/496, deals with co-operation with Op-20-G on cryptanalytic matters. It refers to an 'Appendix B' that 'shows all the major and some of the minor processes involved in the solution of JN-25 at Washington, together with the number of personnel involved'. But Appendix B has been removed from the file copy. Section 9.25 entitled *The Real Secret*, speculates on why Appendix B is missing.

[2]The letter is to be found in the NARA College Park branch in RG38, CNSG Library, Box 19. Fabian notes that the work on JN-25B had kept him too busy for 3 months to make any report at all. Safford was being pushed aside by then.

The bulk of the WW2 signals intelligence records are in the RG457 group at College Park. Some more were found later in the Naval Security Group, Crane, Indiana and moved to NARA in 2000. The Aegean Park Press booklet *NSA Cryptologic Documents* thus does not list the Crane material. However the website www.hnsa.org/doc/nara/ is useful in giving a listing of the Crane material as well as most, but not all, of that in the booklet.

[3]See, for example, Duane Whitlock's *The Silent War against the Japanese Navy* published in *Naval War College Review*, Autumn 1995, vol XLVIII, No 4. The following sentence says it all:

'A vital point that should not be overlooked by historians and students of the war with Japan is the fact that something more than 20 years was required to bring on-line the radio intelligence organization that ultimately gave commanders what was perhaps the greatest strategic and tactical advantage in the history of naval warfare.'

1935 to the Communications Security Group, with other name changes following. It was designated Op-20-G until after the end of WW2. This referred to section G of the naval communications unit (Op-20). Further partitioning followed as the organisation expanded. (See Sect. 18.1.) Thus Op-20-GY was the cryptologic component, Op-20-GYP was the Pacific cryptologic component, and Op-20-GYP-1 was the office working on the principal series of IJN operational ciphers now generically called JN-25. And thus the comprehensive report[4] on procedures used to attack JN-25, as they were in June 1943, was called the *GYP-1 Bible*.

Safford states that the Research Desk had a staff of 7 in 1925. This number increased at the rate of about 10 per year until 1936, when it reached 109. It was at 147 in mid 1940, when rapid growth was authorised, and, by the day of the Pearl Harbor raid, Op-20-G had a staff of 730.

5.2   Agnes Meyer Driscoll

Agnes Meyer, later Mrs Agnes Driscoll, was born in 1889 and graduated from Ohio State University in 1911. She was a successful high school mathematics teacher before enlisting in the USN in 1918. She was initially assigned to the Postal and Cable Censorship Office but was soon transferred to the Code and Signal Section, then responsible for developing codes and ciphers for the USN. After WW1 she rejoined this section as a civilian and spent time both at the Riverbank Laboratory and with Yardley's Cipher Bureau. She then helped develop the 'CM' cipher machine used by the USN in the 1920s. After a brief interlude working with Edward Hebern in an eventually unsuccessful attempt to render his cipher machine secure, she returned to the section as a cryptanalyst at the Research Desk created by Laurance Safford.

Its code-reading effort concentrated on Japanese communications and so Mrs Driscoll's work focused on recovering the regularly changing ciphers and keys used to encrypt code groups from the Red Book. Her contribution to the intelligence gained during the IJN's 1930 fleet exercise was significant. Following her successful involvement with the Blue Book coding system during the 1930s, she managed to break into a new operational code the IJN introduced in 1938, but this lasted less than 12 months. On 1 June 1939 (code changes were usually made on the first day of a month) the 5-digit JN-25A additive cipher was introduced.

Driscoll, assisted by Prescott Currier, had made useful progress on JN-25A by late in 1940, when she was returned to work on German Ciphers. Unlike John Tiltman of the GCCS, they would not have had any contact with research on the Japanese Army cipher introduced in December 1937 and so would have had no reason to anticipate that all its book groups would be multiples of three. Driscoll

---

[4]NARA item RG38, CNSG Library, Boxes 16 & 18, file 3222/65 is *Cryptanalysis of JN-25 (GYP-1 Bible)*. See also Box 116 file 5750/201 *History of Op-20-GY-P*.

and Prescott discovered this almost a year after Tiltman. They must have failed to work out ways to exploit this principal peculiarity of JN-25A. Serious work on the new JN-25B cipher, introduced in December 1940, had to await input from the British.

Various confused accounts of what Driscoll and Currier actually did have appeared in print and on websites. A possible reconstruction is given in Appendix 1 of Chap. 12.

5.3   Advanced DF and Interception Bases in the Pacific

Between 1918 and 1941 the USN steadily developed DF and TA as important aspects of radio interception of (potential) enemy signals. These produced Radio Intelligence in the sense of Sect. 1.13. The number of sites for interception was increased, especially in the Asia-Pacific area, justified as Japanese plans for expansion became more evident. The low-frequency DF network enabled it to monitor various major IJN fleet manoeuvres in the 1930s.

Considerable thought had been put into appropriate sites for interception and processing units. The first was a minor station in Shanghai. Later it was decided to have three major stations in the Pacific. Much less attention was paid to the Atlantic. An interception station was set up on Bainbridge Island (Washington state), just south of Vancouver Island. Another secure base was in Hawaii. By the time of the Pearl Harbor raid, this base, which became known as Hypo, occupied a secure basement in a building in the principal compound of the USN base. This was locally known as the Combat Intelligence Unit. An advance base was in the Philippines. It became known as Cast and was in various sites near Manila.

Plans about a prospective Pacific War against the country usually called Orange—which could mean only one thing—dated back many years. Whereas it had always been thought that Hawaii was safe from invasion, the Philippines, then an American protectorate, was considered to be at considerable risk. The island of Corregidor, at the mouth of Manila Bay, was considered to be a particularly secure fortress for the US Army. In 1938, a bomb-proof tunnel was constructed there for the Cast naval Comint unit. This was occupied well before the outbreak of war. Sadly, the tunnel was destroyed by explosives stored in it when the island was re-captured in 1945.

Of the 730 staff in Op-20-G at the time of the Pearl Harbor raid, around 70 were stationed on Corregidor.

5.4   The October 1940 Instructions

Instructions on 'Orange' cryptanalytical activities were issued by the Acting Chief of Naval Operations in October 1940. Safford incorporated the key parts in SRH-149 (see note 1 again), of which the last paragraph was:

5. It must be borne in mind that the present Orange cryptographic systems may be replaced by new ones immediately on the outbreak of war. Therefore, cryptanalytic intelligence, *per se*, may not be available from that time until after successful attack has been conducted. Meanwhile, enemy information can be obtained from radio intercept and direction finder activities as has been the case during the past year.

It is inappropriate to criticise the lack of concentration on JN-25B in the second half of 1941. It was considered that all codes used by the IJA and IJN were likely to be totally replaced just before any start of hostilities. So working on any particular system might bring in some useful intelligence but otherwise would just amount to a training exercise. If anything, spreading resources over the full range of IJN systems would build up more expertise in cryptanalytical technique.

5.5 Developments in Intelligence Cooperation 1940–1941

Necessity, the mother of invention, is a strong incentive for seeking help. The British, who knew that the European War would totally occupy them for some years, and who were alone (except for Commonwealth support) after the collapse of France, consistently sought ways of involving American help in their struggle. Since the USA was a neutral country pre-Pearl Harbor, all such approaches were conducted with great discretion. Moreover, although President Roosevelt and some of his aides were sympathetic (Roosevelt maintained regular correspondence with Churchill from late 1939 onwards), absolutely no indication of this could be allowed to leak. Roosevelt was running again for the presidency in the November 1940 election at a time when support for an isolationist policy towards Europe remained strong.

The situation in the Pacific was rather different. There was no war there in 1940 and preliminary discussions with possible future allies was possible. Indeed the US Army and Navy started sending junior attachés (not involved in Comint) to Australia around that time.

In May 1940, Op-20-G[5] sent Lieutenant John Lietwiler to Bermuda to inspect some British HF/DF equipment and in July of that year the President sent W. J. ('Wild Bill') Donovan to London to discuss naval intelligence co-operation. Donovan was the head of the predecessor of the CIA. The British DNI, Admiral Godfrey, was in favour of total co-operation.

Progress was made, despite reservations on both sides over the other's capacity to maintain absolute secrecy in intelligence matters. There was a major step forward in August 1940. An American team, led by Rear Admiral Ghormley, met in London

---

[5]NARA item RG457 Box 1123 file 3608 deals with the beginnings of British-US signals intelligence cooperation in 1940–1941. The Lietwiler episode is NARA RG457 SRH-009. John Lietwiler (1908–1978) commanded the Op-20-G unit at Corregidor for part of 1941 and later played an important role in Melbourne as Fabian's deputy at Frumel.

with Admiralty representatives and subsequently were 'observers' at a meeting with the innocuous title of the Anglo-American Standardisation of Arms Committee. A sound basis for future American-British intelligence cooperation was formulated at this meeting.

A follow-up meeting held in Washington later that year produced a further meeting there, over February-March 1941, known as ABC-1. This led to an agreement in principle on far-reaching cooperation between the USA and the UK on all matters relevant to the armed services and their interests in defeating Germany. A particular emphasis was placed on cooperation among the respective naval services, and US liaison officers were also to be exchanged with Australia,[6] Canada and New Zealand. From this time onwards, the importance of naval attachés in diplomatic service was to decline, as direct links were set up between the respective armed services.

The immediate outcome of the August discussions was a move to initiate exchange of Sigint information. On the American side, the US Army was willing to exchange cryptologic (but not cryptographic) knowledge. It had, by October, produced its first Purple machine and several more became available before the end of the year. The British were willing to co-operate fully, except in relation to their work on Enigma traffic. The extreme value of this was seen as requiring complete security.[7] As a result, the Americans sent what has become known as the *Sinkov mission* to Bletchley Park. In reality, it was two missions travelling together in January and February 1941: one from the US Army SIS unit and one from the USN unit Op-20-G. The Army team was Sinkov and Leo Rosen, who had worked on Purple. The USN team comprised Robert Weeks and Prescott Currier, with the latter having worked with Mrs Driscoll on JN-25A. The Americans were shown quite a lot of Bletchley Park and were given documentation of the Enigma. In return, the British were given one[8] Purple machine. Sinkov was able to put experience gained at Bletchley Park to good use at Central Bureau in Brisbane as it expanded during 1943–1944.

---

[6]The NAA file A981 JAP 121 deals with the Japanese Embassy in Australia seeking in March 1941 to introduce military and naval attachés. Permission was not given.

[7]See NARA RG457 A1 9052 Box 940 file 2738 covering co-operation between the SSA and GCCS and TNA HW 14/8 Bletchley Park minutes November 1940.

The NSA website gives access to George F. Howe's *American Signal Intelligence in North-West Africa and Western Europe*, published in 2010 in the United States Cryptologic History series. Its chapter 11 deals with collaboration with the GCCS.

[8]It is possible that the British received a second machine at that time. The GCCS, initially reluctant to inform the Sinkov mission about the Enigma machine, relented rather late in its stay at Bletchley Park. Section 14.2 includes certain paragraphs from Part G of the *Central Bureau Technical Records*. That marked 'A' includes 'A large part of the original work on the Water Transport problem was contributed by CBB, and it was a principal interest for about 18 months, beginning early in 1943'. It is possible, and indeed probable, that Sinkov used knowledge acquired at Bletchley Park about the Polish work on Enigma to help work out the technique for decrypting the Water Transport Code in 1943.

Various published accounts of the Sinkov mission have overlooked the significance of the information about JN-25 acquired from the FECB by the USN.

The Sinkov mission set off shortly after the JN-25A code book had been replaced by the JN-25B. By then, the FECB had used JN-25B5 intercepts to obtain preliminary statistics on the frequency of the common book groups in the JN-25B code book. It had developed efficient techniques for applying this to the new JN-25B6 system. This was exactly the information that was needed to enable Op-20-G to work effectively on that system.

Item HW 67/2 in the British Archives contains a key message of 10 February 1941, a few days after the Sinkov mission reached Bletchley Park, from DNI Godfrey to FECB via the Commander of the British Fleet in Asia, Admiral Phillips:

> IMPORTANT. Your 0844 of 1st January and my 1641 of 5th January. Full exchange of special intelligence material and methods with US authorities should be inaugurated forthwith.

The substance of the matter was recorded by Rudolph Fabian, then a member of the Op-20-G Cast unit in the Philippines and who became commandant of the Frumel unit in Melbourne:

> And we were getting pretty well settled in the tunnel (on Corregidor Island in Manila Bay) when Admiral Sir Tom Phillips—OIC British Eastern Fleet—asked Admiral Hart USN to send a staff down to exchange ideas and get ourselves ready because we were all sure the war was going to come.
> Jeff Dennis was sent to Singapore and brought back a solution of this five number system. (JN-25) When Dennis came back it looked kind of easy to us. So that was when I recommended that we drop everything else to go on to that.
> Jeff Dennis went to FECB from Cast and brought back a solution of the 5-number system. He brought back the solution, how to recover the keys, the daily keys. How the code was made up and a lot of code values (identified book groups).
> And since it was a very heavy volume system, it was the heaviest volume system on the air. I talked with my people and I went back to the CNO and requested permission to drop everything else and go to work on this five number system. And we did pretty well on it. We couldn't do any solid reading but we could pick up phrases like 'enemy submarine'.

This excerpt is taken from an oral history interview with Fabian conducted many years later by the American National Cryptologic Museum. As might be expected, it is somewhat disorganised but the substance comes through. Fabian asked that other units, presumably including that at Pearl Harbor, also concentrate on JN-25. This did not happen before the raid of December 1941. He does not mention the contribution of Cdr Malcolm Burnett of FECB in visiting Cast and helping get it going on JN-25B. This last is noted by another key participant, Wesley 'Ham' Wright, in his oral history (also held by the American National Cryptological Museum):

> The Blue-Book preceded JN-25. In force for about four years. We did pretty good with that. Then they dropped it and went into that numerical thing (JN-25) and we had a hell of a time getting into that. I guess you know we had to call on British help on that one. (On page 38 of his account Wright explains that the exchanges between FECB and Cast on JN-25 were quite valuable.) Malcolm Burnett probably did more on the initial JN-25 (JN-25A) than anyone else, probably more than Aggie (Driscoll). (On page 101) Burnett was a fine linguist who probably did more in breaking JN-25 than anyone else.

Sinkov may not have realised at the time that his mission had opened the way for the USN to obtain essential knowledge about the potential exploitation of JN-25B. This and documentation on the Enigma represented magnificent value for the Purple machine.

Dennis visited Singapore in late February 1941, with the final exchange of information appearing to have come from a visit to Cast at Corregidor by Burnett of FECB in early May. Nave's account cited in Chap. 3 continues:

> This five figure cipher was used by ships and commands throughout the war. FECB flew an officer Malcolm Burnett from Singapore to the US unit at Corregidor to give them the means for breaking and reading this five cipher [digit] messages which they categorised as JN-25. He reported that they were not reading operational messages but relying on their extensive D/F network for information of the Japanese Fleet. This important step enabled Singapore, Corregidor and Honolulu to exchange solutions on a regular basis and greatly speeded up the reading of the important naval traffic. The pads used carried the main Naval traffic giving great depth and therefore had to be changed from time to time. However this presented no great problems with the experience already gained, but a certain delay whilst the change was mastered.

The timing of the exchanges is made apparent by this message[9] of 5 March 1941 from the US Asiatic Fleet to Op-20-G:

> Have received from British following in approximate numbers referring to five numeral system (JN-25B5) effective December to February. Five hundred book values, four thousand subtractor groups, half thousand work sheets with cipher removed and two hundred ninety indicator subtractors for SMS numbers.
>
> Have arranged secure method of exchanging further recoveries by cable.
>
> British employ three officers and 20 clerks on this system alone. They are delaying attack on current cipher until mid-March to accumulate adequate traffic and obtain further book values from preceding period. Due collateral information available here and capacity rapid exchange with English Cavite (the Fabian-Lietwiler team) will assume this system.
>
> As only navy arrangement request Dept forward results to date and technique if considered helpful.

It is clear that the British and the United States started full cooperation in naval Comint in the Pacific just in time. More details on SMS numbers, that is the indicator encryption system then used for JN-25B, are given later.

Whether or not Fabian is correct in claiming credit for switching nearly half of the then USN cryptographic capacity to JN-25B, the fact remains that this was also done just in time and was to be of extreme importance in the early months of the Pacific War.

Thus, Cast joined FECB in working on the principal operational code. The Lietwiler letter reprinted in full in Sect. 10.1 shows that full co-operation did take place.

The USN kept the team at Pearl Harbor working on a difficult code for messages between senior officers. Likewise, once the JN-25B system had been transferred to Cast as its principal responsibility, little work was done in Washington on it.

---

[9]The letter of 5 March 1941 is reproduced photographically by Robert Stinnett in *Day of Deceit*, page 300. The original is in NARA RG38.

By mid-year, there was full collaboration between FECB and Cast, both in terms of exchanging new information on JN-25B and also on exchanging message intercepts. The work of recovering tables of additives was divided up between the two groups. This needed secure communications by radio and these were established. Since Cast was better placed than FECB to intercept JN-25B traffic, this last arrangement was of considerable use to FECB in their attack on this cipher system. Lietwiler's team at Cast made considerable use of a small set of IBM machines, whereas FECB had not been able to utilise similar machinery supplied to them in Singapore. (It seems that this machinery was not used until FECB was relocated to Kilindini in East Africa.)

There was no corresponding cooperation on Japanese Army communications, but at this stage the IJN systems, and particularly JN-25B, were much more important.

As well as the above crucial exchanges of technical information occurring between the Americans and the British code breakers in the Far East, a series of informal intelligence discussions involving American, British, Dutch, Australian and New Zealand personnel, were held late in 1940 and early in 1941. Directed towards establishing better intelligence collaboration in the Far East and Pacific regions, they did result in useful agreements, especially with respect to sharing of naval intelligence.

5.6 Fabian and Friedman on JN-25

In 1996, the National Security Agency brought out a booklet *The Quiet Heroes of the SWP Theater*, edited by Sharon Maneki. The following extract from Fabian's contribution is of considerable interest:

Our business was navy communications. We did everything from beginning to end—decryption, translation and processing. The JN-25 systems that we worked on were very complex. The first thing we had to do was to find the start of the message. Sometimes the beginning of the message would be in the middle of the paragraph or sometimes it would be at the end of the message. We had to find the starting point and then double back. One thing that helped us was that code values were divisible by three.

And so divisibility by three (being multiples of three) turns up again! In the first half of 1942, every piece of assistance was gratefully accepted by the USN cryptanalysts.

Another NSA publication, *The Friedman Legacy: A Tribute to William and Elizebeth Friedman* (reprinted 2006), contains six lectures given to NSA staff by Friedman around 1959. The sixth lecture deals with the period from the end of WW1 to the end of WW2, that is 1919–1945. On page 169 Friedman noted that high level USN communications security in WW2 was quite adequate. Without mentioning

JN-25 explicitly he stated that IJN communications security was quite inadequate[10] and suggested that something like the ambush at Midway in June 1942 was almost inevitable. He commented that the IJN lacked the 'experience and knowledge' to rectify the situation.

---

[10]Chapter 9 identifies the seven principal errors in IJN communications security that were behind the Battle of Midway.

The temptation to draw attention to page 29 of Patrick Beesly's *Room 40* cannot be resisted. Gustav Kleikamp of the German Navy in 1934 wrote a report with the title *Der Einfluss der Funkaufklaerung auf die Seekriegsfuehrung in der Nordsee 1914–1918*. This translates to *The Influence of Radio Intelligence on the Direction of the Naval War in the North Sea, 1914–1918*. Remarkably a copy has survived. It contains the perfectly sound recommendation that the Kriegsmarine (German Navy) should employ experts well in advance of any resumption of war to get its communications security right.

Chapter 6
Developments in Australia

Australia, founded in 1901 as a federation of once self-governing colonies, proceeded to establish its Army and Navy over the next decade. During WW1 the RAN carried out some direction finding activity. But Sigint capability remained low until the late 1930s, when, following strong pressure from Britain, it began to be developed.

6.1 The First Few Steps

In 1917[1] Eric Piesse, the Director of Military Intelligence, recommended that training in Japanese be offered to Army cadets at the Royal Military College (RMC), but that this be disguised by establishing a position in Japanese at the University of Sydney, supported by the Department of Defence. The incumbent lecturer was also required to teach Japanese at the College.

The RMC selected an experienced journalist and Japanese linguist, James Murdoch. He was required to visit Japan regularly in order to gain insight into current Japanese policy directions from his well-placed personal contacts in that country and report this back to the Australian military. Later he recommended to

---

[1] Thus much of the WW2 archival material is stored in Melbourne, with access available upon notice at the NAA, 99 Shiel Street, North Melbourne. In general, there is little pattern in the division of files between the Melbourne NAA, the Canberra NAA and the Australian War Memorial (AWM). At least they all share a common electronic index called *Recordsearch*, accessible through the NAA website.

General reference should be made to the book *Breaking the Codes* by Desmond Ball and David Horner, Allen and Unwin, 1998.

© Springer International Publishing Switzerland 2014
P. Donovan, J. Mack, *Code Breaking in the Pacific*,
DOI 10.1007/978-3-319-08278-3_6

the RAN that Eric Nave[2] be sent to Japan to develop his Japanese. After Murdoch's death in 1921, this arrangement[3] between the university and Defence continued until after the end of WW2.

6.2   The 1919 Jellicoe Report

As First Lord of the Admiralty in the early years of WW1, Lord Jellicoe was, in Churchill's memorable phrase, 'the only man on either side who could lose the war in an afternoon'. Although Jellicoe was eventually eased out of that high office, he had earned immense status and was chosen to examine the need for maintaining the various navies of the British Empire as separate entities. Jellicoe therefore visited Australia, New Zealand and Canada.

His report[4] on the Canadian Navy included the critical phrase:

The war (WW1) has shown the exceeding value of a first-rate naval intelligence organisation.

The section *Intelligence and Coastguard* in Jellicoe's report on the Australian Navy (RAN) advocated the establishment of a Naval Intelligence Department (NID) to deal with activities in the Pacific and Indian Oceans. He also advocated establishment of some sort of coastguard system. He noted:

It is obviously essential, for the purposes of secrecy, that there should be direct communication on intelligence matters between the Admiralty in London and the Navy Office in Melbourne.

He also advocated systematic interception of enemy radio transmissions:

(Direction finding stations) should be erected at convenient positions in the islands and on the mainland of Australia to cover the probable areas of the operation of the enemy, HM

---

[2]Eric Nave is the subject of Sect. 3.3 and the biography (*A Man of Intelligence*, Rosenberg, 2006) by Ian Pfennigwerth. *A Man of Intelligence* complements the earlier book (*The Intrigue Master*, 1995) by Barbara Winter on the wartime DNI, Commander Rupert Long, RAN. Winter makes very clear the importance of Long in the Australian preparation for and effort in WW2.

[3]Jennifer Brewster of Monash University has written a useful paper on *The Teaching of Japanese in Australia 1917–1950*. It is complemented by Colin Funch's *Linguists in Uniform*, 2003.

Canberra NAA file A11093 320/5K5 Part 2 reveals that in May 1944, Roy Booth, the commander of the RAAF component of Central Bureau (Chap. 17), learned that there were four RAF and two RCAF officers in London able to translate Japanese and tried to have them transferred to the RAAF Wireless Units.

Military telegraphic Japanese presented its own problems to potential translators.

[4]Enough of the report on the Canadian Navy is available in the Canberra NAA as A11085 B3C/18. The quotation given in the main text would appear to refer to the Battle of Jutland (Sect. 1.6) and to the confining of the German battleships to the Baltic Sea. Some other material on Jellicoe's visit is in A11085 B3C/7. The first three volumes of the report on the RAN are in the Canberra NAA as CP601/1 BUNDLE 1/3, May–August 1919. The fourth is Melbourne item MP1430/1 NN.

ships and aircraft, it being borne in mind that there must be direct telephonic or telegraphic communication between stations.

(Interception stations) should be installed on the outbreak of war in the direction of probable enemy operations. It is desirable to have special sets made up for this purpose in peace time. This service being of a very secret nature should be entirely under the control and administration of the Intelligence Department.

Some parts of the Jellicoe reports remained more secret than others. The totally secret Part IV states quite explicitly that he believed Japan was the major potential adversary to the British Empire in the Pacific Ocean. Likewise it is clear that the senior Australian military fully agreed with Jellicoe.

Progress on the Jellicoe program was quite slow in the 1920s, but gradually advances were achieved. The British Empire was in fact quite well distributed for the purpose of monitoring Japanese radio. A naval interception station was constructed at the Naden Naval Base in Esquimalt (Vancouver Island, BC) around 1922. The Australian NID was run on a shoestring budget in the 1920s but developed later under Commander Rupert Long.

Although a permanent professional coastguard service in Australia would have been impossibly expensive, it was possible to appoint some part-time coastwatchers. They were given small honorariums and authorised to send telegrams to the NID. This system was later extended to New Guinea and the Solomon Islands where the coastwatchers provided an invaluable supplement to Sigint in the Pacific War.

6.3 Infrastructure in Australia at the End of 1936

The RN and the RAN were in steady communication regarding the uses of wireless interception for obtaining data on the IJN throughout the 1920s and 1930s. The two navies occasionally participated in joint exercises in direction finding and in intercepting messages originating in the Mandated Islands. The net result at the end of 1936 amounted to little improvement within Australia of the infrastructure and personnel needs relevant to interception. There was no activity in the area of message decryption. The latter is not surprising, because the British sought to retain cryptographic expertise within their own GCCS and FECB.

There were, however, two positive developments. Firstly the RAN had maintained a capacity to intercept some Japanese radio traffic for which ongoing training in Kana Morse was needed. Secondly, Australian liaison officers were attached to the FECB.

6.4 Real Expansion: The Period 1937–1941

The Depression of the 1930s had forced the Australian government to slash defence expenditure. The Australian defence forces were able to plan for some expansion from 1937 onwards as more funds became available. By then the UK was placing explicit pressure upon the four Dominions (Australia, Canada, New Zealand, South Africa) to implement effective wireless interception. A meeting in London, in July 1937, was called by the British Director of Military Operations and Intelligence. It was attended by representatives from each Dominion who were informed that the UK had wireless interception in place. This was capable of rapid expansion, but such infrastructure was not the case elsewhere. So:

> It therefore appears to be of the highest importance that the members of the British Commonwealth should consider how best they can gain experience of W/T interception in peace, and prepare schemes for expansion to meet war conditions.

This opinion was later[5] officially conveyed to the Australian government via the British High Commission. But it seems clear that direct RN/RAN exchanges had also occurred, probably via the route advocated by Jellicoe, and that by 1937 planning was at last in place for the construction of radio interception stations, as recommended years earlier by him.

These stations were built at Darwin, Canberra and Jandakot in Western Australia. All three stations had HF/DF capability. Those at Canberra and Darwin were more powerful and had interception capability. The RAN established its Shore Wireless Service (SWS) in 1938. This trained and provided personnel initially to staff its HF/DF stations and then to increase the number of such staff available to meet expanding local commitments. By early 1939 the SWS was able to provide some for service at FECB as trained 'Procedure Y' (Signals Intelligence) personnel.

The RAN Wireless Station Canberra commenced operations in April 1939 and in July the names Belconnen Transmitting Station and Harman Receiving Station were allocated to its two components. The name Harman is a contraction of the names Harvey and Newman, respectively Director and Assistant Director of Signals and Communications at the time. Jack Newman, who had attended a conference in Singapore on wireless intelligence in the Far East in March 1939, was appointed Officer-in-Charge of Harman, and in May of that year he succeeded Harvey as DSC. By late 1940 Newman was in charge of all the RAN shore wireless stations. The title of his position was later changed to Director of Naval Communications (DNC). He played a major role in Naval radio and communications intelligence within Australia throughout WW2.

At its inception the SWS had 23 personnel at Harman. Harman is also notable because, in April 1941, a group of 14 women began duties there as telegraphers. Subsequently 13 of these became the first enlisted personnel in the Women's Royal Australian Naval Service (the WRANS), founded in October 1942.

The naval wireless telegraphy Station HMAS *Coonawarra* at Darwin began operating in September 1939. It continued to provide valuable Sigint information throughout the war. During the bombing of Darwin in 1942, Coonawarra was

---

[5] See NAA Canberra item A1608 O14/1/1.

AWM archive AWM124 5/88 records that an experienced RN Commander, Edmund Harvey, was Director of Signals and Communications in the RAN from May 1937 to May 1939. It is not clear whether Harvey was sent out to upgrade the naval interception facilities in Australia.

Darwin's only radio communications link with the rest of Australia. The first news of the first Japanese air raid on Darwin in February 1942 came in a signal from Coonawarra picked up at the Army's Park Orchards Station near Melbourne.

Initially all the data collected by these RAN stations, including intercepts of messages in various diplomatic and military codes, was sent back to the FECB for analysis and possible reference to GCCS in London. Some local study of the data was made by the few staff with some knowledge of Sigint.[6]

The increase in the availability of this material, and a growing awareness of Australia's almost complete reliance on the FECB and the GCCS for provision of Sigint, began to influence the thinking of Australia's senior military personnel. The declaration of war against Germany by the UK and Australia in September 1939, and the looming threat of war with Japan, accelerated this process, as GCCS became focused on European Sigint work.

Commander Long, the RAN DNI, raised the issue of a *Cryptographic Organisation in Australia* in a memo to Admiral Ragnar Colvin, CNS, in November 1939. Colvin conveyed the substance of Long's memo to Lieutenant General Squires, CGS. Squires had earlier been brought out from the UK to review military organisation and presumably had some knowledge of such activity in the UK. Squires' response, while supportive, was also careful to suggest the extent to which such an activity might be feasible, given the limited resources available. The UK's advice was sought in April 1940. It was decided not to take action until this was received.

Fortunately (since it took some 6 months for the British to respond), it seems that Squires[7] himself may have initiated some 'unofficial' action in January 1940. Then, presumably unknown to the RAN, the Army began supporting a group of four academics at Sydney University. They had volunteered to study the structure of codes and ciphers and to learn some Japanese in order to examine Japanese material. All the messages for examination were provided to them by the Army, using cables and letters intercepted under the 1939 Censorship Act.

This group consisted of Professor T. G. Room, Professor Dale Trendall, Mr. 'Dickie' Lyons, and Mr. 'Aps' Treweek. Room was a mathematician from Cambridge and a former colleague there of Gordon Welchman, one of the Bletchley Park code breakers. Lyons was a fellow mathematician. Trendall was the professor

---

[6]In the last few months of 1941 the task of decrypting JN-25B7 messages was divided between the FECB and Cast. (See Sect. 10.1.) The SWS then had to decrypt message indicators and send intercepts to the appropriate unit. It appears that some limited decryption of JN-25B7 was done in Australia at that time, but more information is lacking.

A letter from Lietwiler to Op-20-G on this matter survives in NARA RG 38 and is reprinted in Timothy Wilford's MA thesis (on ProQuest). NAA Melbourne item MP692/1 559/201/989 notes that by December 1941 Commander Newman was fully occupied with 'W/T intelligence' leaving the other aspects of his job to his deputy.

[7]Squires' involvement is plausible, but the fact remains that around this time he was sent on sick leave with terminal cancer. He had no background in signals intelligence. The original intentions of the academics remain obscure.

of Greek. Treweek also taught Greek and was already a Major in the Sydney University Regiment. Some years before he had decided that war with Japan was inevitable and so started to learn Japanese. By October 1940, internal Army reports indicated that the group was proceeding well and that breaking of a current relatively simple Japanese diplomatic code was expected soon.

The Army's Park Orchards station intercepted Japanese commercial and government radio traffic from mid-1940 onwards. At this time, the Army also formed No. 1 Special Wireless Section in Victoria, to assist in intercepting French and German radio communications.

This unit, renamed the No. 4 Special Wireless Section, was sent to the Middle East as part of the 1st Australian Corps early in 1941. It received additional training[8] at the GCCS Middle Eastern outpost, CBME (see Sect. 3.9), at Heliopolis in Egypt before joining the British force sent to Greece, where it monitored German communications. Later that year it was sent to Syria to assist in the removal of the Vichy French there, and for other work with the GCCS station at Sarafand in Palestine. After the entry of Japan into the War in December, the unit gained training and experience in intercepting Kana Morse. This was of extreme value to Australian Sigint activities after its return home in February 1942.

An incentive for new Navy activity in this area was provided when Eric Nave returned to Australia on sick leave from FECB, Singapore, in February 1940. He was advised, on medical grounds, against any further visits to the tropics, thus ruling out any return to FECB or even a sea voyage back to the UK. The Admiralty reluctantly agreed that he remain with the RAN, for which he had begun a small intelligence operation in Melbourne. Nave collaborated with the FECB in intercepting both radio traffic from the Mandated Islands and also Japanese diplomatic and consular messages. Some traffic analysis and message decryption were carried out. Nave remained sceptical of the viability of this activity owing to the lack of RAN personnel with expertise in Japanese.

The situation began to improve from October 1940 when the British response indicated its support for establishing a cryptographic unit to assist the FECB. In the same month, a *Combined Operational Intelligence Centre* (COIC) was established, modelled along the lines of the joint intelligence committee[9] operating in the UK.

Nave learnt of the Sydney University group's work early in 1941. By May he was able to recommend that its members join his unit in Melbourne. By August 1941, Nave was in command of the newly established *Special Intelligence Bureau* (SIB). This was a joint operation of all three services, with the authority to co-opt civilians.[10]

---

[8]As noted in Sect. 3.9, 'Mic' Sandford of the Australian unit in the Middle East received special training in the contemporary practice of signals intelligence there.

[9]More on the British Operational Intelligence Centres is given in Chap. 7, particularly Note 14.

[10]Of the four Sydney University academics, three—Professors T. G. Room and A. D. Trendall and Mr R. J. Lyons—retained civilian status while the fourth, A. P. Treweek, who was also a Major in the Sydney University Regiment (part of the Militia), was called up by the Army and seconded to the SIB. He was later promoted to the rank of Lieutenant-Colonel. Lyons returned to University

Professor Room and a language expert, Lieutennant 'Jim' Jamieson, were sent to FECB in September 1941 for training in diplomatic cryptanalysis and translation, and to obtain new information on Mandated Islands traffic. As mentioned in the Note 12 of Chap. 3, they also visited the Dutch unit Kamer 14 in Bandoeng (now Bandung), Java.

In July 1941, seven Royal Australian Air Force (RAAF) radio operators, all proficient in ordinary Morse code, together with two Army officers, went through a Kana Morse training program run by Jack Newman[11] of the RAN. Thus more personnel were becoming available for interception of Japanese radio messages.

By the end of 1941 there was a small but effective Sigint organisation in Australia, usefully contributing TA, DF and cryptanalytic information to FECB and GCCS. Of extreme importance to later Allied Sigint activity was the knowledge that interception of IJN messages, including those sent in the JN-25 operational systems, was available at other RAN interception stations. The existence of telegraphic communication between these stations and Melbourne was also very important.

Sigint work continued to be plagued by a shortage[12] of translators of Japanese through WW2. In 1945 the Australian Army was considering recruiting Canadians of Japanese ancestry for translation work.

6.5 Diplomatic Sigint

The work of the Sydney University group on Japanese diplomatic material continued within Nave's unit after its members transferred to Melbourne. By early 1942, after the Pearl Harbor raid, a distinct section[13] with an enlarged membership was

---

duties in the re-organisation of October–November 1942, but the other three continued to make significant contributions to signals intelligence work after then. The SIB had some clerical support in 1941.

[11]The RAAF file A11093 311/236G survives in the Canberra NAA and has been digitised. The RAAF team was sent to Darwin in November 1941 but was then reporting to the Naval Intelligence Department. This was the origin of the No. 1 Wireless Unit usually written and pronounced 1WU.

[12]The NAA Canberra item A432 1940/153 has a memo of 9 May 1938 about the difficulty in obtaining court interpreters of Japanese. The Attorney-General undertook to raise in Cabinet the obvious enough remark: 'Similarly, if Japanese dispatches were to fall into our hands, it seems doubtful if we could even translate them—at all events, with celerity.' The Japanese language itself would be a considerable obstacle to understanding even plain-language military messages.

Canberra NAA file A571 1944/4124 is one of about six surviving records of the need to import suitable 'Nisei' linguists from Canada. In 1944–1945 ever-increasing quantities of Japanese military documents were being captured. Although those of cryptographic significance received the highest priority, the Allied Translator and Interpreter Service (ATIS) had a great deal of work to do.

[13]The genesis, evolution and work of this section is well covered in the book *Breaking Japanese Diplomatic Codes—David Sissons and D Special Section during the Second World War*, edited by Desmond Ball and Keiko Tamura, ANU E-Press 2013.

established, still under the control of the RAN. Headed by Professor Trendall (now a full-time member) and with R. S. Bond (a recent Honours graduate in Classics) brought in on his advice, it was soon strengthened by the fortunate addition of three experienced British officers. Two (C. H. Archer and H. A. Graves) were from the Consular Service and one (the exceptional linguist A. R. V. Cooper) was from the FECB. Some more staff were recruited during 1942. The new section had as its major focus the decryption and reading, jointly with GCCS, of Japanese diplomatic intercepts sent in the FUJI transcription code system, in which it was very successful. Following the Holden agreement (see Sect. 18.8) the unit was transferred to the control of the Australian Army and continued until the end of the War.

6.6   The Coastwatchers

The stimulus to establish an Australian Coast Watching Organisation came[14] in 1919 in an internal RAN memo to the Chief of the Naval Staff. Eventually the Naval Intelligence Department (NID) of the RAN was given approval to build up a volunteer organisation according to an agreed plan. In his foreword to Eric Feldt's *The Coast Watchers*, George Gill, later Official Historian to the RAN, described this as follows:

> In its broad outline, the scheme provided for the appointment of selected personnel, adjacent to or on the coast, who would, in time of war, report instantly any unusual or suspicious happening in their areas, sightings of strange ships or aircraft, floating mines and other matters of defence interest. Appointees included reliable persons such as Postmasters, Police and other Government servants, missionaries at mission stations on little-frequented parts of the coast, pilots of civilian airlines on or near coastal routes; and, in the Territories of Papua, New Guinea and the British Solomon Islands, Patrol Officers and District Officers and other officials, beside planters.

A WW1 Army veteran, Walter Brooksbank of the NID, whose abilities led to his later appointment as Civil Assistant to the DNI, carried out most of the detailed administrative and coordination work required. By the outbreak of war with Germany in September 1939, some 800 coastwatchers were in place and the organisation, now placed under the control of DNI Rupert Long, was ready to function.

Most of the personnel were located on the Australian coast and, for these, the telegraph system provided communication back to Navy Office. For those based in Papua, New Guinea and the Solomons, the only rapid means of communication was radio. The few existing local radio networks were used, but the isolated nature of a number of the coastwatchers required that they be provided with their own 'teleradio' equipment.

---

[14]This memorandum could scarcely have been independent of the recommendation in the Jellicoe report.

Funds for this purpose became available upon the outbreak of the European war. Commander Long appointed Eric Feldt to extend the existing Coast Watching Organisation and to 'place teleradios at strategic points so as to establish a reporting screen and communications network effectively covering the northern and north-eastern approaches to Australia'.

The teleradios supplied to the coastwatchers used special frequencies to minimise detection, with simple codes used for encryption. Some dozen people were needed to move this cumbersome equipment and re-install it anywhere in the difficult jungle terrain. So support of the local tribespeople was essential both for security of personnel and for their relocation when threatened with capture.

During 1942 the Japanese invasion and occupation of numerous sites in New Guinea and the Island Territories forced a rapid change upon the operations of the coastwatchers. Originally civilians peacefully observing and reporting on sea and air movements, many were now in or close to Japanese occupied territory. Although all were offered assistance with evacuation, a number chose to remain active. Those who did so provided invaluable service in helping to organise the safe evacuation of civilian and military personnel. By March 1942 all coastwatchers were given naval service status so as to have them treated as prisoners of war if captured. This did not prevent the torturing and killing of a number of coastwatchers and their local helpers, especially in New Guinea.

By August 1942, the disposition of coastwatchers in the Solomons, from Bougainville down to Guadalcanal, was sufficient for the continual monitoring of Japanese naval and aircraft movements in the region. As there was only limited Sigint available to the USN from June to December 1942, this operation was of paramount importance in the successful defence of the central Solomon Islands.

From late 1942 onwards Feldt could increase coastwatching operations by the deployment of naval and other service personnel, often recruited to assist in reconnaissance of Japanese areas intended for recapture.

Admiral Nimitz described the work of the coastwatchers in alerting Allied commands to enemy ship, troop and plane movements as 'of inestimable value'. Admiral Halsey specifically referred to coastwatcher intelligence as saving Guadalcanal,[15] which then 'had saved the Pacific'.

---

[15]Halsey's comment is given on page 285 of Feldt's *The Coast Watchers*. 'Many appreciative words were spoken by senior officers of the South Pacific command. ... Most treasured were those of Admiral Halsey, who said that the intelligence signalled from Bougainville by [Jack] Read and [Paul] Mason had saved Guadalcanal and Guadalcanal had saved the Pacific.' *Admiral Halsey's Story* praises the Coastwatchers quite enough. Nimitz' comment is in the footnote on page 253 of *The Great Sea War*.

6.7   Developments in New Zealand

By the end of 1913, the NZ[16] government had established five radio stations for maritime telegraphy work. The southernmost one at Arawua, enjoyed excellent reception of signals from the German Pacific islands that later became the Japanese Mandated islands. In particular, it could pick up JN-25 messages emanating from that area in WW2.

Between August and October 1914, NZ Naval Intelligence forwarded transcripts of intercepted radio messages to its counterpart in Melbourne. These had been sent by warships in Admiral von Spee's Eastern Asiatic Squadron. By utilising characteristics of the shipboard operator's 'touch' on his Morse key, and characteristics of the ship's radio transmitter (an early form of radio fingerprinting), the transmitting ship was able to be identified. Further, since the RAN had available to it a copy of the code then in use, these messages could be decrypted and sent on to the Admiralty.

After WW1, the NZ Naval Squadron continued to cooperate with shore-based civilian radio stations and with direction-finding activities. Founded in 1913, it remained a part of the Royal Navy until late in 1941, when the Royal New Zealand Navy (RNZN) was created. As part of its plan to expand its world-wide DF network in the 1930s, the RN provided detailed information to the NZ naval authorities relating to site and equipment choice. Also, the establishment of passenger flights to Australia across the Tasman Sea in the 1930s stimulated the development of tracking stations. Awarua, a station on the South Island, was upgraded and by 1938 was already supplying HF/DF information via the Empire network. For example, in late 1939, it assisted in tracking the *Graf Spee* westwards across the Atlantic from the Cape of Good Hope.

By this time NZ had set up an effective landline communications system connecting all components of its interception system. Soon afterwards this network had been extended to link up with the RAN, FECB, the Admiralty and other British Commonwealth stations.

As noted earlier, immediate connection across a chain of DF stations is essential for effective DF work. Once one station picks up a message from a 'source of interest', then if it can transmit a signal automatically to other stations to lock onto the received frequency, simultaneous fixes of direction enable the source to be located immediately by the intelligence centre receiving the combined data.

When the FECB had sought help from both Australia and NZ in monitoring signals coming from the Japanese Mandated Islands in the Pacific, operators had been trained in Kana Morse to this end. Much information on the main Japanese

---

[16]For the New Zealand story see Desmond Ball, Cliff Lord and Meredith Thatcher *Invaluable Service: The Secret History of New Zealand's Signals Intelligence in Two World Wars*, Waimauku, Resource Books, 2011.

The book reprints the report *New Zealand Naval 'Y', H/F, D/F and Special Intelligence Organisation*, an apparently British report dated 17 December 1942, NARA RG38, Inactive Stations, Box 23.

bases there, the types of warships controlled from there, and other DF plus TA data came from NZ. For example, Commander Newman recorded that radio intelligence data from NZ in early June 1942 led Frumel to infer that radio silence from the IJN ships known to be going to Midway meant that the planned attack was still on track! Intercepted encrypted signals, including those in JN-25, were also passed on for attack, as were all diplomatic cables from Japanese Consular staff that were not routed through Australia. A junior navy officer, Halson Philpott, who had some 15 years' experience in radio work, acted as the wireless intelligence officer. He was then placed in charge of all shore-based wireless intelligence at the start of the Pacific war.

The RN had appointed an officer, F. M. Beasley, to the NZ Navy Office to head up its intelligence section in mid-1941. Shortly after, a NZ Combined Operational Intelligence Centre was established in the Navy Office with Beasley in charge. He held the position of DNI, Wellington. Beasley held this position until 1944, when the tide of war in the Pacific reduced the importance of the New Zealand station to ongoing naval and military operational activities.

By early 1942, a small team had been formed to commence the task of investigating intercepted messages in code. As well as Philpott, the team included James Campbell, professor of mathematics at Victoria University Wellington, and Robert Boulter, the UK Trade Commissioner in Wellington, who had expertise in Japanese. Beasley was able to visit the USN Sigint Station Hypo at this time and on his return reported both to his NZ superiors and to the RAN's DNI in Melbourne, undoubtedly emphasising the new-found importance of JN-25 in the Pacific cipher war.

Throughout the period of intense warfare in the South and South-West Pacific areas, the NZ Sigint organisation, together with its coastwatchers, provided valuable intelligence information to the Allies. For example, it was NZ DF and TA that confirmed the operation of Japanese submarines in the Tasman Sea and off the NSW coast in May 1942.

6.8 The 1947 FRUMEL Commendation

The Recordsearch facility on the NAA website www.naa.gov.au may be used (search for 'Booth Sandford Clark') to locate the file 66/301/232, which may then be read online. It contains the following letter dated 16 April 1947.

The Embassy of the United States of America at Canberra presents its compliments to the Department of External Affairs and has the honor to invite its attention to a letter of commendation from the Secretary of the Navy relating to certain personnel of the Armed Forces of Australia who served in the Fleet Radio Unit, Melbourne, a part of the United States Communications Intelligence Organization, during the period December 7, 1941 to December 12, 1944.

The letter from the Secretary of the Navy states:

> The work of the Fleet Radio Unit, Melbourne (Frumel), was not highly publicised because of its secret nature, but the results obtained were of immeasurable importance in the successful prosecution of the late war. The successful accomplishment of its mission was in no small measure due to the unfailing devotion to duty of the 318 personnel of the Armed Forces of Australia who were attached to that unit and who worked side by side with the United States personnel.
>
> It is requested that you express to the Government of Australia the appreciation of the Navy Department for services of the Australian personnel attached to the Fleet Radio Unit, Melbourne.

The functions of Frumel are variously described as communications intelligence, signals intelligence or special intelligence and may be checked by examining some of the NAA files in the remarkable series B5555. These contain certain Frumel documents assembled by Commander Jack Newman, head of its Australian component.

Chapter 7
Preparedness for Attack?

Much has been written about the raid on Pearl Harbor on 7 December 1941 and why the American forces there were caught unprepared. When it was learnt later that the then current IJN operational cipher (JN-25B) had been under attack since early 1941, claims were made that this activity must have provided foreknowledge of the IJN plans for the raid. Later chapters of this book describe how JN-25B and its successive systems were attacked and thus suggest that it is quite unlikely that such foreknowledge came from JN-25B. Yet three important items of *radio* intelligence (but not *signals* intelligence) were obtained from JN-25B in the weeks preceding the raid. These, when combined with other available intelligence, indicated that there was a very high risk that American territory would be attacked.

7.1 Incompletely Broken Code Books

A bare (that is, not superenciphered) code may be insecure without it being capable of being fully solved. Thus SRH-349 in RG457 in the American National Archives (NARA) comments generally on this issue:

> Reconstruction of a code book is a long laborious process. Each code group needs to be identified singly. The larger the code, the longer is the time needed.'

Thus a decrypted (but incompletely decoded) message may look like 'xxxxx xxxxx submarine xxxxx attack xxxxx xxxxx merchant ship xxxxx xxxxx' and so not supply usable intelligence.

This state of knowledge must be what Alastair Denniston meant by the phrase *skeleton code books* in Chap. 3. It is discussed in Sect. 11.3.

© Springer International Publishing Switzerland 2014
P. Donovan, J. Mack, *Code Breaking in the Pacific*,
DOI 10.1007/978-3-319-08278-3_7

7.2 Aircraft Versus Warships

The Jellicoe Report of 1919 on the Australian Navy (mentioned in Sect. 6.2) includes in Chapter IX(b) of volume II a section entitled *Future Possibilities of the Use of Aircraft against Ships and Submarines*. Jellicoe stated that:

> It also appears that near any fleet anchorage or base a complete system of anti-aircraft defences is essential. Perhaps guns and searchlights can be provided by the ships but an aerodrome or aerodromes close at hand must be part of the defences.

While Jellicoe may not have had a full understanding of the aircraft carriers to be in service 22 years later, he was anything but a fool.

Even though volume II of the Jellicoe Report was confidential, the writings of Hector Bywater (1884–1940) were not. From 1921 onwards he wrote with considerable prescience of a prospective Pacific War, notably in *The Great Pacific War*. He stressed the importance of carrier-based aircraft and surprise attacks.

In 1921 the USN had conducted tests on the possibility of aircraft sinking battleships. This had been done at the urging of General William L. 'Billy' Mitchell (1879–1936). Mitchell's point was demonstrated by the sinking of obsolete battleships by bombers off North Carolina in 1923. Mitchell went on the following year to predict that Japan would initiate a Pacific War by attacking Pearl Harbor and the Philippines by aircraft. Predictions of this type were also made by a Japanese author, Shinsaku Hirata,[1] around 1934. There had been worries expressed[2] about the US Pacific Fleet being inadequately protected at its advance base in Pearl Harbor. However the strength of the aircraft carrier as a weapon against battleships was not yet fully appreciated. The British Wing-Commander H. E. P. Wigglesworth[3] had reported on the defences and intelligence operations in the Far East in 1938 and had noted specifically the vulnerability of battleships to attack by carrier-based aircraft.

The close proximity of air defences to naval bases was thus no accident. To overcome this risk the defensive air cover needed to be on alert and preferably in the air at critical times, such as dawn! This did not happen at either Pearl Harbor or Clark Field.

7.3 The Intelligence Build Up

Collection of intelligence on Japanese military and political activity had continued unabated after the outbreak of the European war. The steady build-up of Japan's armed forces, together with the aggressive tone of its media presentations, and

---

[1] The mention of Hirata is based on that in Layton's *And I was There*, which well deserves its place in the selected bibliography at the end of this book.

[2] Clay Blair's *Silent Victory* on page 49 of volume 1 mentions an objection in 1938 to moving the Pacific Fleet base from California to Hawaii. Blair also points out that long range reconnaissance flights from Pearl Harbor were feasible, even though very wearing if done for months on end.

[3] This report, located at TNA AIR 20/374, is quoted by Richard Aldrich in his *Intelligence and the War against Japan*.

its virtual occupation of northern Indo-China, had led to regular appraisals by the British and others of the situation in South-East Asia and the Pacific prior to mid-1941.

For example, a conference in Singapore in April 1941, attended by senior representatives from Australia, New Zealand, the Netherlands, the UK and the USA, had concluded that the defences of most colonial territories in the Far East were inadequate. The participants considered possible Japanese attacks on Malaya, Singapore, the NEI and the Philippines. Hawaii was not considered part of the Far East and so not discussed. However it was decided that should the United States become involved, the US Pacific Fleet could attack IJN bases in the Japanese Mandates. So the possibility of conflict between Japan and the USA was not entirely ruled out.

Despite the Tripartite Pact, Hitler's invasion of the Soviet Union in June 1941 did not lead to Japan declaring war on the Soviet Union. But Japan began steadily increasing its military strength in Manchuria. This meant that a possible Japanese attack into Siberia could not be eliminated from consideration at that time.

As mentioned in Sect. 2.3, the economic and trade sanctions imposed on Japan in July-August 1941 by the Americans and then the British and Dutch led to the conclusion[4] that a hostile Japanese response was now only a matter of time. The impact of the American oil embargo and freeze of Japanese assets would hit home.

The oil, rubber, mineral and food resources of South-East Asia made Thailand, Malaya, Singapore and Sumatra prime targets. Thailand was the gateway to Burma and Malaya. The conquest of Malaya would expose Singapore and the NEI. Efforts to monitor Japanese preparations for invasion of these countries intensified. The sum total of relevant intelligence from all sources had, by the end of November, identified a 'southern invasion force' assembling off Indo-China. Its progress southwards towards the Gulf of Siam was able to be followed, but the monsoon weather restricted detailed observation. Thus the precise disposition of the forces heading towards the North-East coast of Malaya and the lower South-East coast of Thailand was not known prior to the actual landings.

---

[4] As noted in Sect. 2.4, Malcolm Kennedy noted in his diary on 6 October 1937 that certain embargo proposals then being aired by public figures were extremely dangerous. Arthur McCollum, an intelligence officer in the USN later stationed in Brisbane, set out on 7 October 1940 methods for the USA to incite a war in the Pacific. One of these proposals was to use economic sanctions.

The New York Times on 7 December 2008 drew attention to W. J. Donovan's memorandum to President Roosevelt dated 13 November 1941 found in NARA in 2007. The source is Hans Thoman, the senior German diplomat in the USA.

'In the last analysis, Japan knows that unless the United States agrees to some reasonable terms in the Far East, Japan must face the threat of strangulation, now or later. Should Japan wait until later to prevent this strangulation by the United States, she will be less able to free herself than now, for Germany is now occupying the major attention of the British Empire and the United States.' 'If Japan waits, it will be comparatively easy for the United States to strangle Japan. Japan is therefore forced to strike now, whether she wishes or not.'

Overall, there was adequate intelligence[5] about this southwards Japanese invasion. Should a concurrent Japanese attack on the Philippines, minor American islands and Pearl Harbor have been considered a possibility?

On 31 January 1972 Lester Pearson, Prime Minister of Canada 1963–1968 and once active in Sigint, wrote a considered opinion[6] on the matter to Colonel M. Seymour:

> But the British (and the Americans) seem to have had enough communications and other intelligence at hand at that time to expect that an attack would be made in the near future on Pearl Harbor, quite possibly in early December.

Argument and evidence in support of this conclusion are assembled below. But this book is not intended to be the last word on the Pearl Harbor raid.

7.4   War with the USA?

Negotiations between America and Japan, seeking agreement on conditions that would enable a relaxation or removal of the sanctions imposed on Japan, began in August 1941 and continued into November, with the prospect of success progressively diminishing. The British government was kept informed and was continually consulted about developments. It kept the Dominions[7] in touch with developments and passed on their advice and comments. Both the UK and the USA saw merit in keeping the discussion open, as each wished to avoid confrontation if possible or at least to defer it into 1942. Churchill knew that Hong Kong, Malaya, Singapore and British North Borneo were inadequately defended and wanted British

---

[5]General reference is made to Peter Elphick's book *Far Eastern File: The Intelligence War in the Far East 1930–1945*, Hodder and Stoughton, London, 1997.

[6]The paper by Timothy Wilford entitled *Watching the North Pacific*, published in *Intelligence and National Security* (17(4), 2002, 131–162) draws attention to this letter. The original has survived. Wilford's publications, including his Ottawa MA thesis (2001), on the matter are quite useful.

[7]NAA Canberra file A816 19/304/431 has 340 pages of texts of messages between the future Allies. It has been digitised.

The government of Canada had a meeting of senior officials to discuss how to handle a sudden Japanese attack while the Pearl Harbor raid was in progress! The meeting, and numerous other issues connected with the raid, are discussed by John Bryden in his *Best-Kept Secret: Canadian Secret Intelligence in the Second World War*.

There had been considerable planning for possible war with Japan in Australia. NAA Canberra A1196 3/501/15 is one reference. The Army had to be recalled from the Middle East. The book *Saving Australia* by Bob Wurth (2006) claims that Tatsuo Kawai, the Japanese ambassador in Australia, warned Prime Minister Curtin, with whom he had a friendship, in late November 1941 that war was inevitable. This is quite plausible. However, as the Japanese decision to commence hostilities was made by a very small group of people, it is very unlikely that the ambassador had any useful knowledge of military or naval plans. Likewise General Akin, MacArthur's chief signals officer, took the view from April 1941 that war was inevitable. (His account is in the MacArthur Museum archives.)

supply routes across the Indian Ocean to remain open. All parties knew that the American re-armament program would begin to make an impact from mid-1942 but that the big reinforcing of the USN would not happen until April 1943. Hence, if Japan was to attack, then it should do so earlier rather than later.

The British believed[8] that the existing American naval strength in the Pacific, and especially the Pacific Fleet in Hawaii, would not only discourage Japan from attacking the USA but would dissuade it from undertaking a full-scale invasion to the South. This view appears to have prevailed even though, by late 1941, the IJN had a great supremacy in aircraft carriers in the Pacific and approximate parity with the USN in other warships. But if the IJN had to protect a southern invasion force and provide ongoing cover for further operations, it may not have felt confident of victory against attack by the US Pacific Fleet.

The United States had begun to build up its air force (the USAAF—now the USAF) in the Philippines in late 1941 and its re-armament program was scheduled to further strengthen its forces there. These USAAF planes directly threatened Japanese shipping movements in adjacent waters and various land bases on Formosa (now Taiwan), making the Philippines an obvious target in the event of war. But the American War Plan Orange called for its forces there, if attacked, to concentrate in the Bataan Peninsula north-west of Manila while awaiting reinforcements. The safe arrival of such depended upon the protection of the US Pacific Fleet.

On 27 November, when the pending collapse of negotiations and other intelligence indicated the imminent prospect of war, the Navy Department warned the Commanders-in-Chief of its Pacific and Asiatic Fleets that an aggressive move by Japan, probably directed against the Philippines, Siam (Thailand) or Borneo, might be expected within the next few days. It was reasonable to expect that Admiral Hart in Manila and Admiral Kimmel at Pearl Harbor would have taken some preparatory steps against possible attack. Hart dispersed his small fleet and moved stocks of equipment and fuel from their usual locations. He also organised reconnaissance flights by his long-range seaplanes and co-ordinated with British and Dutch activity. Admiral Kimmel, conscious of an IJN presence in the Mandates, ordered air reconnaissance of the waters around these islands, and also of Wake and Midway, but not of Hawaii.

Army commanders had been given a similar alert. General MacArthur ordered air reconnaissance from his base in Luzon. General Short at Pearl Harbor, who was in command of the USAAF there, had previously decided that internal sabotage was the greatest threat, and made no attempt to increase preparation against external attack.

---

[8] J. M. A. Gwyer's *Grand Strategy* (Volume II, Part I, HMSO London, 1964), particularly pages 280–284, is most useful here. 'Even at that date (early November) few people were really prepared to believe that Japan would risk an open conflict with the United States, or, to speak more exactly, that she would dare to commit her main forces to an operation in the south while the US Pacific Fleet was still in being and able to act offensively against her.' For more on this, see the digitised NAA Canberra file A816 19/304/431 mentioned in Note 7.

The intelligence mentioned above did not indicate the timing of any potential attack other than to show that it was likely to be earlier rather than later. But any surprise raid on Pearl Harbor would have necessitated refuelling of ships at sea in the North Pacific and so would have been difficult in January or February. This was another indication of earlier rather than later. Sunday at dawn was an obvious time for an attack. Moonlight before dawn would have helped in launching planes from aircraft carriers. The moon had been full on Wednesday 3 December 1941.

7.5 Relevant Sigint in November and December

TA and DF throughout November indicated that Japanese merchant shipping was returning to home waters. The pattern of behaviour was quite unusual. Later in that month Commander Nave in Melbourne was obtaining evidence of intensified IJN activity near Truk in the Mandates. Other TA work was going on at FECB, Manila and Pearl Harbor.

In mid-November interception stations in several countries[9] picked up a diplomatic message which became known as the 'winds alert' message:

Regarding the broadcast of a special message in an emergency.
     In case of emergency (danger of our cutting off our diplomatic relations) and the cutting off of international communications, the following warning will be added in the middle of the daily Japanese language short-wave news broadcast:

   (1) In the case of Japan-US relations in danger: HIGASHI NO KAZE AME
        ('east wind, rain');
   (2) Japan-USSR relations: KITA NO KAZE KUMORI
        ('north wind, cloudy');

---

[9]The book *West Wind Clear: Cryptology and the Winds Message Controversy—Documentary History* by Robert Hanyok and David Mowry, published by the NSA Center for Cryptologic History in 2008, and available on the NSA website, is a very useful reference. The 'winds messages' are mentioned in NAA Canberra file A5954 558/1 *Far Eastern Crisis November December 1941. Instructions from Japanese Foreign Office to Posts Overseas*. It is possible that they were a hoax intended to confuse the American and British Intelligence services. If so, they may well have succeeded.

The winds message was an example of an *open code*, that is a piece of plain text to which a special meaning had been assigned. A well-known example of this was the couplet of a French poem by Paul Verlaine (1844–1896) that was widely believed in 1944 to indicate that an Allied invasion of France might be imminent. The relevant text is:

Les sanglots longs
Des violins
De l'automne
Blessent mon cœur
D'une langueur
Monotone.

As a piece of one-time open code this could mean anything. The first half had been broadcast several times well before D-day.

(3) Japan-UK relations: NISHI NO KAZE HARE
     ('west wind, clear').

> This signal will be given in the middle and at the end as a weather forecast and each
> sentence will be repeated twice. When this is heard please destroy all code papers, etc. This
> is as yet to be a completely secret arrangement.
> Forward as urgent intelligence.

There is no agreement that any of the above warnings were subsequently sent and intercepted. Of interest is the fact that the possibility of Japan-US relations being disrupted is placed on an equal footing with the others.

On 29 November, Baron Oshima, Japan's Ambassador in Berlin, reported that von Ribbentrop, the German Foreign Minister, had informed him that, in the event of Japan becoming engaged in a war against the United States, Germany would immediately join in. Tokyo responded the next day, again using the Purple diplomatic cipher (Sects. 4.3 and 8.2):

> Say very secretly to them that there is extreme danger that war may suddenly break out
> between the Anglo-Saxon nations and Japan through some clash of arms and add that the
> time of the breaking out of this war may come quicker than anyone dreams.

An important element here is the use of 'Anglo-Saxon nations' instead of 'Britain'. The phrase must have been interpreted as including the USA.

This message was provided to Churchill and Roosevelt on 1 December. Churchill believed it to be so significant that he incorporated it in his *The Second World War* (Volume III, Chapter XXXI).

Also, on 1 December, intercepts revealed that the IJN had changed its call signs (the parts of radio signals used to identify each of its ships and port facilities). This was unusual in that there had been a change on 1 November and each earlier call sign system had lasted for longer than 1 month. Each change required interception operators to rebuild their lists of IJN radio sources corresponding to each new call sign. This was a routine but time-consuming task. There was an additional complication with this change. Admiral Kimmel's intelligence officer, Commander Layton, told him on that day that there was uncertainty regarding the precise locations of a number of the large IJN carriers, which appeared to be maintaining radio silence. In fact no intercepts from these carriers were picked up prior to the attack on Pearl Harbor.

The third piece of radio intelligence occurred on 4 December when the IJN changed its JN-25 system. This rendered all JN-25 intercepts indecipherable. Such changes usually happened on the first day of a month. (A few days later the change was found to involve only the superencipherment stage of JN-25, with the 'B' code book still in use.)

The total impact of these three IJN actions should have been enough for the USN and the USAAF to raise[10] the level of alertness across all their Pacific bases

---

[10]In particular Admiral Kimmel could have sent most or all of the Pacific Fleet halfway to California keeping radio silence. General Short could have had fully armed planes in the air over

and warships. But the realisation that war was imminent came to London and Washington from different sources. In general, intelligence from all sources should be assessed as a whole.

7.6   The Haruna Code Destruction Messages

Over 1–2 December the Japanese Foreign Office in Tokyo sent coded messages to a number of Japanese consulates and embassies around the world instructing them to destroy all cryptographic material and machines, except a minimum for immediate use, and to advise when this had been done by sending back the single word HARUNA. The message to Washington was sent on 2 December and the response detected on 3 December. The NARA file RG457 SRH-415 has 24 pages of Haruna message material.[11]

It is possible that the Haruna messages were intended to replace the so-called winds message. Be that as it may, such an instruction was to be interpreted as advice of the imminent outbreak of war and not just a temporary break in diplomatic relations. This interpretation was accepted in both London and Washington. Admiral Stark immediately informed Admiral Hart in the Philippines and Admiral Kimmel at Pearl Harbor of these messages.[12] Kimmel did not pass this on to General Short, the

---

Pearl Harbor at dawn each day in December. As pointed out in Note 3, long range reconnaissance flights from Pearl Harbor could have been implemented.

[11]The short opening chapter of Frank Rowlett's *The Story of Magic* is entitled *The Code Destruction Message*. He gives the full text of the translated Magic message to the Washington Embassy and an account of his discussion about its meaning with Colonel Sadtler, who suddenly exclaimed 'Rowlett, do you know what this means? It can mean only one thing, and that is that the Japanese are about to go to war with the United States.' The full text is:

From Tokyo. To Washington. 2 December 1941. #867.

Please destroy by burning all of the codes you have in your office, with the exception of one copy [of] each of the codes being used in conjunction with the machine, the OITE code and the abbreviation code. (This includes other Ministries' codes which you may have in your office.)

[This was sent five days before the Japanese attack. Burning of documents could be observed from outside various Japanese consulates.]

Also in the case of the code machine itself, one set is to be destroyed.

3. Upon completing the above, transmit the one word HARUNA.
4. Use your discretion in disposing of all texts of messages to and from your office, as well as other secret papers.
5. Destroy by burning all of the codes brought to your office by telegraphic courier Kosaka. (Consequently, you need not pursue the instructions contained in my message #860, regarding getting in touch with Mexico.)

This message was sent and decoded more than 48 hours before the raid. The burning of papers was observed at various Japanese consulates soon after its receipt. Its implications could have been made clear to Kimmel and Short in good time.

[12]David Kahn's book *The Codebreakers* (second edition) gives more on this matter.

USAAF commander. Nave in Melbourne predicted that hostilities would commence over the coming weekend. (The 'weekend' east of the International Date Line corresponded to Sunday and Monday west of that line.)

Other information was available. Informed guesses could be made about when Japan would run out of oil and/or foreign currency. The following example (taken from NARA RG80, Box 92, entry 167CC, exhibit 1, item q) has been quoted several times in the literature on the Pearl Harbor raid. It is a message from the British Secret Intelligence Service agent in Manila to his counterpart in Honolulu, dated 3 December 1941:

> We have received considerable intelligence confirming following developments in Indo-China.
>
> A1.   Accelerated Japanese preparation of air fields and railways.
> A2.   Arrival since Nov 10 of additional 100,000 repeat 100,000 troops and considerable quantities fighters, medium bombers, tanks and guns (25 mm).
>
> Estimates of specific quantities have already been telegraphed Washington Nov 21 by American Military Intelligence here.
>   Our considered opinion concludes that Japan envisages early hostilities with Britain and US. Japan does not repeat not intend to attack Russia at present but will act in South.
>   You may inform Chiefs of American Military and Naval Intelligence, Honolulu.

7.7   The Aftermath: Processing and Distribution of Intelligence

The RN had investigated the linking of cryptography and intelligence from 1937, examining the lessons from WW1 and the Spanish Civil War. This resulted[13] in the setting up of an Operational Intelligence Centre (OIC) in London.

The US Army arranged that the Chicago lawyer Alfred McCormack examine what could be learned from the Pearl Harbor raid. It set up a Special Branch of Military Intelligence in May 1942, with a large staff and with McCormack playing a principal part.

The USN took rather longer to reach better practice on pooling and distributing intelligence. For example, General MacArthur had difficulty in getting the full picture[14] after his withdrawal to Australia in 1942. By 1943 the USN was active in the Joint Intelligence Center, Pacific Ocean Area (JICPOA).

---

[13] See page 67 of Mavis Batey's *Dilly: The Man who broke Enigmas* (2009).

[14] The RN had established an Operational Intelligence Centre (OIC) before the war. See Chap. 6. By early 1940, Churchill had in place an effective Joint Intelligence Committee that was responsible for analysis of war-related intelligence material across all services. This Committee was 'the central agency for producing operational intelligence appreciations and for bringing them to the attention of the Prime Minister, the War Cabinet, and the Chiefs of Staff'. There was a Combined Operational Intelligence Centre established in Melbourne well before the Pearl Harbor raid: MacArthur had it moved into his headquarters in 1942.

7.8 The Aftermath: Naval Comint

Since the IJN ships of the raiding force were forbidden to transmit messages once at sea, the JN-25 system change on 4 December 1941 had no direct effect on the Pearl Harbor situation. The change initially rendered the then current intercepts of IJN operational messages totally unreadable. The Allied cryptanalysts would have assumed that JN-25B had been totally replaced. In fact this had not happened. As will be discussed in Sects. 9.15 and 10.8, the changes made on 4 December *minimally enhanced the security of IJN operational messages.* The blunder in not *totally* replacing the JN-25B system early in December 1941 was of enormous importance, setting the IJN on the road to the Coral Sea and Midway and the loss of four of its major aircraft carriers, numerous planes and trained crew in May and June 1942. With Admiral King being the head of the USN, the situation in March, April and May 1942 would be as in Shakespeare's *Henry V* (II, ii, 6 and already quoted at the end of Chap. 3):

> The King hath note of all that they intend
> By interception which they dream not of.

The challenge to the Allied naval commanders was to find a way to exploit this advantage before the inevitable total replacement of JN-25B8.

Appendix 1 The Mackenzie King Diaries

One high-level participant in WW2 left extremely extensive diaries. He was W. L. Mackenzie King, Prime Minister of Canada. The following (slightly corrected) extracts are taken from the Canadian Archives website.

> [27 November 1941]   Read dispatches as to situation between the United States and Japan which seems to have reached a deadlock which the Japanese alone could break. I feel they will attack the Burma Road from Indo-China or perhaps Thailand, but that war in the Pacific is practically inevitable. . . . For I know that with supplies cut off Japan must lose in the end but there will be again an appalling sacrifice of life before she does and a world left more than ever in ashes.
> [30 November 1941]   . . . but Heaven knows when Congress, if at all, would consent to the US going into the war.
> [1 December 1941]   Read to Cabinet all telegrams recently received re Japan. Also my cable sent to Churchill last night.
> [1 December 1941]   Once the US and the world got the impression that it was Britain fighting Japan to save some of her outposts, it would probably result in it being even more difficult than ever to bring the US into the world conflict as a fighting force and without the United States, without the aid of the fighting forces of the US, British forces would be insufficient to cope with those of Japan in the Orient.
> [1 December 1941]   The US is clearly not ready and will not be ready for another 4 or 5 months according to their own communication.

This last point was no doubt evident to the Japanese Government: any delay would be to the detriment of its cause.

[2 December 1941] We all knew that the British Government held the view that without active assistance of the US Britain could not win. Until that assistance came we should, therefore, do everything possible to prevent defeat.

[5 December 1941] The situation is graver than ever and I shall be amazed if, before the week is out, war does not take place in the Far East. Possibly Sunday, as has so often happened, may be the date chosen for the beginning of a fresh war.

The extra risk on Sunday 7 December was thus visible to Mackenzie King.

[7 December 1941] It was an immense relief to my mind, however, to know that their attack had been upon the US in the first instance and the opening shots were not between Great Britain and Japan.

He was not yet aware of the attack upon Malaya having started shortly before the attack on Pearl Harbor. He confirmed that immediate American involvement would be to the immense advantage of Britain and Canada.

[7 December 1941] [Robertson] said that it was understood Germany and Italy might declare war on the United States tonight. This is the most crucial moment in all the world's history but I believe the result will be, in the end, to shorten the war.

The governments of Japan, Germany and Italy did not have to bring the United States into the war at all. War in Asia against the British and Dutch would have been quite enough. Doing so must be one of the biggest blunders of WW2. The consequences of the 1917 Zimmermann telegram had been forgotten. Yet in war madness may produce surprise (Churchill).

Appendix 2   The 1945–1946 Investigation

The 1945–1946 Congressional Investigation into the Pearl Harbor Attack produced 11 volumes of transcripts and numerous volumes of exhibited documents. All this may now be inspected on the web. William D. Mitchell (1874–1955, and not to be confused with the aviator), the first general counsel assisting the investigation, stated (volume 4, page 1585):

I had and still have a definite conviction that the real purpose of this committee was to present facts which would permit a final answer to the basic question *Who was responsible for the failure of our forces at Hawaii to be on the alert and for the admitted failure to use to the best advantage such defensive facilities as were available at Pearl Harbor?*.

Admiral Kimmel, the USN Pacific Commander in 1941, gave evidence that 24 hours notice, but preferably 48, was needed to prepare for an attack. This seems reasonable.

The investigation devoted too much time to the probably non-existent winds message and also to the final 15-part Purple message to the Japanese embassy in Washington telling it to break off relations with the United States soon after the attack on Pearl Harbor.

President Truman (volume 1, page 8) had retained the security on signals intelligence but had made an exception for the investigation. Almost all of the secret information divulged was about the 'Purple' diplomatic code.

The annual budgets for the USN for 1932–1941 are set out on volume 1, page 363. There had been a vast leap in 1941, but the new ships started then were to come into service only in 1943. Indeed a high-level meeting on 3 November 1941 (volume 2, page 650) had concluded that the United States was not yet ready for a major conflict. Admiral Kimmel stated (volume 6, page 2498):

> Japan, at the outbreak of hostilities, had nine aircraft carriers in commission and operating. We had three carriers in the Pacific and these did not have their full quota of planes.

He added that fuel stocks in Hawaii were inadequate in 1941 and there were not enough tankers to keep the fleet going at sea.

The investigation did receive enough evidence as to the general risk of war in the Pacific. In fact warnings had been sent to Admiral Kimmel, and Army (Air Force) General Short in Hawaii. Of interest here is a question asked of General Marshall (volume 3, page 1149):

> Did you think that the first Japanese attack would be at Pearl Harbor?

He replied:

> I did not anticipate that. I thought they would not hazard that.

Captain Alwin Kramer stated (volume 9, page 3959):

> It was the consensus of the opinion of my associates and many of the high officials in Washington that it was very illogical and foolish on the part of Japan to undertake open warfare with the United States, that it was almost inconceivable that they would in view of the fact that it is very likely that they would get everything that they wanted and as they had got in French Indochina and what they wanted south of French Indochina, without any action being taken by the United States.

This differs from the opinion of Admiral Leahy (volume 1, page 348):

> I had thought that war was a likely contingency for several years and I was practically certain in my own mind that it was going to come at some point in the reasonably near future.

These last three quotations scarcely involve signals intelligence.

The evidence of an attack being imminent came from at least five sources:

(1) Traffic analysis had failed to locate certain IJN carriers;
(2) The major IJN operational cipher JN-25, of which much more later, had become indecipherable on 4 December 1941;
(3) The intercepted November instructions in a medium level cipher system to the Japanese consulate in Hawaii to make certain observations at Pearl Harbor;
(4) The 'Haruna' message discussed earlier and the burning of documents at various Japanese consulates;
(5) The 'say very secretly' Purple message quoted by Churchill and discussed earlier.

Evidence had been presented that the weekend was the likely time for any attack. Mackenzie King worked this one out! The investigation did consider (volume 5, page 2108) the *Duty to keep Fleet Commanders informed of Political and Military Developments*. Kimmel and Short had received a general warning of the possible outbreak of war earlier. This may or may not be considered to be enough to require them to act on their own initiative. On the other hand a better central intelligence unit in Washington should have been able to assemble the evidence (3), (4) and (5) and send specific instructions to Kimmel and Short 48 h ahead of the attack.

In retrospect the investigation should also have covered the lack of precautions taken by American forces in the Philippines.

Part II
Technical

Chapter 8
Major Encryption Systems

Several of the cipher machines developed and used as encryption systems before and during WW2 are briefly described, as is a typical additive (cipher) system. Their respective cryptanalyses have both common and distinctive features, resulting in a fundamental difference emerging in relation to the way messages sent in each cipher system could be fully transformed back to the original plain text.

8.1 Cipher Machines

In general, secure cipher machines had to be capable of being adjusted by their operators. There were two different types of adjustment, called here key settings and message settings. Key settings were issued by central management and were implemented simultaneously on a regular basis (often daily) across a complete machine network. Message settings, chosen randomly each time the sending operators had to transmit a message, had to be communicated to the intended recipient(s) in a form that was unintelligible to those outside the network and intercepting the message. The necessary information was sent with the message, sometimes hidden in it, with the word *indicators* being used to describe that part of the message with that function. Security of the indicator against cryptanalysis was of vital importance. The word 'indicators' is used analogously in the later account of additive systems.

8.2 Purple

The Japanese government had used in the 1930s a cipher machine called Type A for the exchange of high level messages with its major embassies. Both the Americans and the British had broken it completely by 1936. The British had used electrical

© Springer International Publishing Switzerland 2014
P. Donovan, J. Mack, *Code Breaking in the Pacific*,
DOI 10.1007/978-3-319-08278-3_8

relays and other devices to build a machine (the 'J' machine) to copy its operation. The cover name Red was introduced by Friedman's SIS for this machine and the traffic generated by it. In 1938 Red decrypts indicated the coming introduction of a new diplomatic cipher machine, Type B. Messages using the new machine began to be intercepted in 1939. The British GCCS did not have the resources to attack it. As mentioned in Sect. 4.3, the SIS took it on as its highest priority. The cover name given for Type B was Purple.[1]

Previous work on the Red machine and its method of operation proved extremely useful in investigating Purple. It used the Latin alphabet and so (presumably) the Japanese characters were transmitted by the Roma-ji system. The Purple machine appeared to use some of the technology of the known Red machine. Instead of using a single encryption system for all 26 letters, it used one encryption system for six letters (which were not always the same) and another for the 20 others. This looks like a gratuitous piece of assistance to enemy cryptanalysts and it was exactly that. Eventually one of the SIS team working on it had the thought that standard telephone switching apparatus might have been used instead of rotors. This was in fact correct. Furthermore, negotiations were carried out in Washington between Japanese diplomats and the State Department with reports being sent back to Tokyo encrypted by the Purple machine. The cryptanalysts were informed of the likely content of these reports and therefore knew what they were looking for in decrypts.

Friedman's report (see Note 1) of October 1940 contains some, but far from all, technical details. This extract illustrates the sophistication of WW2 cryptology. Frank Rowlett was a major player in this achievement.

> 7. In all, the plain texts for parts of some 15 fairly lengthy messages were obtained by the methods indicated above, and these were subjected to the most intensive and exhaustive cryptanalytic studies. To the consternation of the cryptanalysts, it was found that not only was there a complete and absolute absence of any causal repetitions within any single message, no matter how long, or between two messages with different indicators on the same day, but also that when repetitions of three, or occasionally four, cipher letters were found, these never represented the same plain text. In fact, a statistical calculation gave the astonishing result that the number of repetitions actually present in these cryptograms was less than the number to be expected had the letters comprising them been drawn at random out of a hat! Apparently, the machine had with malicious intent—but brilliantly— been constructed to suppress all plain text repetition. Nevertheless, the cryptanalysts had a feeling that this very circumstance would, in the final analysis, prove to be the 'undoing' of the system and mechanism. And so it turned out!

Yes indeed, statistical calculations come into WW2 cryptanalysis!

Team efforts were common in the cryptanalysis of WW2. Friedman ended his report by naming and thanking around 25 members of his staff. This reveals the

---

[1]References here are Frank Rowlett: *The Story of Magic: Memoirs of an American Cryptologic Pioneer*, Aegean Park Press, 1988, and Ronald Clark: *The Man who Broke Purple*, London 1977. The book *Machine Cryptology and Modern Cryptanalysis* by C. Deavours and L. Kruh (Artech, 1985) gives some technical information. But the key document is Friedman's report of 14 October 1940, entitled *Preliminary Historical Report on the Solution of the 'B' Machine*, which is NARA RG457 document SRH-159 and is in the *cryptocellar* section of Frode Weierud's website.

size and capacity of the unit that had been developed since 1930. He thanked Commander Safford of Op-20-G for arranging the construction of a working machine for decrypting Purple traffic. In fact, the American reconstructions of the Purple device have survived to this day while only a few fragments of the Japanese original were ever found.

Purple decrypts, called Magic (as were all decrypts from this section of the SIS), received massive publicity in the various inquiries into the Pearl Harbor raid of 7 December 1941. But messages about IJA and IJN operations were rarely sent by diplomatic ciphers. Information garnered from Purple decrypts was much more useful in the European Theatre of Operations than in the Pacific. Reports from Japanese diplomats in Europe were of considerable assistance to those planning Allied attacks, notably the June 1944 invasion of Normandy.

8.3   The Polish Work on the Early Enigma Machines

Only a brief overview of the Enigma saga is given here. Fuller details are given in Mavis Batey's book *Dilly: The Man Who Broke Enigmas* and elsewhere.

The Enigma encryption machine had been available commercially from the early 1920s. Documents describing its technical details had been lodged with the British Patent Office and no doubt elsewhere. It was a natural enough development. Enigma was an electrical machine that had a typewriter style keyboard with 26 keys, one for each letter, and above that 26 lights, again one for each letter. Plain text would be typed letter by letter into the machine. Then the encryption of each letter was displayed by one of 26 lights being illuminated. The sequence of letters displayed by the lights was copied down and formed the text of the encrypted message to be transmitted by the radio operator. The same procedure was used for decryption.

The main early Enigma machines used three internal rotors and a reflector. The rotors were selected from a set of five provided with the machine. Each rotor contained wiring linking 26 electrical contact points on each side. When linked together the three rotors determined the encryption of each typed letter. Prior to encrypting the next letter, one or more of the rotors was turned automatically and this changed the encryption of that next letter. An extra complication was in the allocation (Ringstellung) of the 26 letters cyclically to the nodes of each rotor: this will not be described here. German military versions included an extra component, a plugboard, which further complicated the encryption and decryption processes.

Security of the Enigma ciphers depended firstly upon keeping the wiring of the rotors secret and secondly upon the enormous number of possible initial configurations. In practice, the choice of rotors, their order in the machine and the plugboard configuration were considered to be keys (in the sense described earlier) and specified by common written daily instructions for each of the various networks. Until late 1938, these written instructions also specified the initial positions (Grundstellung) of the rotors.

Prior to sending a message, the operator was also required to choose a new initial position for each of the rotors. This choice had to be conveyed to the message recipient by including an indicator in the message. Of crucial importance to those attacking the cipher in the above period was that the indicator was encrypted twice.

For example, suppose WDV was specified as the initial rotor positions and the operator randomly chose MHT as the new initial message setting. Having set the rotors to WDV, the letters MHT MHT would be typed in obtaining, say, ZXC HTJ. The rotors were then set to MHT and the message text encrypted. The message transmitted by the radio operator included both encryptions of the indicators, here ZXC HTJ, and the encrypted text.

Enigma was designed to be reversible—the recipient, with machine set to the same initial configuration, typed in ZXC HTJ, obtaining MHT MHT. Now, with the rotors reset to MHT, typing in the received message text would produce the original plain text of the message.

In 1928, Polish military radio interception stations started picking up German Army messages in a new and sophisticated cipher. By 1931 it was decided to employ some young mathematicians to work on the intercepts. Marian Rejewski, and later, Jerzy Rozycki and Henryk Zygalski were chosen on the basis of their performance in a secret course for budding cryptanalysts held a few years earlier. In the same year a German, Hans Thilo Schmidt, approached the French Secret Service. Schmidt eventually provided some documentation of the military Enigma in return for a substantial payment. The Polish Cipher Bureau received copies of this material from the French. Information was also obtained from German Army training messages sent in 1931–1932.

The commercially available civilian Enigma placed the 26 letters A to Z on what might be called the entrance disk in the order used on a standard German typewriter keyboard. It became evident that some different order was being used on the military version. (The GCCS was also working on the new cipher but was unable to determine[2] what that order was and so was stymied.) Rejewski had the initiative to try alphabetical order—and it worked! Had the German military version used a less obvious order the Enigma might never have been broken.

Early in 1933, Rejewski developed a way[3] to exploit the transmissions of the Enigma indicators using the two different encryptions. On a given day it was

---

[2]It is only fair to point out that a female junior member of the staff of GCCS had advocated testing to see whether alphabetical order worked. Some more on this incident is to be found in Mavis Batey's book *Dilly: The Man Who Broke Enigmas*.

NARA item RG457, Box 580, NR 1417 *Tentative Use of Enigma and other Machine Uses* may be found on the Tony Sale web site. It gives information on *Jade* and *Coral*.

[3]The method of using this redundant encryption is too complicated to explain here. Undoubtedly it was a great achievement. The analogous blunder with the IJA cipher 2468 was exploited in a quite different way. See Chap. 14, particularly its Appendix 2.

Cryptologist Abraham Sinkov of the US Army was shown the Enigma and work on deciphering it at Bletchley Park in January and February 1941. It is probable but not certain that this assisted him in assessing the importance of the early stages of finding patterns in the indicators of cipher 2468 at Central Bureau in Brisbane early in 1943.

possible to use the first six letters from a total of 60–80 messages to determine a short list of possible keys for that day.

Up to May 1940, the German Army and Air Force continued to transmit two encryptions of Enigma indicators. Despite some changes to the machine and to operating practices, this gross insecurity remained until then. The German Navy had stopped this practice in May 1937. In May 1940 Bletchley Park had to move on to the *Herivel tip*, which depended upon the prevalence of inertia among enemy operators.

In June 1939, the Polish Secret Service realised that a German invasion of Poland was inevitable. So it invited the GCCS and its French equivalent to send a few appropriate people to learn what was occurring. The GCCS received reproductions of the German military Enigma machine and information about the methods then available to analyse Enigma traffic. After the fall of Poland, and then of France, the work on Enigma continued in Britain with some naval work going (later) to the United States.

Without the Polish work on Enigma, and, prior to that, the information obtained by the French Secret Service from its informant, the GCCS work on Enigma traffic would at best have been running 12–24 months behind its actual timing. All of this would have made it much more difficult for Britain to survive in the period 1940–1943.

The German Naval Enigma was upgraded to have eight wheels available as rotors or reflectors. This made decryption harder, even though the wiring of the wheels was known, mostly as results of captures. Some special equipment needed by Op-20-G for Enigma decryption was made at the Naval Computing Laboratory located in Building 26 of the National Cash Register Company (NCR) in Dayton, Ohio. This facility also made some equipment appropriate for Japanese naval ciphers.

The Melbourne NAA file MP1074/7 contains a message dated 26 April 1941 in which the RAN is seeking British comment on 'the degree of security provided by the Enigma'. Apparently the Dutch[4] military in the NEI used one version of the Enigma and offered one to the Melbourne naval communications room.

8.4   A Remarkable Document

Gilbert Bloch and Ralph Erskine published a paper *Enigma: The Dropping of the Double Encipherment* in *Cryptologia* 10(3), July 1986, 134–141. This reproduced the German Army order that the redundant encryption of Enigma indicators was to

---

[4]The Melbourne file is MP1074/7 1/4/1941 to 30/4/1941, barcode 4169362. This gem is buried among thousands of other high-level secret messages of the WW2 era. The British response may well be in MP1074/7 too, but is better left to the imagination. On page 79 of *The Man who Broke Purple*, Ronald Clark states that Friedman was told before 1930 by the Dutch Army that it was making some use of (civilian) Enigma.

stop effective 1 May 1940. The original text is reproduced photographically. The instruction for encrypting indicators is that it should be done *EINMAL*, that is, *ONCE*. The reasoning behind this decree appears not to have reached the Japanese communications security people. Quite remarkably, the IJA started transmitting two separate encipherments of the indicators of its Water Transport Code from its inception in December 1942. Details of how this was detected and exploited are given in Sects. 14.5–14.9.

8.5   Enciphered Teleprinters

Teleprinters using the 5-bit Baudot code, as briefly described in Sect. 1.11, also provided a transmission system capable of encipherment. In 1917 Gilbert Vernam, an American communications engineer, realised that this code could be used for encryption purposes.[5] Two different quintuples **a** and **b** can be *added* by adding the corresponding bits and taking the remainders upon division by 2. For example:

$$
\begin{array}{ccc}
1 & 0 & 1 \\
1 & 1 & 0 \\
0 + 0 & = & 0 \\
0 & 1 & 1 \\
1 & 1 & 0
\end{array}
$$

illustrates binary addition. Indeed the rules:

$$0 + 0 = 0, \qquad 0 + 1 = 1, \qquad 1 + 0 = 0, \qquad 1 + 1 = 0,$$

are appropriate for electro-mechanical implementation. Vernam's idea was that Baudot code be encrypted by providing both the sender and the recipient with a string of quintuples of bits on a paper tape and sending over the wires (and, later, over the air) the sum of corresponding quintuples. Thus if the message was 'that Baudot code':

---

[5] Steven Bellovin found an old code book in the Library of Congress which describes the invention of the one-time pad by Frank Miller of California some 25 years earlier. See his paper in *Cryptologia* 35(3), July 2011, 203–222. It is not clear whether Miller's idea was ever implemented. It is possible that the idea made its way to Vernam. The principal text in this chapter has been left without incorporating Miller's contribution to additive cipher systems and the one-time pad.

T	H	A	T	sp	B	A	U	D	O	T	sp	C	O	D	E
1	1	0	1	0	1	0	0	0	1	1	0	0	1	0	0
0	0	0	0	0	1	0	0	1	1	0	0	1	1	1	0
0	1	0	0	1	0	0	1	0	0	0	1	1	0	0	0
0	0	1	0	0	0	1	1	0	0	0	0	1	0	0	0
0	0	1	0	0	1	1	1	1	0	0	0	0	0	1	1

and the first 16 quintuples on the encrypting tape were:

1	0	1	1	0	0	1	0	1	1	1	1	1	0	0	1
0	1	1	1	0	0	0	1	1	1	0	0	1	0	1	1
0	1	0	1	0	0	1	1	1	0	1	0	0	0	1	1
1	1	1	0	1	1	1	1	1	1	1	1	0	0	1	1
1	1	0	1	0	0	1	0	1	1	0	1	1	0	1	0

then the transmitted signal is:

0	1	1	0	0	1	1	0	1	0	0	0	1	1	0	1
0	1	1	1	0	1	0	1	0	0	0	0	0	1	0	1
0	0	0	1	1	0	1	0	1	0	1	1	1	0	1	1
1	1	0	0	1	1	0	0	1	1	1	0	1	0	1	1
1	1	1	1	0	1	0	1	0	1	0	0	1	0	0	1

As with Enigma, the decryption and encryption processes coincided.

The Baudot code could be and was represented by holes punched on a paper tape. This was used for signals sent over telegraph wires and printed out mechanically by an appropriate machine. The transmission of the code by radio could also be mechanised. The third row, in which each hole (the sprocket hole) is punched as part of the manufacturing process, is used to guide the punching and reading machinery and is not part of the message. For example, the final string of quintuples given above would yield the tape segment:

Here 'o' is used to denote an unused location for a hole and '•' is used for a hole. A zero (0) on the signal to be transmitted corresponded to a non-punched hole, while a 1 corresponded to a punched hole.

Captain Joseph Mauborgne of the US Army, who later very usefully supported Friedman's unit, now contributed the thought that if the tapes of additives were random and never used again, the encryption would be secure. In fact, the distribution of vast quantities of randomly punched tape proved to be scarcely feasible. This was the first widely recognised combination of use of a one-time pad (OTP) with an additive encipherment. The decimal version of this was heavily used by the Japanese, who did not have a western-style alphabet, from 1937, if not earlier.

What was in fact used for a while in the United States was to have two tapes of additives, one with 1,000 randomly chosen quintuples of bits and one with 999. The transmitting and receiving machines would read one quintuple from each tape at a time. The encryption was formed by adding to the Baudot code of the letter or digit to be transmitted the sum of the two random quintuples being read. The decryption was carried out in the same way. After a letter or digit was processed, each tape would be advanced one space. As 1,000 and 999 have no common factors, the encryption process starts repeating itself only after 999,000 operations. This was not particularly secure. Indeed, in 1919, Friedman[6] managed to break it by examining samples sent to him by the Army Signals Office.

There is another fairly obvious way of enciphering Baudot code without using too much tape. Five different numbers without common factors are chosen. For example, they might well be 41, 31, 29, 26 and 23. Next, five sequences of 0s and 1s are chosen randomly, the first with 41 digits, the second with 31, the third with 29, the fourth with 26 and the fifth with 23. Then a tape with $22,041,682 = 41 \times 31 \times 29 \times 26 \times 23$ of these encrypting additives may be constructed by using the first sequence of 41 over and over again in the first row, the second of 31 similarly for the second row and so on. It might not be necessary to produce tapes at all: the effect could be obtained by having cogwheels with appropriate numbers of teeth carrying out the appropriate electrical trickery.

For example if the five sequences are chosen to be:

```
row 1  (41 bits)  00100  10000  11111  10110  10101  00010  00100  00101  1
row 2  (31 bits)  01000  11000  01000  11010  01100  01001  1
row 3  (29 bits)  00011  00110  00101  00010  11100  0000
row 4  (26 bits)  01101  11000  00111  00110  10002  0
row 5  (23 bits)  01010  01000  00010  01001  110
```

[6]See pages 52–53 of Ronald Clark's *The Man Who Broke Purple*.

the first 75 of these additives would then look like:

1 to 25

0	0	1	0	0	1	0	0	0	0	1	1	1	1	1	1	0	1	1	0	1	0	1	0	1
0	1	0	0	0	1	1	0	0	0	0	1	0	0	0	1	1	0	1	0	0	1	1	0	0
0	0	0	1	1	0	0	1	1	0	0	0	1	0	1	0	0	0	1	0	1	1	1	0	0
0	1	1	0	1	1	1	0	0	0	0	0	1	1	1	0	0	1	1	0	1	0	0	0	1
0	1	0	1	0	0	1	0	0	0	0	0	1	0	0	1	0	0	1	1	1	0	0	1	

26 to 50

0	0	0	1	0	0	0	1	0	0	0	0	1	0	1	1	0	0	1	0	0	1	0	0	0
0	1	0	0	1	1	0	1	0	0	0	1	1	0	0	0	0	1	0	0	0	1	1	0	1
0	0	0	0	0	0	0	1	1	0	0	1	1	0	0	0	1	0	1	0	0	0	1	0	1
0	0	1	1	0	1	1	1	0	0	0	0	0	1	1	1	0	0	1	1	0	1	0	0	0
0	1	0	0	1	0	0	0	0	0	0	1	0	0	1	0	0	1	1	1	0	0	1	0	1

51 to 75

0	1	1	1	1	1	1	0	1	1	0	1	0	1	0	1	0	0	0	1	0	0	0	1	0
0	0	1	1	0	0	0	1	0	0	1	1	0	1	0	0	0	1	1	0	0	0	0	1	0
1	1	0	0	0	0	0	0	0	0	0	1	1	0	0	1	1	0	0	0	1	0	1	0	0
1	0	0	1	1	0	1	1	1	0	0	0	0	0	1	1	1	0	0	1	1	0	1	0	0
0	0	1	0	0	0	0	0	1	0	0	1	0	0	1	1	1	0	0	1	0	1	0	0	

At least if the enemy cryptographers had some idea of what was going on, this by itself would not be much more complicated than the system rejected by Friedman earlier as being inadequately secure.

The Lorenz SZ42 Schlusselzusatz was an enciphering attachment for a conventional teleprinter. It could implement a much more complicated version of such an encryption scheme, using twelve wheels rather than just five. Initially the starting position in the additive was communicated to the intended recipient by indicators transmitted immediately before the encrypted message. The decryption was likewise fully mechanised. This was much more efficient than any machine of Enigma type or any system using code books. However, the equipment was suitable only for major headquarters. A rival machine, the Siemens T52 Geheimschreiber, was also in use by the German military at the time. It had the encryption and teleprinter facilities integrated in the one device.

John Tiltman[7] felt obliged to apologise to William Friedman in 1942 for having to divert his research team from the Japanese Army material to what became a successful attack on the SZ42.[8]

---

[7]Tiltman's letter is mentioned by Paul Gannon on page 146 of his *Colossus: Bletchley Park's Greatest Secret*. The notes refer to TNA files HW 14/24 of 12 December 1941 and HW 14/46 of 29 May 1942.

[8]The person most responsible for this achievement, Bill Tutte, gave an account of his role on 19 June 1998 in the talk *FISH and I*, now widely available. An original *SZ42* is in the library of the Mathematics Department of the University of Uppsala in Sweden. A photograph, by David Kahn, is in *Cryptologia*, 3(4), October 1979, page 210.

Its success depended on one major breach of secure cipher practices and the genius of Tiltman and Bill Tutte (pronounced 'Tut'), a junior mathematician in his team. In essence, a badly trained operator[9] sent two slightly different versions of the one message using the same indicators—the now celebrated HQIBPEXEZMUG— on 30 August 1941. Tiltman was able to deduce from these and a bit of stereotyped material at the beginning what the two versions were and so what was the long stretch of additive that had been generated and used. Eventually Tutte and others in the team were able to work out the structure of the machine used.

Later, after the Germans dispensed altogether with broadcasting indicators and sent substitute information by courier in advance, the GCCS understood the system well enough and was able to devise special machinery[10] to compute the indicators.

From the viewpoint of those studying the cipher war in the Pacific, the work against the encrypted teleprinters is of interest mostly because it produced some highly ingenious special machinery, generally accepted to have been the first electronic calculating devices. The gadgetry needed in the Pacific was tied in with strictly decimal rather than binary calculation and some quite remarkable special-purpose electro-mechanical and electro-optical machinery was built to assist the codebreakers, but that is all.

The major battle of 1943 was the Battle of Kursk, in which the German Army failed to breach a salient that had been very heavily fortified by the Red (Soviet) Army. Harry Hinsley, special assistant to the Director of GCCS in the later years of the war, recorded that Sigint, derived from Enigma and the SZ42, was accumulated at Bletchley Park both in advance of and during this battle and passed on to the Soviets in good time. This is one instance of the great importance of Sigint at key stages of WW2.

Another key use of intelligence occurred in June 1944. As mentioned briefly before, there had been an intensive disinformation campaign to get the German High Command to believe that the then prospective invasion of France would take place near Calais. General Eisenhower received early in June 1944 from decrypted SZ42

---

[9]The key *GYP-1 Bible* or *Cryptanalysis of JN-25* (NARA RG38, CNSG Library, Boxes 16 and 18, 3222/65) comments on another incident on page 9: 'This is an illustration of the fact that errors committed by the enemy afford most valuable information as to the nature of his system of secret communication'.

[10]On page 270 of his 1984 book *Enigma*, W. Kozaczuk quotes Peter Calvocoressi, a senior figure at GCCS in WW2, as saying: 'In order to break a machine cipher, two things are needed: mathematical theory and mechanical aids.' This applies to some of the non-machine ciphers of the era, and in particular to JN-25. The 'Hall weights' of Sect. 15.3 are an example of a mathematical theory.

Stephen Budiansky has written more on this subject. See, for example, *Codebreaking with IBM machines in WW2* in *Cryptologia* 25(4), October 2001, 241–255.

traffic the absolutely vital information[11] that this deception was working. He was thus able to order the planned Normandy invasion to go ahead a few days later.

8.6 The Typex

The Type X or Typex machine was in effect the British version of the commercial Enigma. It had certain technical improvements, for example, it had several alternative sets of rotors, which was a useful and inexpensive extra security feature. In practice, the Typex seems to have been secure.

In general, the internal wiring of its rotors was kept secret.[12] This could have been achieved with the German Naval Enigma, say, by issuing a new set of rotors with a fixed starting date for their use.

Supplies of this machine were initially inadequate. Thus the merchant navy in the Atlantic had to use additive cipher systems which in practice were not always secure.

The RAAF procedures for cipher officers for the later years of WW2 are of some considerable interest. Wing Commander J. P. Lees lectured[13] on cryptographic security for these officers in April 1945. He recommended high grade machine ciphers for long term security while additive cipher systems (to be discussed shortly) had only medium security. He advocated serious planning for any cryptographic system, including consideration of likely traffic loadings. Two key comments on cipher security were given:

---

[11]Much useful information about the defences of Normandy was picked up by decrypting reports sent back to Tokyo by Japanese diplomats using Purple. See Carl Boyd, *Hitler's Japanese Confidant: General Oshima and Magic Intelligence 1941–1945*, University of Kansas Press 1993. Further, the ISK group at Bletchley Park, originally headed by Dilly Knox, was able to read the Enigma traffic of the German intelligence service and so check that all German spies in the UK had been identified and were under the control of the British. Thus the deception campaign was known to be working well.

[12]In 1940 the German Army seized a British Typex machine—broadly similar to the Enigma—without the rotors. A decision was made that allocating serious resources into investigating it would not be justified.

In fact some details of the wiring and other aspects of the Enigma and its rotors were initially obtained by bribery from H.-T. Schmidt and later by physical capture. The Typex instructions issued in Australia—no doubt at the instigation of the British—included having a hammer around to facilitate the destruction of the rotors just prior to being captured. More on this matter is in Sect. 13.7.

[13]The *Lectures at Conference of Chief Cipher Officers* of April 1945 survive in the NAA Canberra as item A705 201/23/453. The master plan for RAAF cipher security was orally explained but is not in the written version. Item A705 201/24/510 makes it clear that additive cipher systems were being used less by the RAAF by July 1944 and even less by April 1945. Item A705 201/28/323 (1945) *Cyphers—procedure re type X indicators* includes on page 46A the quoted comment on insecure encipherment of indicators. The final quote is from A705 201/28/325—*Typex indicators*.

By far the greatest dangers, however, to the security of a cryptographic system are the errors and omissions on the part of operating personnel.

and[14]

The prime requirement for cryptographic security is that the procedure laid down for use of a system shall be followed to the letter completely and without deviation.

Another file comments on possible double encryption of Typex indicators. The following text is memorable:

Response to your signal of 8 Feb 1945. PRACTICE DANGEROUS TO CIPHER SECURITY. General instructions being promulgated.

Another file gives advice (4 March 1945) from the British Air Ministry to the RAAF about Typex indicators:

Any attempt to alter existing indicator system would have serious cryptographic drawbacks since indicator system now used plays very large part in overall security of system. ... Regarding Typex, complications of indicator system are absolutely necessary.

8.7   The American Sigaba or ECM

The USN had been interested in security of communications from around 1920, perhaps reacting to the consequences to the German Navy of insecure communications in WW1. It sought some suitable Electrical Cipher Machine or ECM. The weight or electrical requirements of such a device were not particularly a problem for naval use. It thus monitored the early rotor machines and the development of the Vernam device.

As noted earlier, Friedman had worked out some of the innovations behind such a machine in 1933 and his protegé Frank Rowlett then invented some extra features. Joseph Wenger and Laurance Safford of Op-20-G took up the matter in 1937 and set about getting a working model into production. 'Ham' Wright, later to play a crucial role in the final work of decrypting the JN-25B messages that set up the Battle of Midway, recalled helping with the design in 1938. The machine was examined by the available experts from both the USN and the US Army. In Friedman's phrase, it was way ahead of anything else known. It was in operation before the Pearl Harbor raid.

In addition to its more sophisticated design, the ECM (USN terminology) or Sigaba (Army terminology) had, in common with the Typex, several alternative sets of rotors. This again put it ahead of Enigma.

---

[14]There is great irony here. The JN-25 operators were given written instructions to 'tail' and enough actually did so. This practice greatly weakened the security of various JN-25 systems in 1940–1943. Much more on this is given in Sect. 9.22.

The Sigaba/ECM appears to have been totally successful in defying cryptanalysis in WW2. The reasons for this would appear to be that it was appreciably more sophisticated than the Enigma, yet the latter, particularly the naval version, was very difficult to break. Physical capture of the rotors used in the naval Enigma turned out to be essential for the design of the machinery used against it. Conversely the American machine appears never to have fallen into enemy hands. It was appreciated that a captured Sigaba might have indicated methods of upgrading the security of the Enigma. In fact the machine was not even made available to the British. This created problems but a common machine was produced for communications between the two allies.

The ultimate in Enigma-style encrypting machines appears to have been the KL-7 (KLB-7) made by the NSA. It was phased out when modern electronic calculation became widely available.

8.8   Decryption of a Machine Cipher

Once it is suspected or known that new intercepts are of messages sent using a machine cipher, two possibly formidable tasks confront the cryptanalysts: reconstruction of the machine being used and then determination of its 'operating rules', that is its key and message setting instructions and the data needed to implement them. Thus destruction of cipher machines under threat of capture would be an absolute priority action and so such machines were rarely found in a potentially useful form until the last few months of WW2. Capture of the key setting manuals did occasionally happen (such as with the Naval Enigma in 1942) but the information in them would usually apply fully only for a short period of time. Friedman's team at SIS reconstructed the Purple cipher machine by combining information derived from the breaking of the previous Red machine with sheer cryptanalytic skill.

The GCCS team that broke the German SZ42 system did so by hand cryptanalysis exploiting a serious error made by one operator on one occasion. The next part—finding out how to set a machine so that it can decrypt the text of an intercepted message—usually required the discovery of a procedure that efficiently eliminated possibilities, using whatever cribs were available, until the correct setting was found. This procedure could be mechanised. Running the intercepts through the set machine would now produce the text of the original message. This would then be handed over to expert linguists and service personnel for translation and evaluation.

The essential point to be made here is that once a machine cipher is broken to the stage where a decrypted intercept is produced, the output is the original (but possibly slightly garbled) message text. This is in stark contrast to the process of reading intercepts sent using an additive cipher system, which uses a code book protected by an extra encryption process.

8.9   Additive Cipher Systems

The systems described here used code books whose entries (the book groups) were numeric, that is blocks of digits. For example, the Japanese armed services used 3-digit, 4-digit and 5-digit code books as well as Kana (Japanese syllabary) code books throughout WW2. Thus 3-digit groups were used for local communications between IJA units in New Guinea. As only 1,000 such groups were available, these codes were useful only for simple messages. Several mainline higher level IJA codes used 4-digit groups while the IJN used 5-digit groups in the various JN-25 systems.

Commercial code books, such as the Chinese Telegraphic Code, were usually systematic in their layout. For example, major ports could be listed alphabetically with their corresponding book groups in numerical or alphabetical order. Extra security could (and should) be obtained by allocating book groups randomly, that is using a hatted[15] code.

A confidential message system based simply on constructing texts for transmission by direct use of code book entries is subject to easy attack once it has been used sufficiently frequently. The most commonly used book groups and pairs of book groups would be progressively interpreted by knowledge from intelligence such as TA about the likely subject matter of the messages in which they occurred. Knowledge already obtained about some book groups would help in interpreting others. Infrequently used book groups would, in general, not be interpreted until late in the process or perhaps not interpreted at all.

In the case of a heavily used military system, the weakness of a bare code system could be avoided only by regular change of the code book throughout the entire network. The task of producing new code books and arranging their safe distribution would be daunting.

The most obvious way of providing extra protection for a digit-based code book is to employ an additive table to encrypt the code groups. This combination is called an additive cipher system or just additive system. The encryption process used on the code groups was called superencipherment.

The additive systems[16] considered here consisted of four components, which are now described.

---

[15]See Note 5 of Chap. 1.

[16]The Australian military used additive systems, particularly in the early stages of the Pacific War. Thus NAA Brisbane file BNO407/4 deals with signals between the armed forces in Queensland in 1942. It was agreed to use 'Inter-service Cipher G33/2 with comtab tables' for all transmissions. The NAA Recordsearch index has several entries under *recyphering*, one of the then current words used to describe the use of additives.

The NAA Canberra file A1196 37/501/407 contains the report of a cipher unit set up on Los Negros, an island east of Manus Island, in March 1944 immediately after its recapture. It is noted that 'book cipher was found to be inadequate and the use of machines quickly enabled us to overtake the volume of work'. This confirms what one may well have suspected: additive systems were rather clumsy.

Two relevant files are in the A1196 section of NAA. 12/501/90 reveals that the RAAF was producing additive tables in August 1941. 12/501/133 shows that 12 months later the RAN revealed to the RAAF how tabulators could assist in such work.

8.10 Component 1: The Code Book

Part of a (contrived) hatted code book using 4-digit book groups could appear thus:

cabbage	1783	caldron	7296	camp	5548
cabin	2451	calendar	7428	can	9127
cable	5786	calibre	3731	canal	4146
cake	4472	call	3378	cancel(led)	8426
calamity	7576	camel	1225	candid	2005

As illustrated here, the code book would have a lexicographic list of words and perhaps also phrases appropriate to its intended use. Corresponding book groups are given to the right of each word. These are all different and are all 4-digit groups. As there are only 10,000 4-digit groups available, this code book has at most 10,000 words. The clerk handling a message, such as 'Cable damaged operation cancelled', would begin by looking up the corresponding book groups in the code book and writing them under the corresponding words:

CABLE	DAMAGED	OPERATION	CANCELLED
5786	1221	5346	8426

JN-25, the IJN's principal operational series of cipher systems, was exceptional in using 5-digit book groups. Most of the JN-25 systems had an auxiliary table (A/T) as well. Some 20,000 or more groups would be given second meanings, set out in a second list. Thus in the sequence of groups:

31662 47904 55398 84912 49326 36678 68667 39270 88545 58935
35970 58683 14613 64845 42609 04383 67081 43398 12828 28551

the 88545 might mean 'next two are A/T' and then the interpretation of 58935 and 35970 would be in the A/T.

The capture of documentation for JN-25C9 on Guadalcanal in August 1942 must have helped[17] understanding of later systems in the JN-25 series.

The book *Betrayal at Pearl Harbor* by James Rusbridger and Eric Nave (Summit Books, New York, 1991) contains on pages 83–84 sample pages of a JN-25 code book. The sample of additive table that follows has a 12 × 15 display of additives rather than the 10 × 10 mentioned elsewhere in this chapter. Minor variations like this may be disregarded.

[17]This appears not to be mentioned in the archives, but may well be important. After the capture the USN had a large sample of IJN telegraphic jargon. It also gained some knowledge of the frequency of the more common words and phrases.

8.11   Component 2: The Reverse Book

The second component was the reverse code book. This would list the book groups used in numerical order with the corresponding code words alongside. Thus a few lines would look like:

1217	foreign	1221	damaged	1225	camel
1218	carry	1223	fruit	1227	boat
1220	pocket	1224	strait	1229	bolt

The recipient of a message in this code would look up each of its 4-digit groups in the reverse code book and hence decipher the message. This process, and also the encoding, would be best done on a printed form. Any group not found in the reverse code group would have to be a garble and, if necessary, clarification would have to be sought from the sender.

8.12   Component 3: The Additive Table

This component of the system was also called a table of additives or a table of keys. In book form, it consisted of 100–500 pages, on each of which was printed a 10×10 array of 100 randomly chosen groups. Groups in such tables should be patternless.[18] Occasionally, 120 or 180 groups were printed in a 12×10 or a 15×12 table. This happened with the IJN operational cipher series JN-25 after 1943. For example, a page from an additive table used with a 4-digit code book might look like:

---

[18]In fact the groups in the sample table were selected by letting the computer calculate $\sqrt{\pi} - \sqrt{e}$ to 400 decimal places. Such methods of generating large numbers of groups without following an evident pattern were not available in WW2. Indirect evidence as to how the Japanese manufactured tables of random digits survives. See Note 10 of Chap. 13.

The matter is not far removed from the problem of choosing book groups for code books. Likewise the randomness of the selection of a starting point in the additive table is important. These matters are also discussed in Sect. 13.6.

Page 123

	0	1	2	3	4	5	6	7	8	9
0	4884	2843	4217	7745	4322	4746	9152	7870	8724	4612
1	6859	0840	9240	3899	6300	4929	8070	3518	7237	9417
2	8621	6442	3729	8991	0965	1456	3991	8746	0288	3360
3	3813	9862	9677	9341	7502	3313	8108	7319	4758	4163
4	3839	3165	1507	0845	9639	5445	5001	1202	4021	5531
5	8256	2727	3504	8148	0067	6872	0832	6387	*8363*	*0221*
6	*1907*	*5100*	9671	7081	8582	9571	8480	0690	2411	0503
7	2931	5542	1699	3517	4816	7249	8091	1505	4469	2625
8	8205	4831	5602	5346	4786	4528	9948	7633	4686	2177
9	7347	8954	6055	9835	8859	6783	7362	1560	2411	9387

The numbers in bold face to the left of or above the printed groups are called row or column co-ordinates respectively. In certain circumstances, these may be better used in a randomly chosen order such as:

$$\textbf{4 \quad 1 \quad 6 \quad 9 \quad 8 \quad 7 \quad 0 \quad 5 \quad 2 \quad 3}$$

varying from page to page. If there were (say) twelve columns, the enemy cryptanalyst might be held up by column co-ordinates like:

$$\textbf{69 \quad 04 \quad 15 \quad 75 \quad 98 \quad 23 \quad 42 \quad 95 \quad 54 \quad 56 \quad 30 \quad 11}$$

The analogous comment applies to row co-ordinates. This happened with superencryption for JN-25 in 1944. During WW2 the IJA and IJN used numerous variants of the concept of an additive system.

The clerk processing the original message first converted its plain text into book groups. The example presented in the previous section is repeated in the first two lines of the following display. The clerk was then supposed to choose at random a group from an accompanying additive table. Suppose that this table contained the above page 123 and that 8363, the group at line 5, column 8 on page 123 is chosen. It and the following groups (italicized above) are then written under the code groups for the message as below:

	CABLE	DAMAGED	OPERATION	CANCELLED
book group	5786	1221	5346	8426
additive	8363	0221	1907	5100
GAT	3049	1442	6243	3526

The final row in the example is obtained by false addition[19] of the previous two rows: one adds each digit in the second row to the corresponding digit in the third row and then disregards the initial 1 if the sum exceeds 9. In other words there is no 'carrying' in the false addition process. Thus, the false sum $6 + 7$ is not 13 but just 3. The groups in the last row, obtained by false addition of the two groups above, are to be transmitted. The acronym GATs was used for groups as transmitted. The message actually transmitted should therefore include the sequence of GATs 3049 1442 6243 3526. It is likely to include some prefix[20] (then called the discriminant) stating which cipher system was in use. There should also be another prefix, the call sign, stating the sender of the message. There might also be a prefix identifying the intended recipient. There may be a group which just gives the number of groups in the message, this being a simple check in a rather tedious process. Information telling the recipient exactly what part of the additive table has been used to encipher its book groups is an essential part of each message and this is discussed below.

The first page of the sixth additive table for JN-25B (JN-25B6) follows. It was used from February to July 1941, and is one of tens of thousands of such pages reconstructed (or partly reconstructed) by Allied cryptanalysts.

---

[19]George Aspden, formerly of CBB, recalled seeing some captured encrypting forms in 1945. They had four lines with the functions shown in the example.

The decryption process needs 'false subtraction':

GAT	3049	1442	6243	3526
additive	*8363*	*0221*	*1907*	*5100*
book group	5786	1221	5346	8426

An extra complication called 'Op-20-G Usage' is needed in Chap. 15. In decrypting IJN ciphers the 'false negatives' of the original additive table were stored, but not the table itself. The decryption process then took the form:

GAT	3049	1442	6243	3526
neg. add.	*2747*	*0889*	*9103*	*5900*
book group	5786	1221	5346	8426

The motivation was to replace false subtraction by false addition, less prone to error.

[20]The reference is page 3 of part G of the *CBTR*. 'To indicate the method of encipherment and materials in use, a four-figure group, the *discriminant*, preceded the message. This was originally unenciphered, but from August (19)44 was enciphered with materials set aside for that purpose. ... Encipherment of discriminants was part of a general scheme to give all traffic the same external features.' The early naval JN-25 systems were readily distinguished by the practice of transmitting its messages in groups of five digits. However from late 1942 onwards there were JN-25 code books being used with up to five different additive tables concurrently and so discriminants were needed. The word 'channel' was the jargon for any one of the combination of code book and additive table.

It was possible to transmit a 4-digit system with four groups of five representing five groups of four. This would have been prone to error.

	0	1	2	3	4	5	6	7	8	9
0	81417	25240	47079	53584	63401	14698	02802	96362	50644	25281
1	36434	10017	73582	24697	03492	88536	28868	80785	35875	63903
2	45942	97036	05685	30425	61478	46305	30923	14616	50683	75380
3	77978	29488	60506	91512	88559	64577	90380	54578	84421	09620
4	82579	35303	11250	57915	30518	45590	27157	92454	16688	46910
5	96457	41910	74591	03499	59864	23299	49465	35586	79456	31065
6	21778	60559	52468	79545	46867	89515	12377	82021	96558	67863
7	59536	01354	35564	41873	61922	16105	68891	19803	03408	35388
8	77389	31994	86027	73056	96908	70394	52582	01247	24269	19088
9	69802	18342	91044	51450	41074	64092	75932	11021	61761	34036

From the viewpoint of the Allied cryptanalytic units, the entries in the pages of the additive table had to be identified efficiently and speedily. The above page was one of 500, so the entire JN-25B6 table contained 50,000 entries and hence 250,000 random digits. Earlier tables, such as JN-25A1, had 30,000 entries and so 150,000 random digits. The sheer magnitude of the task focused attention on the need to automate some or all of the process. Thus, the special devices described in later chapters for working out additives for the extremely important JN-25 and JN-11 series of systems were immensely valuable.

8.13 Component 4: The Rules

Any message transmitted using an additive system must contain a critical item, the precise location of the starting point in the additive table chosen to encrypt the first group. Analogously to the terminology for machine ciphers, the GAT or GATs in the transmitted message that were used for this purpose are called the indicator group(s) or just indicators.

If the recipient and sender had identification numbers 2483 and 8563 respectively, cipher system 3699 was being used, there were 4 groups in the message and the date was 7 February the message might be transmitted in a standard form something like:

2483	8563	3699	0123	0058	3049	1442	6243	3526	0702.
recip.	send.	disc.	page	place	text	text	text	text	date.

Here the 'page' and 'place' groups 0123 0058 form the indicator—that is, page 123, row 5, column 8. It is obvious that an indicator of this simple form, appearing in all messages, would soon be identified by a codebreaker for what it is—a gratis piece of extremely useful information. As remarked in the final section of this chapter, *indicator systems should be seen as being particularly important and no effort should be spared in concealing, encoding and encrypting them.*

In an additive system, therefore, the *rules*, setting out the method for describing the starting point of the additive table in an encoded, enciphered and/or disguised way, are very important. Various examples of rules are to be found in this book.

A minor weakness is also visible in the example above. Granted that several ciphers were in use simultaneously, it was essential to have the discriminant (system identification) included in the message. The discriminant was transmitted without encryption by the IJA in 1942–1943, so gratuitously making the task of the Allied cryptanalysts somewhat easier.

8.14   A Distinction in Additive Table Usage

When IJN operators reached the end of a page in a table of additives they moved on to the top left group of the next page. This practice opened the way to tailing (Sect. 9.22) and so turned out to be quite important.

With some exceptions, the IJA followed the other reasonable convention.[21] They went back to the top left of a page of additive when the bottom right was reached. Thus if the first of 26 additives used was that on row 8, column 7 of the additive table page 123 (as shown on p111 above), the groups used would be those italicized in the rows as given below (where row **0** follows row **9**).

	0	1	2	3	4	5	6	7	8	9
8	8205	4831	5602	5346	4786	4528	9948	*7633*	*4686*	*2177*
9	*7347*	*8954*	*6055*	*9835*	*8859*	*6783*	*7362*	*1560*	*2411*	*9387*
0	*4884*	*2843*	*4217*	*7745*	*4322*	*4746*	*9152*	*7870*	*8724*	*4612*
1	*6859*	*0840*	*9240*	3899	6300	4929	8070	3518	7237	9417

8.15   General Comments

Superencipherment predates WW1. The German WW1 HVB code used a book and a second stage encipherment that was relatively easily broken, as demonstrated by Dr Wheatley in his report (see Note 4 of Chap. 1).

---

[21] The *CBTR*, part G, page 9, makes it clear that the group used after the last group on a page of additives was usually the first group on that same page. In effect each page in a book of additives was yet another separate cipher. For the naval additive systems of the JN-25 series the group used after the last group on a page was the first group on the next page. This turned out to be quite important: see the discussion of *tailing* in Sect. 9.22. In view of the vast number of enciphering systems used for various purposes in different parts of the Pacific Theatre of operations in 1941–1945, it is difficult to determine which practice was more common.

Some later JN-25 codebooks were used with up to five different tables of additives at once. These were called 'channels'.

It may be desirable, especially when there are multiple exchanges between two stations, to insert a message number into each message. However, because this may well be inferred, it may give the interceptors a little information that they may not otherwise have had. Indeed, enciphering such details by the same table of additives as is used for the main text may well just give the enemy cryptanalysts potentially useful information[22] gratuitously. Using the unenciphered date and time of transmission for reference is more secure.

8.16 Non-primary Decryption

In the case, say, of a 4-digit system, if any one fixed group, say 1357, is (false) added to each book group and (false) subtracted from each additive, the same GATs are produced and so, since the transmitted messages do not change, the cryptanalysts cannot observe any difference. This ambiguity does not preclude progress. As soon as the enemy cryptanalysts are confident that a single word in the code book, say CABLE, has been identified, the unknown book group for this word may be taken to be 0000. Later, a book group differing from that for CABLE by 0660 may be identified as OPERATION. Then 0660 is called the relative or non-primary book group for OPERATION. Likewise, given time, lots of intercepts and lots of work, all other non-primary book groups are well determined. The primary (original) book groups cannot be deduced in general, but may not be needed. In the early work around May 1943 on the very significant Water Transport Code, the phrase MESSAGE BEGINS was initially[23] taken to have book group 0000.

---

[22]This particular blunder did happen in WW1. See page 14 of Patrick Beesly's *Room 40*. This is confirmed by A. G. Denniston's comment 'The Germans, whose folly was greater than our stupidity, reciphered the numbers of the messages thus offering the simplest and surest entrée into their reciphering tables.' See document DENN1/2 at Churchill College, Cambridge, or Robin Denniston's *Thirty Secret Years* (2007).

The *CBTR* (Part G, page 2) notes that this happened in WW2 IJA codes. It comments 'Since the serials (serial numbers) ran consecutively they were predictable and therefore of great use in extracting keys'.

The RN had adopted the practice of referring to a message sent at, say, 08.48 (Greenwich mean time) on the tenth day of a month as 0848Z/10. The month would be mentioned only if there was ambiguity.

[23]The discovery that 6666 should be taken to be the WTC book group for 'message begins' was analogous to the discovery by Mrs Driscoll that the primary book groups for JN-25A were multiples of 3. See Appendix 1 of Chap. 12. See also Chap. 14.

The very important JN-25 series of naval ciphers, discussed in great detail in this book, is an exception[24] to these remarks, in that it was possible to identify primary book groups.

8.17 The One-Time Pad

An interesting comment on the July 1942 British practice with additive cipher systems may be found in the Australian War Memorial Research Centre.[25] This is an instruction manual for the British Inter-Service Code and must have originated in the production section of the GCCS. It would thus have been created by people with at least indirect experience in handling Japanese codes such as JN-25. It urges the user to use the code with a one-time pad (OTP) whenever possible. Such a pad had pages which would look like this:

----

[24]The point here is that by adding one suitable fixed group to all the non-primary book groups in a JN-25 code book all the book groups obtained would be multiples of three. The 'one suitable fixed group' that achieved this result would be unique. (Highly contrived exceptions to this last remark are to be found in Appendix 4 of Chap. 10.)

Suppose that the following are 20 book groups in a 5-digit additive cipher. They are in fact all *scanning*, that is the sums of their digits are all multiples of three.

80169 83997 92208 32949 17580 64995 00933 35604 23985 96423
93294 19515 58113 07320 09774 78765 35655 50502 11367 78492

The enemy cryptanalyst cannot read the code book, but may be able to calculate the (false) difference between any pair of these *primary* groups. It is convenient to store this information by selecting one such book group and assigning a specific (but arbitrary) group to it. For example, it may be convenient to take the first to be $40697 = 80169 - 40572$. Then, to preserve the known differences, the 20 non-primary book groups are each obtainable from the corresponding primary book group by subtraction of 40572. So they are:

40697 43425 52736 92477 77018 24423 60461 95132 83413 56951
53722 79043 18641 67858 69202 38293 95183 10030 71895 38920

The recovery of the first batch of 20 groups from the second needs either extra information (see Chap. 12) or special inspiration.

[25]The reference is AWM124 4/71. Patrick Beesly's book *Very Special Intelligence* gives more information on the use of these codes by the RN. The manual in the AWM makes it clear that by July 1942 the RN was urging the use of one-time pads whenever possible to protect the security both of the message to hand and other messages.

The NAA item A1606 I8/1 Part 2 contains a message dated 22 June 1949 from Britain offering to supply one-time pads for communications between all British Commonwealth countries. By then the quantity of confidential traffic would have been much less than in WW2.

OTP SQ8357 PAGE 191.

```
31783 72451 95782 24472 57576 17296 17428 83731 33378 43343 25548
79127 24146 12005 38446 69299 82307 58652 42960 70029 40612 29518
44944 06005 04510 90391 48215 26616 75340 14994 43781 71148 70565
97479 99893 12174 86959 99318 92862 69354 52027 83295 29735 26121
16075 39265 41933 20552 58615 79330 50815 53235 75347 41272 21592
90196 26864 02235 22551 68061 06936 49515 83274 29211 79615 66575
06673 07507 74778 20954 52378 28094 67271 34941 26301 06033 58697
89381 17935 65505 07571 16384 76546 61295 81262 97531 04061 86185
42652 37031 96475 14019 40132 68117 70993 22243 35278 14685 99510
83566 42047 87219 42177 12886 93182 15531 82689 68976 30263 53593
```

When a random OTP is used (only once) as an additive table the outcome is random and so cannot be exploited by enemy cryptanalysts. Another way of looking at this is that the various methods for attacking additive ciphers all depend upon comparing GATs which have been encrypted by the same additive.

Small numbers of each OTP would be prepared with one supplied to the future sender and the other to the future recipient(s). The page would be used in the same way as a chosen part of an additive table but would be destroyed after just one use.[26] Thus the transmitted message may as well begin with something like OTP SQ8357 PG191 to assist the intended recipient. Provided the OTP itself has been generated totally unsystematically, an enemy cryptanalyst cannot make any use of the page number. However it would be good practice to encrypt OTP SQ8357 PG191 to encourage the enemy to waste resources attempting to break into the message. The use of the OTP has the two advantages of being (almost) beyond decryption and of being incapable of being used to compromise the underlying code. The disadvantage is that much more effort has to go into preparing, distributing and storing adequate numbers of OTPs.

Abe Sinkov[27] recalled that the 3-digit codes used locally by the IJA in New Guinea in 1942–1943 were enciphered by do-it-yourself one-time pads. These were occasionally captured, but the Australian and American troops fighting in New Guinea in 1942 and 1943 received relatively little information from direct breaking of their opponents' local codes. Eventually it was decided that work on the mainline ciphers would be more profitable. (The terminology was that the mainline codes were those used between headquarters and local codes were those used within a division or smaller unit. Local codes tended to be sent on low power and thus were harder to intercept. The Water Transport Code (frequently called 2468 after its discriminant) was a mainline code.)

---

[26]Protocols would be needed to prevent re-use of OTP pages.

[27]The Sinkov oral history interview is in the National Cryptologic Museum oral history series. See page 31 of the document. Note that the account of the Sio capture given there is somewhat inaccurate. The *CBTR* has a section on these 3-digit systems.

8.18   The Security of Additive Systems

The special methods devised for attacking JN-25 additive systems and the much harder processes needed for more general additive systems are described later. Even when the 'additive is stripped off', that is the enciphering is removed, the task of recovering the code book remains. Some comments[28] made by Professor Room in the *CBTR* in 1945 confirmed that a well-designed and correctly used additive system (for example, most of those used by the IJA) could defy WW2 codebreakers:

It seems that at least one of four conditions is needed to solve one of these codes:

(1) Additive starting-point indicator system should be such that messages can be set in depth from it alone.
(2) A body of stereotyped messages for which the additive book is used very badly, with several sets of messages starting at the same point.
(3) The code book must be heavily patterned and these patterns are clear from the differences of groups in the same column when the messages have been set in depth.
(4) Enough material must be captured to give a fair sample of book groups and possible stereotypes.

Although Room was fully aware of the special features of the peculiarities of the JN-25 series of naval ciphers, it would appear that he had been required not to disclose anything about them in the *CBTR*.

8.19   Maxims about Cipher Security

The cipher wars in Europe and the Atlantic in the era being considered are now well documented and widely understood. They thus provide a list of maxims[29] about

---

[28]The quote is from Part D of the *CBTR*. The text was probably written by Professor Room or at least with his active involvement. The word '3-digit' has been dropped from the quoted text as the comments apply with equal force to 4-digit and 5-digit systems. The *CBTR* may be overstating the difficulty in breaking into a properly used additive cipher.

Possibility (3) in the *CBTR* extract: 'The code book must be heavily patterned and these patterns are clear from the differences of groups in the same column when the messages have been set in depth' is examined in detail in Appendix 1 of Chap. 12.

Stephen Budiansky's meritorious *Battle of Wits* gives another account of additive cipher systems. However its main emphasis is on the European Theatre of Operations.

Those attacking the Enigma tended to use 'cribs', that is common words or phrases that could be anticipated in a message, in attempts to recover the original setting. This could have been thwarted by the simple device of spelling such crib words badly, such as by randomly inserting the letters 'Q' and 'Y'.

[29]Auguste Kerckhoffs in his 1883 paper *Cryptologie Militaire* (already mentioned in Chap. 1) sets out in Part II six desiderata for military cryptography. By 1937 greater cryptographic sophistication had rather changed the requirements, but four of the six will be given here: (1) the system should be practically, if not theoretically, indecipherable; (4) it should be compatible with telegraphic

secure pre-electronic cipher practice. These maxims seem to apply equally to the non-machine ciphers of the Pacific Theatre.

The following list is supplied with the addition of a few references:

Partial changes to a possibly broken cipher system tend to be inadequate. The enemy may[30] be able to use knowledge of the unchanged part of the new system to identify the changes. Periodic complete changes are much better, particularly for additive cipher systems.

Training sessions for new operators should be conducted via cable rather than over the air waves. This prevents gross blunders in operator practice reaching the enemy and giving away information on the system in use.

Indicator systems should be recognised as being particularly important and no effort should be spared in concealing, encoding and encrypting them.

Any easy method of complicating the task of the enemy while not throwing much extra work on one's own staff should be adopted.

Experienced and sophisticated specialists should be used to design secure communications procedures and (Tiltman[31]) take full responsibility for security.

Random choices required to be made by operators should be made with dice, playing cards[32] or some other such mechanism. This must be enforced.

---

transmission; (5) it should be portable; and (6) it should be easily used. Chapter 13 applies the maxims of this chapter to the additive ciphers prevalent in Japanese communications in WW2.

[30]Bengt Beckman's book *Codebreakers* gives on page 54 a Swedish report on exploiting Russian errors with an additive cipher system in 1940: the Russian failure to change a superenciphering (additive) table when a code book was replaced saved the Swedish team several months. Other errors noted in that report resemble those listed in this book.

Chapter 9 discusses the partial change in the JN-25B system made in December 1941.

[31]This point is to be found in Tiltman's *Reminiscences*, which have survived in NARA College Park RG457 as in Box 1417, NR 4632. It is fully compatible with the views of Wing-Commander Lees. The need to have appropriate personnel designing cipher systems was noted by Kerckhoffs at the end of Part II of his *Cryptographie Militaire* paper and is repeated from Chap. 1 for emphasis: *Do not use new cryptographic systems invented by people without the necessary experience.*

And indeed it is repeated by William Friedman on page 157 of the NSA booklet *The Friedman Legacy*. 'Cryptographic invention must be guided by technically qualified cryptanalytic people.'

Of course if the specialists are not experienced and sophisticated and instruct operators to 'tail' (Chap. 9) then security may be non-existent.

On page 116 of her *Dilly: The Man Who Broke Enigmas* Mavis Batey notes that 'the biggest gift the Italians gave us was that they insisted on their operators spelling out full stops as XALTX'. Experienced security people would have stopped that error.

[32]If operators in the field need to select letters randomly from the 26-letter alphabet, they can use a common pack of 52 playing cards with each letter written on the face of two cards. The pack can then be shuffled and a card selected. The *Herivel tip* used at one stage to work out Enigma indicators depended on no such randomising device being used regularly.

Never transmit two separate encryptions of the same plain text.[33] In particular, *never transmit two separate encryptions* of the indicators, and *never transmit an encryption* of a plain language text available[34] to the other side.

Diligently monitor enemy documents, announcements and actions for any evidence[35] that your own systems are being read.

Communications with units exposed to capture should be in special[36] systems not in general use.

---

[33]This maxim is the central theme of Chap. 14. Stereotyping is included here: it is an insecure cipher practice.

[34]In particular it was not appropriate for the Japanese Embassy in Washington to use Purple to report on discussions with the US State Department and particularly not on documents emanating from the Department.

[35]Breaches in the security of Allied Comint are discussed in Chap. 19. These were not identified by the Axis. Chapter 16 mentions how the German Navy gave away the secret that it was reading the cipher system used by the Atlantic convoys by putting current information from that source in an Enigma message.

It has been argued that the Japanese Army and Navy leadership in WW2 put little faith in any aspect of intelligence and so would not have insisted on this precaution.

[36]For example the RAAF 14 W/T Unit on Los Negros, east of Manus Island, in 1944 had Typex machines with special rotors. See NAA Canberra item A1196 37/501/407. The US Army and Navy were very careful with Sigaba/ECM machines for another very valid reason: they did not want to present the Germans with technology that could be used to upgrade Enigma.

Chapter 9
JN-25 and Its Cryptanalysis

The JN-25 series of IJN operational codes was the most important source of Allied Sigint throughout the Pacific War. From early 1942, intelligence supplied by this source was significant in determining the operational decisions made by the Allied commanders. This intelligence was supported by some derived from other IJN systems, but only after mid-1943 was significant intelligence obtained from reading IJA intercepts. This chapter explains how the weaknesses in the structure and use of the early versions of JN-25 were successfully exploited by Allied codebreakers.

9.1 The Overall Importance of JN-25

The role played by JN-25 Sigint in the decisions to bring on the Battles of the Coral Sea and Midway is well known. Its role as an ongoing major source of Sigint is illustrated by the following two extracts from Archival records:

(1) from the *GYP-1 Bible*[1] written in July 1943:

> JN-25 is the largest of the Japanese naval cryptographic systems. From sixty to seventy-five percent of each day's radio intercepts are in this system. The system is used by all men of war, shore stations, commands and bureaus for the administration of the Navy. It is, therefore, a fruitful source of information.

---

[1]NARA RG 38. CNSG Library. Boxes 16 and 18. File 3222/65. In general the USN WW2 cryptological archives are in Record Group 38 in NARA. The reports on JN-25 form a high proportion of the technical material there.

© Springer International Publishing Switzerland 2014
P. Donovan, J. Mack, *Code Breaking in the Pacific*,
DOI 10.1007/978-3-319-08278-3_9

(2) from a circular[2] dated 4 September 1944 written by the Director of the GCCS.

> The loss of JN-25 over so long a period has deprived us of about 60 % of normal
> strategic intelligence.

Striking evidence of the extent to which signals and other intelligence had revealed the state of the IJN by mid-1944 is to be found in a 90-page document *Order of Battle of the Combined Fleet*. The copy[3] in the NAA has a cover page in attractive Japanese characters after which the details are remorselessly displayed in English typescript and American spelling.

Early work done by GCCS in London and at the FECB and by the USN (especially by Station Cast) was of crucial importance in laying the foundation for the rapid development of the Allies' ability to obtain Sigint from JN-25 in the early months of 1942. A further crucial factor was the decision taken in January 1941 by the American and British governments to share Sigint knowledge and skills. Thus the period from June 1939 to the end of 1942 is given particular attention in this account. Apart from a later Chap. 15 explaining the Scanning Distribution and Chaps. 14 and 16 on the important IJA Water Transport Code, little attention is given in this book to post-1942 developments.

9.2   The JN-25 Series 1939–1943

The name *JN-25* and the systematic notation used by the USN to describe its successive versions was introduced in 1942 and applied retrospectively back to June 1939 for this series of additive cipher systems. The code books were named A, B, C etc in order, while the additive (enciphering) tables were numbered 1, 2, 3, etc. These were changed more frequently than code books. For example, the original

---

[2]The report by Sir Edward Travis is TNA HW 14/142. A report there dated 29 September 1944 states that this current problem with JN-25 'is now having the attention of Brigadier Tiltman and the Research Section. As you will remember, Brigadier Tiltman broke the cipher in 1939 and we are hoping this re-inforcement of our strength will do the trick'. In fact, JN-25N62 was never broken.

A quick examination of the list of WW2 signals intelligence material in the American National Archives confirms that JN-25 was the principal series of ciphers of interest.

[3]NAA Canberra item A10909 2 may be read via Recordsearch and the web. The spelling reveals its American origins. It would not have been compiled by Frumel alone. Some of its contents would have come from using captured code books, such as the JN-25C material captured on Guadalcanal in August 1942, to decrypt old JN-25 messages.

Unfortunately no work has been done to compare A10909 2 with Japanese archival records of the IJN in 1944.

A code book was used with five enciphering tables producing JN-25A1, JN-25A2, JN-25A3, JN-25A4 and JN-25A5. The chronology for the period under study is:

JN-25A1	1 June 1939	to	31 August 1939.
JN-25A2	1 September 1939	to	30 November 1939.
JN-25A3	1 December 1939	to	31 May 1940.
JN-25A4	1 June 1940	to	30 September 1940.
JN-25A5	1 October 1940	to	30 November 1940.
JN-25B5	1 December 1940	to	31 January 1941.
JN-25B6	1 February 1941	to	31 July 1941.
JN-25B7	1 August 1941	to	3 December 1941.
JN-25B8	4 December 1941	to	27 May 1942.
JN-25C9	28 May 1942	to	14 August 1942.

In the above chronology, two items should be noted for later reference: the fifth additive table continued in use for 2 months after the new B code book was introduced and the change from the seventh additive table to the eighth, with the same B code book, occurred rather irregularly on the fourth day of the month rather than the first. The change of 4 December 1941 was mentioned in Sect. 7.5 as suggesting that some attack somewhere might be imminent.

In view of the amount of misleading material published on the breaking of JN-25A and JN-25B, the following account makes specific reference to a most informative collection of messages on JN-25, file HW 8/102 in the British National Archives. This GCCS file, called *Early Signals on JN-25*, records exchanges between GCCS and Captain Shaw at FECB in the period June 1939 to July 1942.

9.3   The First Step in Cryptanalysis

More on the decryption of additive systems is given in Chap. 12.

If cryptanalysts, working on an unknown cipher system, are able to determine that it is in fact an additive system, then, as previously noted, they face a formidable task. The content of any intercepted message can only be even partially found if meanings have confidently been attached to some of the book groups in the message. Since the book groups are hidden from view by the additive table encipherment, progress can only be made by successfully recovering entries in the table of additives. If the rules for the indicators of the system can be discovered, so that the starting point for additive table use in each message can be determined, then this can be used as a basis for attack. If cryptanalysts are unable to work out the rules for the indicators, then the additive system may successfully defy repeated attack, as was the case with major Japanese Army systems used in WW2. It is unlikely that the indicator system rules can be worked out without access to a large number of intercepted messages.

Once the starting point information is available for a large number of messages, it should be possible to determine the form of the additive table used—its number of

pages and the arrangement of entries on each page. With this data, then the next step
is to place together all the GATs calculated from a single entry in the additive table.
A useful way of proceeding is to use specially printed large sheets of paper, one for
each page of the additive table. Suppose it is known that each page contained 100
additives. This is too many to record on one side of a sheet and so both sides would
be used, with 50 columns on each side. Each message to be recorded on this sheet
would have its data entered on a single line, and the first side of the sheet would
look something like:

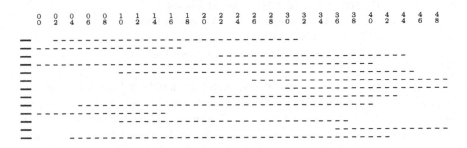

Here the longer and heavier dashes at the left represent details of the message (date,
time, source and the like). Each of the other dashes represents a GAT, *with GATs
constructed with the same additive in vertical alignment.* There are 50 columns of
GATs and the sheet of paper must provide 15 mm width for each vertical column.
The reverse side of the sheet would be similar in form, but headed by:

$$5\ 5\ 5\ 5\ 5\ 6\ 6\ 6\ 6\ 6\ 7\ 7\ 7\ 7\ 7\ 8\ 8\ 8\ 8\ 8\ 9\ 9\ 9\ 9\ 9$$
$$0\ 2\ 4\ 6\ 8\ 0\ 2\ 4\ 6\ 8\ 0\ 2\ 4\ 6\ 8\ 0\ 2\ 4\ 6\ 8\ 0\ 2\ 4\ 6\ 8$$

Thus one sheet would be used for GATs calculated from the one page of additive table. If there were 50,000 additive groups spread out over 500 pages, the
cryptanalysts would need 500 of these pieces of paper with lines specially printed.
Considerable clerical work would be involved. Over time more intercepts would
build up, hopefully producing enough to work on.

British Archives item HW 43/34 describes the 'Tiltman forms' used at Bletchley
Park in 1944–1945.

*Intercepts displayed as above, that is with GATs constructed with the same
additive being in the one vertical column, were said to be written in depth.* The
word 'depth' was also used for the number of GATs in a column, which was also
called a 'depth'. Thus the first column, headed by 00, has a depth of 3. Additives
had to be recovered one at a time, using the evidence supplied by GATs aligned as
above in sufficient depth.

The obvious question is what should be done when enough columns have enough
depth. It will be seen that *there were special methods available for the JN-25 and
JN-11 additive systems but additive recovery in most other systems was much
harder.*

9.4 The Breaking of JN-25 in 1939

JN-25A1, the first code of the JN-25 series with the first table of additives, was introduced on 1 June 1939. The FECB, then in Hong Kong, was able to intercept substantial numbers of messages in what was initially called 'the five-figure cipher'. Advice[4] was sought from the GCCS. Denniston advised the FECB not to try to research the new cipher but instead just to send the intercepts back to GCCS on paper in the secure bag. Fortunately for the future Allies, this happened a few months before the war in Europe broke out, and Denniston was able to allocate the very experienced John Tiltman to the analysis of JN-25. Tiltman was assisted by three[5] other very capable men.

The team led by Tiltman made its first serious breakthroughs in August and early September 1939. Their experience with Japanese naval and military codes would have made them suspect from the start that JN-25 was an additive system. This assumption would have had to be checked as the research developed. GCCS was able to report that it had broken into the indicator system that communicated the starting point in the table of additives. On Wednesday 9 August the GCCS telegraphed to the FECB:

> We are now concentrating on the 5-figure naval cypher but no opinion can yet be given as to when it is likely to become legible. At present, it is considered that all investigation may still most suitably be carried out in London. Any development which might affect this policy will be telegraphed immediately.

On Thursday 24 August the GCCS telegraphed:

> Satisfactory progress is being made with the 5-digit subtractor cypher but no opinion can yet be given as to when it is likely to become legible.

The notification of the first major breakthrough came on Thursday 7 September 1939, 6 days after the invasion of Poland, the declaration of war, and the move of GCCS to Bletchley Park:

> 5-figure method can now be described. One subtractor of 30,000 5-figure groups arranged on 300 pages. 10 starting points on each page 10 groups apart numbered consecutively

---

[4]TNA HW8/102, page 1.

[5]The three were Richard Pritchard, Malcolm Burnett and Neil Barham.

Apparently Pritchard was an Army officer who had been with GCCS for about ten years. His speciality was German Army material. He was commandant of a short-lived GCCS unit based in France before the German invasion.

Malcolm Burnett was a naval officer who also had substantial experience with GCCS. Some information on his career, including a wedding photograph, is given by Ian Pfennigwerth in his book *A Man of Intelligence*. Burnett was later to be instrumental in getting FECB and then Op-20-G up to speed in working on JN-25.

Neil Barham was a naval officer working for GCCS and involved with IJN ciphers at various stages.

Curiously the commendation that GCCS Director Alistair Denniston wrote about Tiltman (29 April 1940, now in TNA HW 14/4) does not mention his work on JN-25.

from 0 to 9. Pages numbered consecutively 002, 005, 008, 011, etc. up to 899. Starting point indicator gives page and line in first four figures of indicator group. Function of fifth figure uncertain. Indicators are covered by daily 5-digit subtractor. Indicators for all days from 1st June to 19 August have been equated. In many cases subtractor has been stripped from first 6 groups of messages. Place names appear to be grouped together—not known whether book is otherwise alphabetical. It is intended that Barham and Burnett will take this cypher to Singapore when form of book understood. No repetition no more material is required here.

This message implies that the indicator encryption for JN-25A1 had been determined and that, from this, the structure of the first additive table had been determined. The fact that some GATs in some messages had been stripped of additives implies that sufficient depth for some entries in the additive table had been obtained for these entries to be determined and so for the underlying book groups to be exposed.

Given the relatively small total number of JN-25A1 messages received by GCCS from FECB, this implies that some operators using the new system were regularly choosing the same or adjacent starting points in the additive table rather than deliberately spreading these out.

Note that the above message does not describe exactly how the encrypted indicator GAT was identified and decrypted. Nor does it give any account of the nature of the decrypted book groups, except to suggest a 'patterning' had been observed in the code book. This information was too valuable and perhaps too complex to be explained in a message. Malcolm Burnett was a RN officer and cryptologist who had assisted Tiltman in the breaking process. Burnett was sent out alone to FECB in September 1939, by then re-located in Singapore. All that could be done was for FECB to allocate some resources[6] to JN-25A while keeping other people working on diplomatic material.

9.5   Tiltman and the Indicators of JN-25A1

The first task was the breaking of the indicator system. This needed identification of the indicator. Fortunately, as noted in the GCCS file, the second GAT in messages in the batches sent from the FECB was always the same as the second last. The last

---

[6]Nave, by then developing serious health problems, did some work on JN-25A at FECB. Autobiographical notes written by Nave, which are now preserved in the Australian War Memorial, indicate that he believed that JN-25A was insecure due to the patterns in the allocation of book groups in the code book. This may refer to the 'multiples of three' patterns, which he mentions elsewhere, or to lesser patterns or (quite likely) both.

Deidre Macpherson's book *Betty Archdale* (2002) describes how H. E. Archdale, a former captain of the English women's cricket team and later headmistress of Abbotsleigh school in Sydney, led a group of young women out to Singapore to do interception and clerical work necessary for breaking into JN-25A and JN-25B. The key letter of 5 March 1941 from the Asian Fleet to Op-20-G and quoted in Sect. 5.5 and also in this chapter mentions them: 'British employ three officers and 20 clerks on this system alone.'

The FECB still worked on diplomatic systems in 1941. An interesting report by Professor Room survives in the British Archives.

GAT obviously represented the time of transmission. A bit of traffic analysis would have revealed that the first three digits of the first GAT identified the sender. The other two digits in the first group could be recognised as being the part number and number of parts—both useful when long messages were broken up for security reasons. So the second group presumably was the indicator.

That was the easy part. Tiltman would then have copied out together the indicators of the 40 or 50 messages for each day and looked for patterns. And there was a pattern: each day only nine of the ten digits occurred in the first place of what was believed to be the indicator GAT.

The system was that the additive table had 300 pages, beginning with a page numbered 002, and then with successive pages having numbers jumping by 3: 005, 008, 011, ..., 890, 893, 896, 899. Each page had 10 permissible starting-places: these were labelled in order from 0 up to 9. Thus, starting place 7 on page 356 was denoted by the group 3567. This was padded out on the right by an extra[7] digit to produced a 5-digit group, such as 35672. The operators were issued with a special table of indicator encryption groups, one for each day. The indicator additives for June 1939 would have looked something like:

01.06	25626	02.06	79474	03.06	45749	04.06	32130	05.06	27486
05.06	81661	07.06	15743	08.06	58836	09.06	31979	10.06	72162
11.06	81552	12.06	65407	13.06	03741	14.06	25159	15.06	94977
16.06	53759	17.06	77244	18.06	82878	19.06	66198	20.06	91029
21.06	33938	22.06	19005	23.06	52867	24.06	44064	25.06	39966
26.06	73514	27.06	81383	28.06	59838	29.06	92801	30.06	04373

For example, if starting-place 7 on page 356 was being used for a message sent on 17 June 1939, the transmitted indicator group would have been the false sum 35672 + 77244, that is 02816. (The significance, if any, of the fifth digit was minimal. It may be safely disregarded here.) The additives tabulated above are taken from the USN records in NARA RG38 Box 29, save that the fifth digits are randomly selected for the purposes of this book.

Let us consider[8] an example. Suppose the message:

A MIKADO MORE HUMANE THAN HIROHITO NEVER EXISTED IN JAPAN. HE IS CERTAINLY RECKONED TO BE A TRUE PHILANTHROPIST.

---

[7]The function, if any, of this digit was not discovered by either the GCCS or Op-20-G at the time. However it appears not to have been a *null*, that is a randomly chosen digit inserted as padding and hopefully also to confuse the other side.

[8]Here one more complication is overlooked. Messages would usually be cut in two and transmitted with the second piece coming before the first piece. So the message in the text might well end up as: NEVER EXISTED IN JAPAN HE IS CERTAINLY RECKONED TO BE A TRUE PHILANTHROPIST START A MIKADO MORE HUMANE THAN HIROHITO.

has been encoded using the following 18 book groups:

	36264	60138	31716	14955	27471	00069	19809	63411	64920
13428	44158	30081	47559	77880	02223	33663	75756	73326	

and is to be sent on 16 June 1939.

Next a page in the additive table and a starting-point on that page are chosen randomly, providing the following additives to be used in order with the above book groups:

	64913	56220	43844	54565	58391	08795	55474	74309	31360
08382	75592	88613	67277	93749	28335	01261	70548	84003	

Supposing that the indicator of this starting place is 18237, it is placed at both the beginning and the end of the original sequence of book groups to give the 20 groups:

18237	36264	60138	31716	14955	27471	00069	19809	63411	64920
13428	44158	30081	47559	77880	02223	33663	75756	73326	18237

The clerk then looks up the indicator additive for the day. Suppose it is 57183. This is placed at both the beginning and end of the chosen string of additive groups:

57183	64913	56220	43844	54565	58391	08795	55474	74309	31360
08382	75592	88613	67277	93749	28335	01261	70548	84003	57183

The final step is to calculate the false sums of the groups in corresponding positions in the last two arrays:

65310	90177	16358	74550	68410	75762	08754	64273	37710	95280
11700	19640	18694	04726	60529	20558	34824	45294	57329	65310

These are the GATs. Extra groups would be prefixed to show the originator and recipient. An unenciphered group specifying the date and time of transmission would be placed at the end.

An account of how the system underlying these indicators probably was determined may be found in Peter Donovan's paper.[9] The interpretations of the fourth digit would have needed some stripping of messages encrypted with some of the more heavily used pages of additives. More on stripping techniques is given in Sect. 12.2.

If, instead of using so simple a sequence of page numbers, the IJN had used a randomly produced one, such as 002, 003, 005, 007, 008, 011, 013, 018, 019, 023, 026, 033, 035, 036, 042, 043, 046, 053, 055, 056, 057, 066, 067, 070, 071, 073, 081, 085, 086, 087, 091, 097, 098, 109, etc, the Tiltman team would have taken somewhat longer to find it. In fact the situation would have been similar to that of recovering the code book of an additive cipher: the solution is not quite unique. After August 1939, the GCCS would have attached lesser priority to possible trouble with Japan, because of the war in Europe. Also, it is not difficult to produce more complicated indicator encipherment systems[10] that would not have been broken so readily. As previously noted, the simplicity of the system used was the first blunder in the JN-25 saga, allowing the GCCS to break through its first defence.

It was rather to be expected that the new JN-25A1 system would be poorly used—most new systems in that era were. In particular, the various pages of the additive table received quite unequal use. Presumably those in the middle of the book would be used most frequently. Intercepts derived from these relatively few pages would be available to help the GCCS team break into the page co-ordinates problem. It would have been reasonable to expect the fourth and fifth digits of the indicator GAT to determine the page co-ordinates of the starting-position. Some considerable work similar to that described in Chap. 12 would have revealed that this was not the case. Tiltman and his team would have had to search for GATs that occurred in different intercepts corresponding to the same page, and use these to work out common differences. In time they would have put intercepts in considerable depth and finally deduced what the starting-place indicator system had to be.

The first two JN-25 systems used this fairly simple (and thus insecure) method of encrypting indicators. Moreover, the practice of *tailing*, whereby a cipher clerk would use consecutive entries in his additive table for a sequence of messages,

---

[9]Peter Donovan's paper *The Indicators of Japanese Ciphers 2468, 7890 and JN-25A2* was published in the journal *Cryptologia*, July 2006. It does not mention how the reading of JN-25A1 and JN-25A2 indicators made possible the detection of the practice of tailing, which became most useful in breaking later indicator systems.

[10]In view of what would be at stake, it would have been appropriate to create a really difficult indicator system. For example, it would have been feasible to pad out the indicator 3567 by a null (always 2 in the examples) in varying positions, such as on the left on the first 6 days of the month making 23567, in the second place on the next 6 days of the month marking 32567, in the middle on the next 6 days of the month making 35267, etc. And the encryption could be achieved by adding to the padded indicator group the group for the day and the third encrypted group of the message itself. It is unlikely that the Tiltman team would have broken such a system using only a couple of thousand intercepts.

was observed. The FECB obtained information as to which IJN cipher clerks were tailing and this contributed to making both this and the next generation of indicator encryption insecure.

Already two major blunders committed by the IJN with its JN-25 series may be identified:

(1) The indicator system for JN-25A1 (and, in fact, also for JN-25A2) was insecure. The later IJN practice of 'tailing' seriously weakened stronger indicator systems.
(2) Excessive use was made of a few pages of additives in the early stages of using JN-25A1.

9.6   Multiples of Three

It is a quite well-known piece of elementary mathematics that a number is a multiple[11] of three (or, equivalently, is divisible by three; the jargon was *scanning* or *scannable by three*) if and only if the sum of its digits is a multiple of three. For example, the following 15 groups (considered as 5-digit numbers):

$$
\begin{array}{ccccc}
06343 & 07995 & 11772 & 12939 & 25703 \\
29505 & 35410 & 43729 & 46542 & 54474 \\
57895 & 67828 & 74456 & 93372 & 99076
\end{array}
$$

yield the digit sums shown in the corresponding places in the table below.

$$
\begin{array}{ccccc}
16 \times & 30 \checkmark & 18 \checkmark & 24 \checkmark & 17 \times \\
21 \checkmark & 13 \times & 25 \times & 21 \checkmark & 24 \checkmark \\
34 \times & 31 \times & 26 \times & 24 \checkmark & 31 \times
\end{array}
$$

Here and later $\checkmark$ is used to denote *scanning*.

In practice, testing hundreds of thousands of groups for divisibility by three is quite tedious and tiring. In the quotation given in Sect. 3.7 and repeated later in this chapter, Alan Turing mentioned that, if the digits in the groups are read off digits

---

[11]Likewise *a number leaves a remainder of one (two) on division by three if and only if the sum of its digits leaves a remainder of one (two) on division by three.* These are useful in understanding the process of stripping at least the earlier versions of the code JN-11 and the process of stripping the indicators of JN-25A1 and JN-25A2.

displayed on the curved surfaces of rotatable discs, there is a method to facilitate this process. The digits are given coloured backgrounds as follows:

Digits	Colour
0, 3, 6, 9	White
1, 4, 7	Red
2, 5, 8	Blue

Then a group is divisible by three if and only if the number of red digits differs from the number of blue digits by 0 or 3. This may be checked by eye without calculation.

Now suppose the same group, say 56789, is (false) subtracted from each of the 15 in the last display and divisibility by three is checked again. The output is:

50664 ✓	51216 ✓	65093 ×	66250 ×	79024 ×
73826 ×	89731 ×	97040 ×	90863 ×	08795 ×
01116 ✓	11149 ×	28777 ×	47693 ×	43397 ×

Here three of the 15 sums are divisible by three. As the probability of a randomly chosen group being divisible by three is about ⅓ (actually .33334), three out of 15 is low but scarcely surprising.

If instead the group 49578 is (false) subtracted from each of the original 15 groups, and divisibility checked, the output is:

67875 ✓	68427 ✓	72204 ✓	73461 ✓	86235 ✓
80037 ✓	96942 ✓	04251 ✓	07074 ✓	15906 ✓
18327 ✓	28350 ✓	35988 ✓	54804 ✓	50508 ✓

If instead the group 48567 is taken, the result is:

68886 ✓	69438 ✓	73215 ✓	74472 ✓	87246 ✓
81048 ✓	97953 ✓	05262 ✓	08085 ✓	16917 ✓
19338 ✓	29361 ✓	36999 ✓	55815 ✓	51519 ✓

Each of the 30 groups displayed in the last two arrays has *the sum of its digits divisible by 3* and so is a multiple of 3.

An electronic study of the effect of subtracting each of the other 99,997 5-digit groups from each of the original 15 shows that 49578 and 48567 are the only two such that all 15 false sums are multiples of 3. But if the initial 15 groups are replaced by:

16343	06995	11872	12939	25713
29595	35411	43728	56542	54474
58895	66828	74556	93272	99086

then there is no group that (false) subtracted from each of the 15 gives groups each of which has the sum of its digits a multiple of 3.

9.7   Tiltman and the JN-25A Code Book

The pagination system[12] used for the JN-25A1 additives was somewhat reminiscent of the first lines of one of the best-known poems in the English language, Coleridge's *Ancient Mariner*:

> It is an ancient Mariner,
>   And he stoppeth one of three.

Twelve months before working on JN-25, Tiltman had discovered that the book groups of an additive cipher used by the Japanese Army in China were exclusively multiples of three. The use of every third 3-digit number as a page number in the first additive table for JN-25A must have reminded him of this practice.

Ronald Whelan, deputy head of the IBM card processing unit at Bletchley Park for much of WW2, wrote an account[13] years later of his experiences, including the following most interesting paragraph:

> Some of these special procedures were required to be prepared for Brigadier Tiltman. He made a point of returning to us following the successful outcome of his labours, so he could personally thank the operators for the work they had done for him. Also he always made light of his break-in into the cipher, sometimes saying he didn't quite know how he had done it, that he had just had a 'hunch' and found that it had worked. It was always a great pleasure working for him.

It is extremely likely that this 'hunch' refers to Tiltman's work on JN-25A1. The hunch would then have been that the more modern Japanese mariners had followed the practice he detected in the earlier Japanese Army cipher system. This was closely related to the method of allocating page numbers to the table of additives. One is led to suspect that Tiltman's great achievement was not the hunch of August 1939 about JN-25A1 but instead the hunch of 1938 about this otherwise insignificant army cipher system. (See the British Archives item HW 67/3.)

Tiltman had to check whether whenever about 13 GATs a, b, c, d, e, f, g, h, i, j, k, l, m were in depth, there existed a group x such that each of the (false) differences $a - x$, $b - x$, etc, had digits adding up to a multiple of 3. A crude calculation shows that this is quite improbable: for a random choice of x the probability of getting all of these sums with this property would appear to be the 13th power of $1/3$. Even trying the 100,000 permissible values of x would appear to be unlikely to find one that worked. The GCCS file HW 8/102 on the JN-25A1 project includes a message (9 August 1939) to FECB stating that 'we are all concentrating on the 5-figure naval cipher'. Presumably, Tiltman had 'borrowed' all available staff to search for such

---

[12]Op-20-G appears to have interpreted the page numbers as 001, 002, . . ., 299, 300 and to have brought in the rule 'add 1 and divide by 3' to get page numbers from the indicator. Of course neither GCCS nor Op-20-G could see a copy of the original additive table. There is no surviving evidence that the GCCS numbering did give Tiltman the hunch mentioned by Whelan 12 lines further on in the main text, but it seems likely that this was the case.

[13]The Whelan account of tabulators at Bletchley Park is in TNA HW 25/22.
One can only speculate about the 'special procedures' that were carried out for Tiltman.

an **x** for the various large sets of GATs in depth. Checking the 100,000 possibilities one by one consumes much time. Even though the process can be made somewhat more efficient, it is still very slow. Prolonged boring calculation would be needed for each case.

It should be noted that for the Japanese Army system introduced in December 1937 the groups consisted of four digits rather than of five. As there would be only 10,000 values of **x** to check for each such set, once the indicator system had been broken the process would take one tenth of the time.

After about ten GCCS staff had worked on these sets of GATs in depth for a week or two, sufficient evidence had been found for Tiltman's hunch to be accepted: the JN-25A code book used only 'multiples of three' as book groups. Here again a crude calculation of the probabilities is helpful. Suppose that instead of 13 groups **a** to **m** being known to be in depth the GCCS people had 23 such groups **a** to **w**. Then the probability of there being a group **x** such that each of $a - x, b - x, \ldots, w - x$ are all multiples of three is about .0001%. Once this has been found for a few sets of GATs in depths of 20 or more it is in practice certain that a genuine phenomenon has been identified.

Exploitation of this phenomenon by Allied cryptanalysts is described in some detail later in this book. The following is confirmed as the third major blunder made by the IJN with JN-25:

(3) The limitation that only scanning groups be used as book groups in the various versions of JN-25 (with one exception) greatly accelerated the process of 'stripping' GATs to obtain additives and book groups and reduced the depth required to do so.

The matter of how to exploit the fact that all book groups were multiples of three must have been discussed with the mathematical mafia at Bletchley Park. The only reference to this is a comment in a report[14] by Alan Turing, which deserves to be quoted again:

---

[14]This was reprinted in *Cryptologia* in 2000 and is available in facsimile on the Turing web site. The report was mostly on the machinery being made at the NCR plant in Dayton, Ohio, to handle German submarine Enigma codes.

Mavis Batey on page 96 of *Dilly* draws attention to a meeting held at Bletchley Park on 1 November 1939 involving Knox, Twinn, Jeffreys, Turing and Welchman. Equipment needed from BTM for work on Enigma was discussed. Travis then fixed up the purchase. This may have led to the same people designing the first JN-25 decryption device.

Carrying out large numbers of arithmetical operations in one's head is tiring and prone to error. In practice some machine would be essential for the processes described in this chapter and Chap. 10.

The British GCHQ, successor to the GCCS, cannot find documentation of the Letchworth device. It is difficult to understand why Turing, Welchman and others did not think up this use of colours while they were at it.

Cryptanalysis of JN-25 (July 1943) includes on page 231 a sketch of the Dayton version of the machine and states that various others had been tried earlier. This sketch closely resembles the device known as the 'fruit machine' and preserved in the Wenger Display at the Naval Station in Pensacola, Florida. Instructions for its use are given on pages 230–233 and 501–502 of the above-mentioned document.

SUBTRACTOR MACHINE. At Dayton we also saw a machine for aiding one in the recovery of subtractor groups when messages have been set in depth. It enables one to set up all the cipher groups in a column of the material, and to add subtractor groups to them all simultaneously. By having the digits coloured white, red or blue according to the remainders they leave on division by 3 it is possible to check quickly whether the resulting book groups have digits adding up to a multiple of 3 as they should with the cipher to which they will apply it most. A rather similar machine was made by Letchworth for us in early 1940, and, although not nearly so convenient as this model, has been used quite a lot I believe.

The above quote confirms that the work on JN-25A was being carried out at a site, namely FECB, not easily accessible to Turing. It confirms that advice on this matter was given by Turing and his associates and that it worked. It is not at all clear why there was a delay of 4 months between the commencement of work on JN-25A and the involvement of Turing.

9.8   The Probabilities

The rather crude probability argument used previously shows that if 15 or more groups a, b, etc, are randomly selected, it is unlikely that there would be any 'potential decrypting' group x, that is a group such that all of $a - x$, $b - x$, etc, are scanning. The probabilities are quite different if a, b, etc are obtained by randomly selecting *scanning* groups A, B, etc, and another group z and setting $a = A + z$, $b = B + z$, etc. These latter probabilities are relevant to the analysis of a column of JN-25 GATs in depth. Considerable (totally anachronistic) computer experiment reveals that the probability that such a 'column' of GATs in the depths shown will have a *unique* decrypting group x as above is as set out in the second line of the following table. The third line of the table gives the average number of potential decrypting groups. In about 90 % of the samples, the number of decrypting groups is at most that shown in the fourth row of the table.

depth	7	8	9	10	11	12	16	20
Prob. unique	0 %	0 %	1 %	3 %	6 %	11 %	42 %	68 %
average	148	73.2	38.4	21.7	13.3	8.6	2.7	1.6
90 % level	318	164	87	49	30	19	5	2

Some samples give many more decrypting groups than might have been expected from knowledge of these averages. Appendix 4 of Chap. 10 gives contrived columns with remarkable numbers of decrypting groups.

Turing in his report *The Applications of Probability to Cryptography* (1941) compared certain work in cryptology to searching for a needle (the correct decryption) in a haystack. Knowledge that all of the decrypted JN-25 groups scan greatly reduces the size of the haystacks, particularly if the depth is reasonably large.

The processes called 'Phase 2' and 'Phase 3' below use knowledge of the commonly occurring book groups to find likely decrypting additives without wasting too much time on possibly numerous alternatives. The cryptologists needed efficient methods to decide whether a candidate decryption was likely[15] to be correct. This very important issue is discussed in Appendices 2 and 3 of Chap. 10.

9.9 Phase 1: Getting Started

The cryptologist attacking a new JN-25 system would begin with the columns of GATs in greatest depth (preferably at least 17) and then find all such decrypting groups for each. An undoubtedly laborious process is available for this task. The 100,000 5-digit groups are tried one by one until one which works is found. Methods were developed to accelerate this process. A principal archival item is a report[16] by Robert Ely dated 24 January 1944 which begins:

> Given that by indicator analysis we can line traffic up in depth, and that we know the code groups scan by three, but we have no data as to the meanings or relative frequency of the underlying groups. Required: To find for a given depth (column) the additive(s) whose application will make the resultant groups scan.

Ely analysed the methods then in use and concluded that mechanisation might well be possible using the technology already developed. Instead of reporting on the Ely document or earlier accounts of methods used for initial decryption of JN-25, Sect. 15.6 sets out how quite sophisticated methods might well have worked.

The initial process will be called Phase 1 and would be carried out on a new JN-25 system to the minimum extent necessary to enable other methods (Phases 2 and 3) of stripping to work. It is not feasible for use in recovering even a significant proportion of all the 50,000 groups in an additive table of 500 pages each containing 100 entries.

---

[15] A 1945 reference to the effect that at least a sprinkling of the JN-25B7 and JN-25B8 additives had been evaluated incorrectly by Op-20-G is NARA RG457 Entry A1 9032, Box 578, file 1391.

This file refers to a proposed exercise of reworking the decryptions of JN-25B7 and JN-25B8 intercepts using the improved method discussed in Appendices 2 and 3 of Chap. 10. Presumably the motivation was to obtain more complete and accurate decryptions and decodings of more intercepts and so determine whether the Pearl Harbor raid could have been anticipated. Not much came of this. In any case Chap. 7 showed that more than enough information about the risk of such an attack in December 1941 was available at that time.

[16] The Ely report is to be found in NARA RG38, CNSG Box 7, item 2500/7 and also RG457 Box 600, item 1551. This refers to an earlier report by Lt. (jg) Tompkins dated 14 November 1942. The GCCS file on JN-25 (TNA HW 8/102) contains a terse summary dated 5 September 1942 of a method developed by Thomas Room and Richard Lyons in Melbourne.

Ely's use of the word 'additive' may cause some confusion here. The matter has little intrinsic importance but has been explained in Note 19 of Chap. 8.

More sophisticated methods of handling Phase 1 would consist of taking various columns of depths 14 or more, selecting those with relatively few decrypting groups and making optimal use of the information they provided together.

In August 1939 the Tiltman team began work on JN-25A1. Inadequately trained IJN cipher clerks had tended to choose one of a few obvious places in the additive table to start encrypting several[17] messages. So a supply of columns of depths 20 or more was available. Most of these would have had unique decryptions and so provided an initial list of book groups in use. The exercise, for which Tiltman co-opted all available GCCS staff, would also have confirmed that his hunch that all book groups were scanning was probably correct. There would have been no difficulty starting on a list of frequently used book groups. The correct decryptions of other columns of depths 15 and upwards would then have included groups already on this list. Incorrect decryptions would have produced random scanning groups, few of which would have been on the list. So most of these columns could have been decrypted and the list of frequently used book groups improved. Quite obviously the strategy would be to work with the columns of greatest depth first. As searching for all decryptions of a given column of low depth is quite laborious, Tiltman would have switched to Phase 2 (below) or Phase 3 (Sect. 10.2) as soon as possible.[18]

Suppose now that a new JN-25 system is found to have weak indicators. A few depths of 15 or more would confirm the continuation of the use of only multiples of three. If only depths of 19 or less are available in quantity, a new strategy for identifying book groups taken randomly from intercepted signals must be found before Phase 2 and/or Phase 3 methods can be used. In these the occurrences of previously identified common book groups are a good guide to determining the correct decryption.

In fact computer simulation experiments, based on information about the JN-25L codebook (1944) preserved in the British Archives as item HW 43/34, provide the following rather startling insight:

If a JN-25 system has a code book using only multiples of three, has a weak indicator system, and has the frequencies of occurrence of its common book groups comparable with those of JN-25L, then around 1,000 GATs in columns with average depth around 13 (and none in depths of less then 11) will yield on average correct decryptions for about 90 % of the GATs.

This demonstrates the magnitude of the insecurity of the JN-25 series of ciphers!

In general, an important feature of any serious attack on an additive system is regularly updating the statistics on the frequency of occurrence of the various progressively identified book groups. This will yield gradually more definitive lists of the 100–250 most commonly used book groups and the frequency of their occurrence. Once this has been worked out the cryptologist may well wish to check the decryptions obtained by Phase 1 methods.

---

[17]The Barham *Japanese Ciphers—Notes* (in TNA ADM 223/496) state that the Tiltman team, of which he was a member, was able to obtain intercepts in depths of up to 20. This would not have happened if the starting places used for JN-25A1 messages had been chosen by lot!

[18]When JN-25A5 was replaced by JN-25B5 the FECB cryptologists had already worked out some of additive table 5. They could thus effortlessly decrypt some of the JN-25B5 messages and thus get initial statistics on the most commonly used book groups in code book B. When additive table 5 was replaced by additive table 6 the British and American teams could use these statistics. *Thus Phase 1 was by-passed for JN-25B.*

9.10   Phase 2: Using Known Common Book Groups

Appendices 2 and 3 of Chap. 10 explain the theoretical basis for Phases 2 and 3. In particular, almost every column of GATs in depth of 6 or more will, when decrypted, yield at least two common book groups.

The following example illustrates the Phase 2 method. Suppose nine JN-25 messages have been put in depth and that one column (typeset here as a row) consists of the GATs

> 60621   31945   16464   54959   98825   63959   53626   43861   03930

which therefore were all obtained from book groups—all multiples of three— by (false) addition of the same unknown 5-digit group. Suppose that previous work with this system has yielded ten common[19] book groups, which in order of frequency are

> (1) 04518   (2) 13854   (3) 45297   (4) 59094   (5) 49242
> (6) 56127   (7) 89112   (8) 80784   (9) 24495   (10) 71469

The first step is testing whether the first GAT, 60621, is obtained from the most common book group, 04518, by (false) addition of the appropriate entry in the table of additives. If this is the case, the additive has to be:

$$60621 - 04518 = 66113$$

and so 66113 may be subtracted from each of the nine GATs:

GATs	60621	31945	16464	54959	98825	63959	53626	43861	03930
ADD	66113	66113	66113	66113	66113	66113	66113	66113	66113
DIFF	04518	75832	70351	98846	32712	07846	97513	87758	47827

If this were the correct stripping, all of these differences should be multiples of three. So this is checked with the following outcome:

DIFF   04518 √   75832 ×   70351 ×   98846 ×   32712 √   07846 ×   97513 ×   87758 ×   47827 ×

---

[19]The *GYP-1 Bible* states on page 48 and elsewhere that collateral information in what is in the messages being used to recover additives may influence the decision about which common groups should be tried. Of necessity parts of this book oversimplify some of the issues. Thus in practice garbled GATs were quite common and so the decrypter may well have been happy to accept eight groups out of nine being divisible by three as good enough.

Next the same book group, 04518, is tried successively in the other eight positions. None of these yields a line of differences that are all scanning.

Next, the second most common book group, 13854, is tried in various positions. Eventually it is tested as a prospective book group from which the seventh GAT is obtained by use of the additive $53626 - 13854 = 40872$. The calculations are then:

GATs	60621	31945	16464	54959	98825	63959	53626	43861	03930
ADD	40872	40872	40872	40872	40872	40872	40872	40872	40872
DIFF	20859	91173	76692	14187	58053	23187	13854	03099	63168

A check for divisibility by three yields:

DIFF	20859 ✓	91173 ✓	76692 ✓	14187 ✓	58053 ✓	23187 ✓	13854 ✓	03099 ✓	63168 ✓

which includes 13854 and so is unlikely to have happened by chance.

Thus more book groups are known and so the statistics of common book groups could be adjusted. (In practice, this would happen only after a significant amount of new data had become available.) The missing additive has been found with relatively little labour. And now attention may be turned to the next set of nine GATs in depth. If an exercise of this type gets nowhere, it may be sensible to defer working on that particular set of GATs until additional common book groups are known or more messages in depth are obtained. If all else fails it would be possible to try the exhaustive search on a few sets of GATs, but this should be done as little as possible.

The following table has the above calculations set out in columns rather than in rows to show the stripping of the nine given JN-25 GATs in depth. Experience eventually showed that stripping devices based on columns were more user-friendly. It is evident that such calculations are much better done mechanically. It is inefficient to return to the initial depth (column) of nine groups for each step. Instead the group that needs to be added to each group in one column to produce the next column is worked out and then the next column calculated. The italicized are used to show where the common book groups are being tried. On a special-purpose device, flags are needed for this function.

One stops at the column in which all the groups are scanning and notes that it may be one of the about 38 possible decryptions and so may not be correct. But if there are two (or, much better, several) known common book groups in it the probability that it is correct is much higher. As already noted in the section on *The Probabilities*, this very important issue is left to Appendices 2 and 3 of Chap. 10.

```
60621 √    04518 √    33294 √    58775 ×    10280 ×
31945 ×    75832 ×    04518 √    29099 ×    81504 √
16464 √    50351 ×    89037 √    04518 √    66023 ×
54959 ×    98846 ×    27522 √    42003 √    04518 √
98825 × → 32712 √ → 61498 × → 86979 √ → 48484 ×
63959 ×    07846 ×    36522 √    51003 √    13518 √
53626 ×    97513 ×    26299 ×    41770 ×    03285 √
43861 ×    87758 ×    16434 √    31915 ×    93420 √
03930 √    47827 ×    76503 √    91084 ×    53599 ×

           76314 √    01280 ×    11513 ×    21378 √
           47638 ×    72504 √    82837 ×    92692 ×
           22157 ×    57023 ×    67356 √    77111 ×
           60642 √    95518 ×    05841 √    15606 √
         → 04518 √ → 39484 × → 49717 × → 59572 ×
           79642 ×    04518 √    14841 √    24606 √
           69319 ×    94285 ×    04518 √    14373 √
           59554 ×    84420 √    94753 ×    04518 √
           19623 √    44599 ×    54822 √    64687 ×

           61209 √    13854 √    42530 ×    67011 √
           32523 √    84178 ×    13854 √    38335 ×
           17042 ×    69697 ×    98373 √    13854 √
           55537 ×    07182 √    36868 ×    51349 ×
         → 99403 × → 41058 √ → 70734 √ → 95215 ×
           64537 ×    16182 √    45868 ×    60349 ×
           54204 √    06859 ×    35535 √    50016 √
           44449 ×    96094 ×    25770 √    40251 √
           04518 √    56163 √    85849 ×    00320 ×

           29526 √    85650 √    10526 ×    20859 √
           90840 √    56974 ×    81840 √    91173 √
           75369 √    31493 ×    66369 √    76692 √
           13854 √    79988 ×    04854 √    14187 √
         → 57720 √ → 13854 √ → 48720 √ → 58053 √
           22854 √    88988 ×    13854 √    23187 √
           12521 ×    78655 ×    03521 ×    13854 √
           02766 √    68890 ×    93766 ×    03099 √
           62835 √    28969 ×    53835 √    63168 √
```

9.11   Patterning and the Use of Multiple Book Groups

The IJN further simplified the breaking of JN-25A by committing two more blunders.

There were regularities in the code book for the numbers from 1 to 999. Thus 273 had the book value 27324, where the final '24' is double the sum of the digits. The months had code values 01005, 02010, etc, with December having code value 12060. The days of the month were encoded as 01302 for the first day, 02304 for the second, etc, with the 31st having code value 31362. All of these values are multiples of three.

Each of these conventions poses grave risks to the security of the system. For example, suppose that an enemy cryptologist has worked out relative (non-primary) code values (Sect. 8.16) for certain days of the month. Even when some fixed group, say 78251, is added to some of the values 01302, 02304, 03306, etc, a pattern will quickly become evident. This is likely to lead to quick discovery of the general fact that all code values appear to be multiples of three. So the fourth major error was:

(4) Patterning in the JN-25A code book helped to obtain correct meanings for some of its entries.

The comment made by Professor Room in the *CBTR*, and already quoted in Sect. 8.18 is relevant here. At least one of four conditions had to be satisfied to make an additive system vulnerable. Two of these were:

(1) Additive starting-point indicator system should be such that messages can be set in depth from it alone.
(3) The code book must be heavily patterned and these patterns are clear from the differences of groups in the same column when the messages have been set in depth.

All JN-25 systems would have been much harder to attack if their code books had provided multiple book groups for commonly used texts and if, when such were available, operators had been required to choose one of the available groups randomly each time such a common word was being encoded. This was not done. More on this matter is in Sect. 13.3.

This provides the next major blunder made by the IJN in its use of the JN-25 system:

(5) Failing to have enough alternative book groups for commonly used words and phrases and failing to insist that operators use dice or similar devices to determine which alternative to use.

It is not obvious just how many alternative book groups a common work or phrase needs to preserve security. Modern computer experiments can work this out but the IJN did not have such capabilities at the time.

9.12 Early American Work on JN-25

Op-20-G (Sects. 5.1 and 5.2) became involved with JN-25 in November 1939. Agnes Driscoll, an Op-20-G veteran, and Prescott Currier, later to be one of the two USN representatives in the 'Sinkov mission' to Bletchley Park early in 1941, started work on JN-25A. An almost contemporary study[20] of the role of captured documents in the decline and fall of JN-25 states that:

> The break finally came some time in the early (North) fall of 1940. The fact that additives could be recovered only at the beginning of dispatches led to the hypothesis that these beginning groups contained code groups representing numbers. Someone remembered that in an old four-numeral code, which we had stolen from a consulate, the Japanese had generated their code groups for numbers according to a regular pattern. When this old 'S' code, as it had been designated, was obtained from the files, it was found that the four-numeral group for each number was composed of the number itself plus a regular multiplying factor.

This work revealed the book groups for all numbers from 0 to 999 in JN-25A. It also yielded access[21] to the primary book groups and so inevitably exposed the phenomenon that each book group had digits which summed to a multiple of three.

A summary account of Op-20-G work on JN-25A1 and JN-25A2 may be found in NARA RG38 Box 33. It is named RIP 74-A, dated 25 January 1941, and entitled 'AN Cipher No. 1'. ('AN' was the identifier given by Op-20-G to the JN-25 system prior to the introduction of the 'JN' notation.) It contains:

(a) a description of the sequence of items in a message text—each GAT is a 5-numeral group, the first gives the sender's identity, the part number and the

---

[20]The item is in NARA RG38, CNSG Library Box 115, 5750/119. Although it was written soon after the war ended, the author could not guarantee completeness of the account of the use of captured documents by Op-20-G. There is some considerable difference between documents captured in conflict and those obtained by sleight of hand or theft earlier but the material on the 'S' code was of considerable importance.

It is not clear whether Tiltman had memories of this same code as that called 'S' by Op-20-G.

Stephen Budiansky's paper in the April 2000 issue of *Cryptologia* refers to archival material to the effect that Op-20-G was decoding some JN-25A messages in October 1940 and expected to be reading it more or less in full by April 1941. This hope died when JN-25A was replaced by JN-25B on 1 December 1940. Budiansky has written another extremely relevant paper on the early American work on JN-25. This is *Too Late for Pearl Harbor*, published in the *Proceedings of the US Naval Institute*, December 1999.

Jeff Bray interviewed the veteran Arthur Pelletier in 1994 and obtained information supporting this account of Agnes Driscoll's role. This is published as a footnote on page xx of Bray's version of SRH-009, *Ultra in the Atlantic*, Aegean Park Press.

Stephen Budiansky in his paper in *Cryptologia* (April 2000) draws attention to the *History of GYP-1* in RG38 of NARA. The reference is CNSG 5750/202, pages 14–18. The American break into JN-25A was indeed as late as September 1940.

[21]In Sect. 14.2 some text is quoted from Part G of the *CBTR* about how the corresponding discovery was made for the Water Transport Code. The process appears to be quite similar to the Driscoll 'break' on JN-25A.

total number of parts while the second is the key indicator. The second last part is the key indicator repeated and the last gives the date and time of transmission. It is noted that the message text will contain at most 15 groups.

(b) a list of the message key indicators for each day that JN-25A1 was used, plus instructions for using these to obtain from the key indicator the true key (starting point) used for that message.

(c) instructions for using the starting point in the additive table for stripping the message GATs, whenever the relevant additive table entry had been previously found.

(d) a list of the 300 10 × 10 tables in the first JN-25 additive table, showing the additives recovered to date on each page.

It does not explicitly state that all book groups were multiples of three, but a decrypted message text exhibits that property.

9.13   The Switch from JN-25A5 to JN-25B5

The JN-25A code book was replaced by a new code book (JN-25B) on 1 December 1940, but the fifth table of additives continued in use for another 2 months. Thus JN-25A5 was replaced by JN-25B5. The significance of errors of this type was appreciated by Op-20-G later[22]:

> It had always been realised that by simultaneously changing all three elements—code, cipher and rules—the Japanese could set up a system which could be broken only with extreme difficulty and after much loss of time.

The switch from JN-25A to JN-25B was noted[23] by Neil Barham, who had worked with Tiltman and Burnett in the first examination of JN-25A in July and August 1939:

> The Japanese introduced a new codebook but, unfortunately for them, retained in use the current reciphering system and indicator system. These had already been solved in some positions and new code groups were discovered immediately. But for this mistake on the part of the Japanese the form of the book may have taken a couple of months to discover.

All this describes a two-stage blunder. The switch from JN-25A4 to JN-25A5 did not change the code book and, in particular, did not change the most common book groups. It has already been noted that the stripping process makes heavy use of

---

[22]NARA RG38, CNSG Box 115, 5750/119, page 12, quotes this from page 100 of the Op-20-GYP history.

The rules involved in some Japanese naval and army ciphers could be quite complicated. Thus the Jamieson report (B5554 in the NAA) mentioned in Note 46 below gives some minor IJN systems which had seven different variations, one for each day of the week.

[23]This is taken from the Barham report previously cited in Note 17. It was found courtesy of the book *Action this Day* by Michael Smith and Ralph Erskine (Bantam Press, London, 2001).

these. Thus at least some of the more heavily used parts of the fifth table of additives could be determined. When the 'A' code book was replaced by the 'B' code book 2 months later, these pieces of the additive table could be used to decrypt some intercepts. Even as few as some hundreds of decrypted groups would suffice to give usable information on which were the commonest book groups in JN-25B.[24]

This was the sixth blunder made by the IJN in its use of the JN-25 series:

(6) The switch from JN-25A5 to JN-25B5 (continuing with the fifth table of additives) enabled enough common JN-25B book groups to be identified very quickly.

It was immediately evident that the sum of digits of JN-25B book groups appeared to be all multiples of three. No special hunch was needed here. A single decrypted JN-25B intercept would have been enough to show that the use of multiples of three was continuing. This repeated blunder would now be anticipated in each new version in the JN-25 series by the more optimistic Allied cryptanalysts and would be one of the first objects of researching each new version. See Sect. 15.7.

At the time JN-25A was replaced, it was estimated that its code book would have been substantially recovered in another four to 6 months more. Instead JN-25A provided the British at FECB with a very useful training exercise and, by indirectly providing information on the most common JN-25B book groups, opened the way for a serious attack upon JN-25B.

9.14 Co-operation Between FECB and Op-20-G

This section inevitably repeats parts of Sect. 5.5.

---

[24]The paper by Peter Donovan in *Cryptologia*, October 2004, gives some reasoning why, with JN-25 having book groups which were known to be multiples of 3, extra tricks were available for the cryptologists when either

- the additive table was changed but not the code book, or:
- the code book was changed but not the additive table.

Changes of both types happened with JN-25B.

In fact this chapter, Chaps. 10 and 15 reveal that the use of only multiples of three as JN-25 book groups was appreciably more pernicious than as suggested in the 2004 paper.

Once again attention is drawn to the Swedish experience with a Russian additive cipher in 1940. Bengt Beckman, on page 54 of *Codebreakers*, quotes a contemporary report to the effect that the Russian failure to change both code book and additive together saved the Swedish unit 'several months'.

'But for this mistake on the part of the Japanese the form of the book may have taken a couple of months to discover.' (This is from the Barham quotation given earlier, see Note 23 above.) Would JN-25 have been broken in time for the Coral Sea and Midway battles if there had not been an overlap in additive tables for JN-25A and JN-25B?

Early work on JN-25 carried out in Washington had been transferred to the advanced station Cast in the Philippines. An oral history interview[25] with Rudolph Fabian, the head of Op-20-G Station Cast in Manila during 1941, is now highly relevant:

> Jeff Dennis went to FECB from Cast and brought back a solution of the 5-number system [JN-25B]. He brought back the solution, how to recover the keys, the daily keys. How the code was made up and a lot of code values. And since it was a very heavy system, it was the heaviest volume system on the air. I talked with my people and I went back to the CNO and requested permission to drop everything else and go to work on this five number system.
>
> And we did pretty well on it. We couldn't do any solid reading but we could pick up phrases like 'enemy submarine'.

This does not mention a subsequent visit to Cast made a little later by Malcolm Burnett of FECB. Captain 'Ham' Wright, in another National Cryptological Museum oral history, does mention this later visit and confirms that either or both visits enabled the Americans to learn how to process JN-25B efficiently.[26]

A key USN document of 5 March 1941 is repeated from Sect. 5.5:

> Have received from British following in approximate numbers referring to five numeral system (JN-25B5) effective December to February. Five hundred book values, four thousand subtractor groups, half thousand work sheets with cipher removed and two hundred ninety indicator subtractors for SMS numbers.
>
> Have arranged secure method of exchanging further recoveries by cable.
>
> British employ three officers and 20 clerks on this system alone. They are delaying attack on current cipher until mid-March to accumulate adequate traffic and obtain further book values from preceding period. Due collateral information available here and capacity rapid exchange with English Cavite (the Fabian-Lietwiler team) will assume this system.

---

[25]The Fabian interview is one of about 35 oral history records available from the National Cryptological Museum.

The NCM oral history is somewhat different from the version printed in the booklet *The Quiet Heroes of the South-West Pacific Theater: An Oral History of the Men and Women of CBB and Frumel* edited by Sharon Maneki. In particular, Fabian devoted some considerable time to explaining how having the book group digits summing to a multiple of three was exploited. This had been crudely blacked out in the original version available from the NCM. An uncensored copy has now been released by the NSA. The paper by Peter Donovan (*Cryptologia* 28(4), 2004, pages 325–340) appears to be the first reconstruction of this key aspect.

The NAA Melbourne file MP1185/8 1937/2/415 shows that Nave in March 1941 confirmed that the '5-figure code' (JN-25B) and two other diplomatic codes were the only Japanese codes intercepted in reasonably large quantity at FECB.

[26]The timely transfer of key technical information is rather analogous to that of 27 July 1939, when a GCCS delegation, including Dilly Knox, was in Poland meeting the Polish cryptanalysts. Marian Rejewski was asked *Quel est le QWERTZU?*, that is in which order the 26 letters were channelled into the military Enigma machine. (The standard German typewriter has QWERTZU on the left of the top row of letters whereas English keyboards have QWERTY there.) After being told he responded *Nous avons le QWERTZU—nous marchons ensemble!*, that is 'we have the QWERTZU—we will walk together'. In Singapore Op-20-G acquired information about how to exploit the use of multiples of three in the JN-25B code book and so could walk with FECB.

Chapter 6 of Mavis Batey's *Dilly: The Man Who Broke Enigmas* gives much information on the July 1939 meeting.

> As only navy arrangement request Dept forward results to date and technique if
> considered helpful.

It seems clear that 'Dept' (the Navy Department, including Op-20-G) was not able
to help with techniques.

Fabian is stating that his team recognised that JN-25B had acquired enormous
importance. Obtaining the allocation of resources to its breaking becomes one of
his major achievements. He also recalls attempting unsuccessfully to direct more
USN resources elsewhere to working on this cipher system. But the USN had no
foreknowledge that the JN-25 systems would continue in use after any onset of war.
The decision to keep the cryptologic team at Pearl Harbor working on another cipher
system throughout most of 1941 was rational, but unfortunate. However its leader,
Joe Rochefort, was well informed about the techniques used for JN-25B.

Thus in the first half of 1941 the number of people working on JN-25B increased
from around 40 to around 100. These included interception staff, traffic analysts
and other support staff. Cast and FECB were co-operating and developed a secure
OTP radio link[27] in April 1941. Fabian revealed in his interview that there was a
useful two-way interchange of information. This would have included page after
page of additive and (much more slowly) book group after book group. Duane
Whitlock,[28] a traffic analyst at Cast, makes it clear that traffic analysis was useful
in interpreting book groups. Neil Barham of FECB has been quoted[29] as saying that
about 3000 book groups were known by 1 December 1941. Here 'known' cannot
mean 'decoded' but simply 'having been observed in a stripped (deciphered) JN-
25B message'. This is broadly compatible with a statement made by Commander
Jack Newman, Director of Signals and Communications for the RAN. And so there
would have been a reasonably accurate list of the most frequently occurring 100 or
150 book groups.

9.15   The Evolution of JN-25B

With JN-25B, a new major complication was introduced, which is described as
follows in the USN Radio Intelligence Publication RIP 80 of 1 September 1943:

> Many code groups (below 69999) have an additional 'Auxiliary Table' (A/T) meaning,
> which is to be used if so indicated by an appropriate group in the normal table. The Auxiliary
> Table contains Roman letters, digraphs, and high frequency trigraphs, kana, units and
> activities, place and proper names, movement report vocabulary, punctuation and numerals.

---

[27]Evidence of the existence of the OTP link is to be found on page 20 of Robert Benson's *History
of US Comint in WW2* published by the Center for Cryptologic History of the NSA.

[28]The Whitlock interview is in the same Cryptologic Museum series as that with Fabian.

[29]Barham is quoted on page 96 of Michael Smith's *The Emperor's Codes*. Newman's account is
available in the NAA Melbourne file B5555 2.

The fifth additive table was replaced on 1 February 1941 to produce JN-25B6. This continued until the end of July 1941. The method of enciphering the indicator was somewhat changed. More changes were made with the introduction of a new (seventh) table of additives on 1 August 1941. This continued in effect until 3 December 1941. JN-25B7 has considerable interest. As was learnt much later, had enough resources been made available to work steadily on it since its introduction, some further information relevant to the coming Pearl Harbor raid might have emerged.

The seventh additive table was replaced by an eighth one (producing JN-25B8) on 4 December 1941, only a few days before the Pearl Harbor raid. The fact that the change of additives was not implemented tidily on the first day of the month should have suggested that trouble was brewing.

All previous work on JN-25 would have amounted to a valuable training exercise but no more had the IJN on 4 December 1941 followed the first maxim given in Sect. 8.19. The relevant quotation[30] is:

It had always been realised that by simultaneously changing all three elements—code, cipher and rules—the Japanese could set up a system which could be broken only with extreme difficulty and after much loss of time.

Thus, when JN-25B suddenly became totally unreadable on that day, the anticipated imminent outbreak of war would have led the Allies to expect a complete change— new code book, additive table and rules. But some 15 h after the Pearl Harbor raid a JN-25 message was intercepted in which every GAT was a multiple of three. This most likely happened because the operator forgot to add on groups from the additive table. But several groups in this message were known common JN-25B groups. This immediately aroused suspicions that, even though it had been anticipated that the code book would have been totally replaced immediately prior to the commencement of hostilities, this had not happened. Another such plain code JN-25B message was intercepted about 6 days later. This confirmed that the old 'B' code book was still being used. That was the seventh major blunder made by the IJN in its early used of JN-25:

(7)   The failure to change the JN-25B code book early in December 1941 when the switch from JN-25B7 to JN-25B8 was made. The JN-25B code book was used for nearly 18 months, which was far too long for a heavily used code.

Laurance Safford later described[31] this as a most significant blunder. It most certainly reveals gross overconfidence by the IJN in its cryptographic capacities.

---

[30]The quoted text is taken from NARA RG38 CNSG Library Box 115, 5750/119. See also page 100 of the history of Op-20-GYP in the SRH series at NARA.

[31]The source is Laurance Safford's *A Brief History of Communications Intelligence in the United States*, SRH-149 in NARA RG457. An edited version is published by Aegean Park Press under the title *U.S. Naval Communications Intelligence Activities*. Although there are some errors in matters of detail, Safford makes the key point: 'If—and it is a big if—the Japanese Navy had changed the code book along with the cipher additives on 1 December 1941, there is no telling how badly the War in the Pacific would have gone for Australia and the U.S. or how well for the Japanese in the middle stages.' (The correct date was 4 December 1941, not 1 December 1941.)

This blunder by the IJN was one of the major historical factors that contributed to the actual outcomes of the Battle for Australia, the Battle of Midway and (to a lesser extent) the struggle for Guadalcanal. In fact, Safford quotes the key message from 'COM 16' to 'OPNAV INFO CINCAF' of 15 December 1941 referring to intercepts in plain code on both 8 and 13 December and the all-important sentences:

CODE REMAINS UNCHANGED
WILL SEND RECOVERIES THIS SYSTEM IF YOU DESIRE

Captain Shaw, the head of FECB sent an even more encouraging message to GCCS at this time stating that about 3 months further work would see major breaks into JN-25B. Note that the computations needed for this process were done on sequences of about 10–40 groups at a time out of a total of 50,000 groups in the eighth additive table. It required numerous staff but the method was clear.

In summary, the successive changes of additive tables used with the JN-25B code book minimally enhanced security but considerably enhanced the building up (by the prospective Allies) of knowledge of book groups. In turn, this information enabled a reasonable knowledge of the most commonly used book groups and their relative frequencies of occurrence to be available for immediate attack on constructing entries in the new additive table.

9.16 Joe Rochefort

Commander Joseph Rochefort (1898–1976) has become the hero of what is seen as the greatest cryptanalytic achievement of the USN. Yet Rochefort was rewarded for his contribution by being transferred to an insignificant position elsewhere. It was left to President Reagan to make amends in 1986 by presenting the appropriate award to Rochefort's children. The new (2011) major NSA building in Hawaii was named after Rochefort. His name has been included in the National Security Agency's Hall of Honor while the character assassins are forgotten.

Rochefort had served 6 months in WW1 before becoming a career naval officer. He was one of several key early figures recruited to what became Op-20-G by Laurance Safford. The selection appears to have been based on solving puzzles set in a magazine circulating in the USN. He learned the elements of cryptanalysis from Safford and Agnes Driscoll. When Safford was sent on a tour of duty in 1926, Rochefort took over as the head of Op-20-G. He had language training in Japan in 1929–1932, at the same time as Edwin Layton. In the 1930s he mostly served at sea but had some experience as an intelligence officer.

Fabian elsewhere states that there was an unenciphered message intercepted from a ship under attack near the Philippines some 12 h after the raid on Pearl Harbor. It would have been sensible to wait until several depths of 12 or more of new system messages were available and then test them to see if the known common JN-25B book groups were being used. If this was the case it would then be responsible to announce that CODE REMAINS UNCHANGED. However the second unencrypted message must have removed most doubt.

The paper[32] *How Op-20-G got rid of Joe Rochefort* by Frederick Parker describes how lesser people in the Navy Office in Washington managed to get Rochefort side-lined a few months after the Battle of Midway. It suffices here to quote the following paragraph from a letter from Captain Joseph Redman to Admiral Horne (June 1942):

> Strong people should be in strong places and I do not believe the Pacific organisation is strong because the administration is weak in so far as Radio Intelligence is concerned. I believe that a senior officer trained in Radio Intelligence should head up these units rather than one whose background is in Japanese language.

This conspicuously fails to note that Rochefort had been in charge of Op-20-G in the early days while Safford had been required to serve on a ship and that he had just brought off one of the greatest intelligence coups ever.

After the Battle of Midway, Admiral Nimitz had recommended Rochefort for the award of a Distinguished Service Medal, and General MacArthur had recommended a similar award to Rudolph Fabian, head of Frumel. (The acronym Frumel, from Fleet Radio Unit Melbourne, was eventually adopted for the Melbourne unit described later in this chapter.) Both were unsuccessful.

9.17   JN-25B in Hawaii

Soon after receipt of the first bare code intercept, the Hawaii Sigint unit must have been provisionally ordered to concentrate on JN-25 intercepts. The second such intercept would have led to management confirming this decision. Once adequate depths of JN-25B8 intercepts had been built up, the correctness of this decision would have been confirmed repeatedly.

Rochefort had developed a first-class team of over 40 people, including linguists and intelligence officers, in the Combat Intelligence Unit. The last language officers to visit Japan before the outbreak of war were included. It operated in the basement, commonly called the dungeon, of an administration building at the Pearl Harbor base.

For over ten years the USN had been committing far more resources into Op-20-G than the US Army put into its SIS. This practice paid off immediately after the raid on Pearl Harbor, when it became clear that JN-25B had not been discarded

---

[32]Parker's paper is in *Cryptologia* 24 (July 2000) 212–234. Parker also wrote two booklets *Pearl Harbor Revisited: USN Communications Intelligence 1924–1941* and *A Priceless Advantage: US Naval Communications Intelligence and the Battles of Coral Sea, Midway and the Aleutians.* Both were published by the Center for Cryptologic History, National Security Agency. All three of Parker's publications are highly recommended.

SRH-268 in NARA RG457 is given in www.history.navy.gov/library/online/srh268.htm and contains the full text of Redman's letter.

See also Eliott Carlson, *Joe Rochefort's War*, Naval Institute Press, 2011.

but was merely being used with a changed additive table. So Rochefort's unit was available to be switched to what had become the only game in town.

The forced withdrawals of FECB and Cast increased the relative importance of the Hawaiian unit. Although the stripping of the additives from the intercepts was not a fundamental problem, sufficiently many plain language code words had to be recovered individually from the flood of intercepts. As explained in Sect. 11.5, it was possible to use IBM cards to locate past usages of any particular book group and thus make the task feasible.

In fact the sinking of the battleships at anchor nearby made available the extra staff needed for the dungeon. Musicians from the band of the USS *California* were co-opted into working out additives, punching cards and stripping the additive from intercepted GATs.

The recovery of enough of the code book accelerated as the weeks wore on. Thus the quality of the resultant intelligence improved. A final triumph was the breaking of a somewhat complicated system used to encode dates in JN-25B. The date encryption system for JN-25A had easily yielded to cryptanalysts. (Having any such system at all was another classic blunder: a random allocation of book groups for a good range of future and recent past dates would have been safer.)

The following list of principal members of Rochefort's unit in the lead-up to the Battle of Midway is due to Forrest Biard.[33]

Joe Rochefort	Eddie Layton	Joe Finnegan	Red Lasswell
Tom Dyer	Ham Wright	Tom Huckins	John Williams
Jack Holtwick	John Roenigk	Allyn Cole	Art Benedict
Gil Slonim	John Bromley	Banks Holcomb	'and others'.

---

[33]One reference is Graydon Lewis: *Setting the Record Straight on Midway* published in *Cryptologia*, volume 22(2), April 1998, 99–101. This debunks theories that JN-25B was broken by means of recovering the code book from the wreck of the submarine I-124, sunk off Darwin early in 1942. The book *Sensuikan I-124* by Tom Lewis on this submarine contains several high-quality photographs of it. Of course the first-hand accounts of Biard and others, coupled with overwhelming documentary evidence, make the truth of the matter quite clear. See also Biard's paper *The Breaking of Japanese Naval Codes: Pre-Pearl Harbor to Midway* in *Cryptologia* 30(2) (April 2006) 151–158.

The myth of JN-25B being broken courtesy of the I-124 is attributed in Ken Kotani's *Japanese Intelligence in World War II*, Osprey Publishing, Oxford, 2009 to a semi-published account of Commander Motonao Samejima. See page 87 of Kotani's book, where it is stated that 'in the lead-up to the Battle of Midway, it was clear that the IJN's operational codes had been compromised, since the Allies had obtained the code books from the I-go 124 submarine sunk to the north of Australia in January 1942'.

In an interview given in 1969 and preserved in the US Naval Institute, Rochefort explains how his unit was dependent on IBM cards and associated machinery. His colleague Thomas Dyer had been responsible for putting together this early large-scale cryptanalytical data processing system.

Donald 'Mac' Showers, the 'last man standing' of Rochefort's team, has recorded his experiences on a DVD prepared and sold by Shoestring Educational Productions of San Diego. He mentions various key figures in the Dungeon and confirms that JN-25B was not providing useful intelligence until late February 1942. This DVD is strongly recommended.

Biard later emphasised that the key figures were Rochefort, Finnegan, Lasswell, Dyer and Wright. He stated that without any one of these men the Midway coup would not have happened and that the USN might well have lost the war in both the Pacific and Atlantic Oceans. He included Jasper Holmes in another list.

9.18   Emergency Re-location

The FECB had to be evacuated from Singapore within a few weeks of the Japanese landing in Malaya. It withdrew initially to Ceylon and then to Kilindini in East Africa (Kenya). Its role[34] during 1942 was much reduced. However in the critical first 4 months it did some useful work on JN-25B in Ceylon. In 1943, under the name of HMS *Anderson* it was re-established in Ceylon as a serious participant in naval codebreaking.

Station Cast, which had been moved from the mainland to a purpose-built tunnel on the heavily fortified island of Corregidor in Manila Bay, had to be evacuated somewhat later.

The New Zealand[35] DNI, Lieutenant-Commander F. M. Beasley RN, had intended to visit Pearl Harbor in mid-December 1941 to consult with the British DNI Admiral Godfrey. The onset of war had resulted in the Clipper seaplane services being cancelled. So it was arranged that he travel in January 1942 instead. His mission was to try to improve the quality and quantity of American naval intelligence reaching New Zealand and also Australia. Beasley was interviewed by Admiral Nimitz and others on the defence capacity of Fiji and various other islands under the influence of New Zealand. He then spent an afternoon and the following morning in discussions with intelligence staff, including the senior cryptologist, Commander Rochefort, and the fleet intelligence officer Commander Layton. His report notes that he discussed TA, DF and Y intelligence. Beasley was taken on a brief inspection of sunken battleships and then embarked on a particularly harrowing return flight to Auckland. He wrote out reports on 27 January (the day he returned), not giving any details on TA, DF or Y intelligence and then rushed off to Melbourne to discuss these matters. So on 28 or 29 January Commander Long, the Director of

---

[34]The report by Fabian to Safford of June 1942 is found in NARA RG38, Crane NSG Library Box 19, and explains his difficulty in working out the intended functions of British naval signals intelligence units in Colombo and Kilindini. At that time Frumel was sending them progress work on JN-25 by means of a one-time pad supplied by Nave. Some of the unreadable OTP messages have survived in the Melbourne NAA.

[35]This paragraph is based on item N2 30/68/3 in the New Zealand Archives. This most valuable source was found by Alan Bath and is mentioned in his *Tracking the Axis Enemy: The Triumph of Anglo-American Naval Intelligence*, University of Kansas Press, 1998. However Bath does not place great importance upon its relationship with spreading the news about how JN-25B could be broken.

An unexpected security consequence of the Beasley visit is discussed in Sect. 19.3.

Naval Intelligence (DNI) for the RAN, his colleagues Newman and Nave and the Chief of the Naval Staff would have been informed that JN-25B8 was extremely vulnerable and the transfer of the Cast team to Melbourne was under consideration. The boost to the local Sigint operation that would stem from the latter would have ensured their full support for it.

The report written by Fabian to Safford in June 1942 (see Note 27) states:

> When things got from bad to worse in the Philippines area and when CinCAF went South John (Lietwiler) and I had a long conversation regarding plans for the future. It appeared to both of us that it would be a good idea not to lose the whole unit in one crack—and we consequently proposed to CinCAF that he consider getting a nucleus unit plus equipment aboard a submarine Java bound. He agreed and so directed and we left on the night of 5 February. Incidentally Cominch's directive to evacuate us was received just before we left and we were happy to see ourselves so strongly backed.

The message from Washington has survived elsewhere[36]:

> Since the withdrawal of Singapore CI unit to Colombo, communication intelligence organization under your command is of such importance to successful prosecution of war in Far East that special effort should be made to preserve its continuity.

Fabian was in the first of three evacuation groups. Amazingly all Op-20-G personnel were successfully evacuated from Corregidor. After a short stay in Java his group landed in Perth on 3 March 1942, and then went by train to Melbourne.

The negotiations between the USN and the RAN have survived. The Australian Legation in Washington cabled[37] to the Chief of the Naval Staff on 19 February 1942:

> USN had total of 8 officers and 68 men specially trained in W/T Intelligence work in the Philippines at outbreak of Japanese War. Included in this number are translators in Japanese, cryptographers, W/T Operators trained to read Japanese transmission and clerical staff trained in analysis of intercepted messages. It is understood that about 20 of them have already been evacuated to Java and steps are under way to evacuate the remainder. Admiral Glassford has suggested to the Navy Department that some and in certain circumstances all of this party should proceed to Australia and it is probable that the Navy Department will approve.

---

[36]The invaluable Aegean Park Press *Cryptographic Series* includes a version of SRH-180 edited by Sheila Carlisle. The original is available in NARA RG457. An internal Op-20-G memorandum of 23 January 1942 on this matter is mentioned on page 53 and the quoted text of the cable message is on page 54.

The second evacuation group from Cast was ordered on 12 March 1942 to report to the senior naval officer 'for further transfer to Melbourne, New South Wales, Australia'. Likewise Fabian's group initially had been to Exmouth Gulf hoping to catch a train to Melbourne. They all got there in the end.

Those interested in such things may find some amusement in Melbourne NAA file B5555 14, which is a 'collection of comic relief' assembled by Commander Newman. Some more were accumulated in the data base used for this book.

[37]The message of 19 February 1942 may be found in the Research Centre of the Australian War Memorial, AWM124 4/132. The response of 22 February 1942 is in the NAA Canberra file A816 43/302/18.

The response was:

> US personnel referred to would be a valuable addition to W/T and Special Intelligence
> organization which now exists in Australia and we would be glad of services of any who
> can be made available.

Joseph Wenger of Op-20-G wrote the essential point in a report[38] of 1945:

> In Melbourne, a junction was made with a joint British-Australian naval, military and
> civilian unit which was found to be in operation there. The resultant combined unit evolved
> into what was essentially a US-Australian naval unit [Frumel] under US control and
> leadership.

However Wenger appears to have been unaware that Layton, Rochefort, Beasley and
Long would have considered this relocation back in January 1942.

9.19   The Decryption and Decoding of JN-25B8

By March 1942 JN-25B was starting to become readable. One early piece of
information from it was preserved[39] by Commander Newman and covers the first
Japanese landings in West New Guinea in late March 1942. Perhaps General Akin
is correct in saying that credit for the turning point here should be given to the
American soldiers who sacrificed so much in a delaying action in Bataan province,
north of Manila. All told it was a major factor in the coming Battle for Australia.

Although JN-25B had not been read usefully before the Pearl Harbor raid, in
March, April and May 1942 the hard work was rewarded. Another quote from the
SRH series[40] is appropriate:

---

[38]The report is dated 8 October 1945 and is now SRH-197 in NARA RG457. It is reprinted by
Aegean Press in booklet 65 of its *Cryptographic Series*; see page 63.

[39]The Newman material is in the Melbourne NAA as B5555 1. Earlier material is to be found in
NARA.

It is of interest to compare a report sent by Captain Shaw of the FECB to GCCS in mid-
December 1941. The source is TNA HW 8/102. The text is 'New naval recyphering tables
introduced 4th December, books and system unchanged. Progress expected to be similar to last
period of 4 months, viz tables broken into after a month and all the 1,000 indicator subtractors and
about 10,000 recyphering groups solved after 3 months.' In fact there was massive disruption at
FECB and Cast but reinforcements were provided at the Pearl Harbor Combat Intelligence Unit,
later called *Hypo*. The process of handling JN-25 was well understood by December 1941.

The NAA Canberra file A6923 SI/2 includes on digital page 249 a report of 5 April 1942 for
the Deputy Chief of the General Staff containing the cryptic sentence 'It cannot be sufficiently
stressed that Japanese Naval traffic is totally different in character from the Army'. As both the
Japanese Army and the IJN used additive cipher systems, this would appear to be asserting that
it had become possible to handle JN-25B (by exploiting its flaws) but high-level Army codes
remained unreadable.

[40]SRH-020, *Narrative Combat Intelligence Center JICPOA* may be consulted in NARA RG457
but is also available in the book *Listening to the Enemy: Key Documents on the Role of*

Just before the Battle of Midway there was a major Japanese code change. The period of darkness, when the major Japanese code was not being read, extended well past the initial stages of the Guadalcanal campaign. After being accustomed to reading the enemy's mail, the denial of this information was acutely felt. . . . After the Battle of the Coral Sea most of his major moves were disclosed through this source alone.

Throughout April and May it was becoming clear that the continued use of JN-25B was most unlikely to last indefinitely. So the Allied leadership must have wanted to exploit it while this was still possible.[41] Indeed, a message from General MacArthur to General Marshall of April 1942 makes this point. And here lies part (but only part) of the rationale behind the USN being prepared to put so much at stake in the Battle of the Coral Sea and the Battle of Midway. (Another element of the rationale was the foreknowledge that the IJN was not sending all of its fleet to Midway.)

9.20   JN-25C and JN-25D

The new JN-25C was introduced on 27 May 1942. The GCCS file (HW 8/102) contains a message to Op-20-G of 3 June 1942 asking whether the change to JN-25 made on 27 May was just of the additive or involved both additive and code book. The response on 4 June 1942 stated that:

Change of 5 numeral system involves new additive table and new code book. Indicator system is similar and new code has DIVISIBILITY by 3.

JN-25C had a short career from 28 May 1942 to 14 August 1942. Because there was only one table of additives, the system is also called JN-25C9. It has been suggested that the early switch to JN-25D was caused by fears of compromise following the Battle of Midway. However it is much more likely that JN-25C was dumped once it was realised that the US Marine Corps had probably captured much cryptographical material on Guadalcanal soon after the American landing there on 7 August 1942. In fact, this had happened[42] and the material included a JN-25C

Communications Intelligence in the War with Japan by Ronald Spector. This source notes that call sign ciphers became more difficult from the introduction of JN-25C.

[41]Despite all the Sigint, the decision to risk so much at Midway must have been taken with some trepidation. The appropriate quotation from Shakespeare is to be found in *Julius Caesar* (iv 3 218):

There is a tide in the affairs of men
Which, taken at the flood, leads on to fortune,
Omitted, all the voyage of their life
Is bound in shallows and in miseries,
On such a full sea we are now afloat.
And we must take the current when it serves
Or lose the ventures before us.

[42]The reference is NARA RG38, CNSG Library, Box 115, item 5750/119. In fact the JN-25C code book was recovered from a fire. See also RG38, Box 42 RIP 83 for the JN-25C code book.

code book. By that time some progress had been made on it but the cipher system was replaced a week later.

Samples of the original JN-25C code book survive in NARA RG38. No doubt it, coupled with intercepted messages covering eleven weeks, gave very useful intelligence on the IJN.

The good news about JN-25D was that it continued the use of multiples of three!

Around October 1942 something referred to as JN-25XX was starting to gain prominence. On 4 December 1942 Op-20-G reported[43] that JN-25 was now known to be divided into at least four and possibly five systems. Of these, JN-25D10 carried 65 % of the traffic. By then the USN had awarded cryptanalysis of JN-25 the extreme priority it clearly warranted and so an adequately staffed special unit (Op-20-GYP1) was in place in Washington.

9.21   The GYP-1 Bible of July 1943

Mention has already been made of *Cryptanalysis of JN-25*, or the *GYP-1 Bible*. This was written in mid-1943 and is a very detailed account of the entire operation that turned intercepted messages into book group form. The size of the task was immense: by July 1943, 3,500 messages per day had to be handled. Substantial duplication would have reduced this to around 1,500 different intercepts. The documentation explains how additives were recovered and how priorities were set. So on page 67 one may read:

> For example convoy control reports, giving estimated noon positions for several days in advance of receipt, are almost always placed on additive recovery priority by GZ.

The 'estimated noon positions for several days in advance' were exactly what submarines needed for the remarkable successes discussed in Chap. 16. Here 'GZ' was the jargon for Op-20-GZ, the translation unit of Op-20-G. The difficulties of translation are stated on page 240:

> The actual text of the clear messages in JN-25 is terse and entirely innocent of the elaborate circumlocutions of polite Japanese. Full inflexion of verbs and adjectives is practically never observed. Honorific expressions are absent. It frequently happens that even after the full text of a message has been recovered only an expert in naval text can render it . . . into plain English text.

Cryptanalysis of JN-25 is a remarkable account of the work of a team of hundreds of people in Op-20-GY-1, later Op-20-GYP-1. It sets out detailed procedures for working out literally hundreds of thousands of additives. The basic procedures (described in its chapters VIII and IX) systematically exploited the practice of using

---

[43]TNA HW8/102, page 24.

The different systems were generally called 'channels'. They had the same code book but different additive tables. The less commonly used channels tended to be secure enough.

only multiples of three as book groups. Intercepts with garbled indicators were not thrown into a too-hard basket but instead received special processing that enabled about 10 % recovery. It is clear that substantial thought had gone into finding the optimal organisation for a very complicated exercise. The decision to centralise all additive recovery in Washington is very understandable.

9.22 Tailing and Indicator Recovery

According to the *GYP-1 Bible* (page 191) it was common for JN-25 messages from the same location to *tail*, that is the additives used for each message would follow immediately after those used for the previous message. As noted in Sect. 8.14, it was the practice with JN-25 to use the first additive of the next page immediately after the last additive on a page. Furthermore, with JN-25D and some later systems, it was the practice to send with each message encrypted indicators for the positions of both the first and last additives used. This practice is dangerously close to being redundant encryption, discussed in Chap. 14.

The *GYP-1 Bible* states that:

> Although tailing is responsible for most of our (indicator) key recoveries, the Japanese apparently believe that tailing contributes to security by insuring that the same text additives are not used repeatedly by one sender. As a matter of fact, copies of Japanese code books in our possession reveal that Japanese senders are instructed to tail.

(JN-25C9 system material was captured on Guadalcanal in August 1942.)

This tailing[44] practice, when taken with having encrypted indicators for the first and last additive used, produced pairs of encrypted indicators, being that for the last additive of the previous intercepted message and the first additive of the current message. These were adjacent in the table of additives. If the encryption system was anything like that used for the indicators of system 2468, the result would be potentially very insecure. The last quote from the *GYP-1 Bible* reveals that it was in practice totally insecure. Thus it is scarcely surprising that tailing was so useful to Op-20-GY. The *GYP-1 Bible* names a naval unit based in Rabaul that ran a single solid tailing series and so contributed handsomely to the reading of JN-25 traffic.

The *GYP-1 Bible* refers to a common practice of *trailing*, defined to mean 'approximate tailing'. It would have been convenient to copy the additive onto the message forms (see *CBTR* paragraph B) in advance. In practice this was not pernicious.

---

[44]Michael Smith's *The Emperor's Codes* mentions on page 159 how Tiltman in March 1942 managed to break into the Japanese Military Attaché Code—an additive system—by putting a complete run of messages from an active tailer into depth by simply counting the groups used! The point is he would have guessed that the number of groups in the additive table was 5,000, 10,000 or 20,000 and so obtaining depth from a long run is easy.

For example, the indicator system for JN-25B7 is taken from RIP (Radio Intelligence Publication) 74-E of 1 December 1942, which survives in NARA RG38, Box 33. Its historical interest lies in the fact that it was only minimally changed when the switch to JN-25B8 was made, and so the decryption priority for Op-20-G was obvious. The sample message would look like:

```
75800   75811   67316   21672   G01   G02   G03   G04   G05   G06   G07   G08
G09     G10     G11     G12     G13   G14   G15   G16   G17   G18   G19   27150
```

Here G01 is the first message GAT, G02 is the second, and so on. The 27150 at the end gives the time of transmission: the 27th day of the month at time 1,500 (to the nearest 10 min). All the groups have five digits. The first three digits of the first group are repeated as the first three digits of the second group and are called the *SMS*. The SMS may be any 3-digit group from 001 to 999. The fourth and fifth digits of the first group were used to designate the part when a long message has to be broken up. Very wisely these were not encrypted. The fourth and fifth digits of the second group are the same odd digit whose purpose may well have been just padding. The third and fourth groups are two encrypted forms of the indicator group.

The pages of the additive table were numbered from 001 to 500 with 100 additive groups on each page. These were given row and column co-ordinates in the straightforward way and so were numbered from 00 to 99. The *clear* indicator group was obtained by (true not false) doubling the page number and (true, not false) subtracting 1 or 2 and thus obtaining a 3-digit group between 000 and 999 to which the row and column co-ordinates were adjoined on the right. For example, starting with page 346, row 7, column 3, one doubles to obtain 692, subtracts to get 691 or 690 and so obtains clear group 69173 or 69073.

The operator was given an *SMS table* with 999 pairs of randomly chosen indicator additive groups. Thus if line 758 of the SMS were:

```
758         08243         62509
```

the two encrypted versions of the plain indicator 69173 using this SMS line are 67316 and 21672. This double encryption is not of itself the cause of the insecurity. The point here is that given a few known reliable tailers and preferably also a known operator who has a strong preference for a few pages of the additive table, reconstruction of the SMS table is quite easy.

It is natural to speculate that Commander Malcolm Burnett was responsible for breaking into this indicator system. The details of how to handle it appear to have been given by FECB to Cast. The quotation from Wesley ('Ham') Wright (Sect. 5.5) praising Burnett's contribution is most likely to be referring to breaking the indicators of JN-25A and JN-25B. In fact Tiltman broke the simpler system used for the earlier JN-25A systems and this would have made tailing obvious. Burnett would have worked out how to exploit tailing in breaking second generation indicator systems.

Professor Room wrote the following comment (already quoted) about IJA Mainline traffic on page 6 of Part F of the *CBTR*.

> They did not regard code books as having any great protective value in themselves. They must have known that once depth can be recognised, messages with any sort of disguise can be stripped and read and the code book broken. Hence they staked their security on the concealment of depth. As explained elsewhere, in doing so they set our cryptographers some major problems and narrowly missed making their traffic completely unreadable.

For JN-25 in 1939–1943, the 'concealment of depth', that is the indicator system, was ineffective. As shown in this chapter, the process was rendered much easier by exploitation of the use of multiples of three as book groups. All that was left to protect the messages were a sequence of long and complicated code books. By 1942 Op-20-G had the resources, including cribs and TA, to do quite a lot of reconstruction of these.

9.23   Later Versions of JN-25

Later in the JN-25 saga, the indicator systems became much more difficult. This was not an intractable problem in 1939–1943. However the trend is clear enough from the following extract from RIP 7F:

> This cipher system consists of 25 indicator additive tables and 500 text additive tables. The indicator additive tables are made up of 999 SMS numbers with two 5-digit groups of indicator additives for each SMS number.

9.24   JN-11

The additive system JN-11 was introduced early in 1943 for merchant shipping other than that controlled by the Japanese Army. The reason was that the previous code was not trusted and that the use of diplomatic codes was throwing an excessive burden on that network. It ran through a series of mutations which are listed in the Jamieson Report,[45] item B5554 in the National Archives of Australia. A digital version is available on its Recordsearch system. The following account disregards various complications introduced for JN-11 late in 1944.

Although it was not of the same importance as JN-25, it—from the viewpoint of the Allied cryptanalysts—nicely complemented the Water Transport Code 2468. Both were broken, though for quite different reasons. Both were extremely useful

---

[45]NAA Melbourne item B5554, known as the Jamieson report, gives a general account of minor IJN ciphers. The section on JN-11 includes a description of the Ransuuban, a substitute for an additive table. Another reference is page 152 of Norman Scott's paper *Japanese Naval Ciphers 1943–1945*, published in *Cryptologia* 21(2) (1997).

in getting submarines in the right place at the right time. As already noted, the common Japanese practice of sending anticipated noon positions of convoys over the air waves was extremely helpful to the other side. Broken intercepts in JN-25 also helped at times in tracking convoys. See Chap. 16.

The remarkable discovery made early in analysing JN-11A, the first version of JN-11, was that it was a 4-digit additive cipher in which each book group was one more than a multiple of three. Thus there were only 3333 book groups available, these being:

$$0001\ 0004\ 0007\ 0010\ 0013\ 0016\ \dots\ 9991\ 9994\ 9997$$

Thus once a method had been developed for stripping enciphered intercepts back to the book groups, the task of reconstructing the code book was much simpler than the corresponding task for JN-25. The latter had around eight times as many code words.

Yet a method had been developed already. The method for stripping JN-25 intercepts applied with minimal adjustment to those in JN-11A.

9.25   The Real Secret

There is some evidence that from 1939 onwards the insecurity of using only multiples of three as book groups was considered to be a very high-level secret. For example, the 1943 report on *Collaboration of British and U.S. Radio Intelligence* made by Malcolm Burnett contained an appendix on methods[46] used in Washington

---

[46]The following report by Hugh Foss was found by Ralph Erskine in record group HW 37 in the British Archives. It shows how sophisticated JN-25 decrypting became:

'The text cards and additive masters are then put through an NC-4 machine which has been specially devised for Op-20-G and may be new to Freeborn.

It has two hoppers for master (read) and detail (punch) cards respectively. In this case the additives are masters and text cards details. It compares five columns of the master card with five columns of the detail card (cell number and last two digits of page in this case) and if they are different it takes the detail card and passes it through unpunched. It would do the same for the master, if it were lower, but in this case the detail cards are from a complete series, only the masters can be missing. Incidentally the machine can compare anything up to ten columns. When the master and detail cards are the same it takes both cards, punches the five columns of the additive group onto the detail card (cols 76–80), adds the additive group to the text group and thus gets the code group which it also punches on the detail card (cols 31–35). Then it works out whether the code group is divisible by 3 and if so, column 25 of the detail card is punched with the 11 punch. The detail hopper can be used without the master hopper so that two five-column fields of one card can be added and punched on it. Though this machine can only deal with 5 columns at a time it is used in preference to a multiplier, except when there are a large number of columns to be added. The reason is that this machine works at the rate of 100 cards per minute providing all detail and masters match, or only one feed is used, and the multiplier (type 601) can only handle 25 per minute.

to decrypt JN-25: this is missing in the copy in the British National Archives (ADM 223/496). See Note 1 of Chap. 5. As the principal methods used a few months later are described in the *GYP-1 Bible*, the Burnett document is now of limited importance. There are a few other references to the importance of the use of multiples of three. But some of these, such as the Lietwiler letter quoted in Sect. 10.1, mention it implicitly rather than explicitly.

The text cards are next sorted on column 25 into two decks accordingly to whether or not they have the code groups divisible by three. The deck with scanning groups is then sorted into numerical order (cols 35–31) and collated to verify the sorting.

All three decks:

1. headings
2. with scanning code groups
3. with code groups that don't scan or are absent for lack of additives

are then taken over to GS 1-A where a special unit is waiting to receive them.'

Chapter 10
Using Common Book Groups

By late 1941, sufficiently many book groups in the JN-25B code book had been found to enable the compilation of an accurate list of those most commonly found in decrypted messages. The list would be regularly updated as more GATs were decrypted. A new method of attacking the problem of finding additive table entries was now based on the information in this list. This chapter describes the successive refinements of this method that included the application of Bayesian ideas in devising scoring systems for choosing the correct decryption from a small number of possibilities.

10.1 The Lietwiler Letter

John Lietwiler, head of the Cast Sigint unit at Corregidor, wrote a letter[1] on 16 November 1941 to Leigh Peake of Op-20-GY Washington. It throws considerable light on the state of play with JN-25 three weeks before the Pearl Harbor raid and the direction of attack in the lead-up to the Battle of Midway.

---

[1] The Lietwiler letter is to be found in NARA College Park in RG38, Crane Inactive Stations, 3200/11. It has been made widely accessible by being incorporated in the appendix to the 2001 Ottawa MA thesis of Timothy Wilford, *Pearl Harbor Redefined: USN Radio Intelligence in 1941*, which is included in the ProQuest electronic data base. Wilford later furthered his studies and obtained a PhD degree from that university. That thesis is also on ProQuest and is of interest. Some of the writings of Wilford's thesis adviser, Brian Villa (see below), are also quite relevant. Duane Whitlock in an oral history interview with the National Cryptological Museum explains that Lietwiler was intended to be Fabian's relief but Fabian refused to be relieved. So Fabian, no longer commandant, was evacuated from Manila ahead of Lietwiler and, reaching Melbourne first, became commandant of the relocated Sigint team.

This letter is mentioned by Wilford in *Intelligence and National Security*, December 2002 in Note 20 of his paper and by Phillip Jacobsen in *Cryptologia*, 27(3), July 2003, 193–205.

© Springer International Publishing Switzerland 2014
P. Donovan, J. Mack, *Code Breaking in the Pacific*,
DOI 10.1007/978-3-319-08278-3_10

The novelty has worn off the JEEP IV[2] now and it is seldom in use. I believe the reason is mainly the time required to make a set up. We have pretty well proved that it takes longer to set up the ten groups and run the check by machine than to do it by hand. Also there are about 10 people working at one time and only one machine. A further reason is a new system of attack we have put in use. Using the 400 high frequency groups we have compiled a table of 24,000 differences. When we are stuck on a column now we take any likely looking group and subtract it from every other group in the column. The reciprocals of these differences are also written down which gives the differences of every group in the column from the master group. By reference to the table, the groups which produce these differences are found and tried in the proper spots, i.e., on the master group in the case of the original column and on the columnar group in the case of the reciprocals. Two days ago I saw MYERS (A. E. Myers, Jr.) walk right across the first 20 columns of a sheet using this method almost exclusively. In view of this I do not believe we want a new JEEP IV.

I do not understand your paragraph about 'whether we found the true key or simply applied the key subtractor'.

We have stopped work on the period 1 February to 31 July (JN-25B6) as it is all we can do to keep up with the current period. We are reading enough current traffic to keep two translators very busy, i.e., with their code recovery efforts, etc., included. In this connection, I certainly wish you could see your way clear to drop the ancient history side of this cipher and work with us on each current system as it comes up. With Singapore, we have adopted a system of exchanging block numbers to prevent duplication. We have more or less given them a free hand in selecting the cipher blocks they tackle on account of their more limited traffic.

I am going to be very frank about the code recoveries you have sent us so far. The translators say they are of no help, as we already have the confirmed ones and those which we do not have are usually proven wrong. They note that in one case one group we misspelled when we sent it in was copied exactly and sent back, and several values on which we were mistaken and have since corrected were sent back to us as confirmed (i.e., the original wrong value). Let me urge you once more to have someone check on this work.

Please do not adopt the microfilm system for traffic. It would be practically impossible for us to handle our volume in this way. The air freight system I think is an excellent idea. I regret the delay as much as you do, but the Issuing Officer here assures me he is doing the best he can. This also applies to the delay in our station reports. I will be glad to send them by the air freight also if you authorise this plan. I suppose some allotment will be set up to cover the cost.

Sadly Peter Donovan in his first paper on JN-25 (in *Cryptologia* 28(4), October 2004, 325–340) made no reference to it. The most recent paper about the revisionist line appears to be *Signals Intelligence and Pearl Harbor: The State of the Question*, by Brian Villa and Timothy Wilford. It appeared in the journal *Intelligence and National Security*, volume 21, issue 4, August 2006. John Zimmerman's paper *Pearl Harbor Revisionism: Robert Stinnett's 'Day of Deceit'* in *Intelligence and National Security* vol 17, June 2002, 127–146 should be studied very carefully before any conclusion is made about the letter. Chapters 9 and 10 of this book appear to give the first account of the technical matters being raised by Lietwiler.

[2] Other correspondence from Lietwiler and copied by Wilford (Note 1) give some hints about the JEEP IV machine. SRH-355 *Naval Security Group History to World War II* (NARA RG457) states on page 433: 'Also reported was the arrival of LT Hess, a USNR officer, from the Navy Department, bringing a device known as JEEP IV. This evidently was some sort of mechanical device used for recovering additives (or subtractors) which formed the cipher key for encrypting messages in the Operations code (later JN-25).'

Section 9.2 gives the background information on the changes in JN-25 code books and additive tables in 1941.

The jargon used here needs some explanation. The word 'block' must mean a page of the table of additives. Rudolph Fabian mentioned in an oral history interview with the NSA how for each block a large sheet of paper was used to record all the intercepts encrypted using that block. A 'code recovery' means working out the word or phrase corresponding to a particular book group. 'Singapore' refers to the FECB. The 'reciprocal' of a group **a** has to be the false difference $00000 - \mathbf{a}$. For example, the reciprocal of 12345 is 98765. Note that if (say) 12345 occurs as the difference $\mathbf{x} - \mathbf{y}$ of two GATs $\mathbf{x}$ and $\mathbf{y}$, then its reciprocal 98765 is the difference $\mathbf{y} - \mathbf{x}$.

Handwritten annotations on the original letter confirm that Lietwiler was under a misapprehension in believing that there was work on JN-25B going on in Washington at the time.

In his letter, Lietwiler described a method of stripping messages encrypted by 'unrecovered blocks' (pages of additive not yet fully worked out) and placed in depth of ten. Someone in Washington has annotated the letter with the words 'same as our new technique'. As ideas would have been shared with the FECB team, it may also have contributed.[3] The method used '400 high frequency (book) groups', which will produce a total of $400 \times 399 = 159{,}600$ differences, about half of which would be minor differences in the sense of Appendix 1. It is not clear[4] how Lietwiler's 24,000 were chosen for his table. One way would be to just take the 156 most common groups to obtain 24,180 differences. Another way would be to take the differences between the 400 most common groups in which at least one of the 32 most common groups occurs.

Whatever actually happened in Cast in 1941, it would have been obvious that almost halving the number of entries in the table of differences was a sensible, if not essential, step. Appendix 1 explains how this was done.

10.2   The Phase 3 JN-25 Stripping Method

The selected common differences would be punched on IBM cards as in the table below. Here 52728, 31803 and 08712 are taken to be frequently occurring groups and used to produce the six cards punched below. On each card the third group is the false difference of the first and the second.

---

[3] Section 5.5 explains the extent to which the USN used ideas from the FECB.

[4] In an oral history interview with the National Cryptologic Museum, 'Ham' Wright says that in early December 1941 Rochefort's Combat Intelligence Unit at Pearl Harbor had only the 100 most frequently used groups in order of frequency. Evidently the CIU had to use these to prepare its own table of differences.

52728 31803 21925
52728 08712 54016
31803 08712 33191
31803 52728 89185
08712 52728 56094
08712 31803 77919

The 24,000 cards[5] punched thus would then be sorted mechanically and printed out in numerical order of the third groups. This would produce the required table. In fact the office (in a tunnel on Corregidor!) described by Lietwiler would have needed ten copies of the table, which could have been produced by running the tabulator, set up to serve as a printer, ten times. It would not matter if occasionally the same difference arises as the difference of two or more pairs of frequently occurring groups. However two conditions need to be satisfied, with the second being far more important than the first:

- there are 25,000–40,000 differences in the table, and
- usually when eight or nine messages, independently encrypted using the same portion of the additive table, are placed in depth, each column contains at least two GATs encrypted from book groups in the list of 150 (or perhaps 100) most commonly observed book groups.

The first condition helps speed the process along: when calculating differences one does not want too many to be common differences by chance. The second amounts to the requirement that the frequently occurring book groups occur sufficiently frequently. Lietwiler in effect asserted that this was the case for the usage of JN-25B at that time.

We now consider an example of stripping nine[6] GATs assumed to be in depth. The GATs are 32815, 30763, 35252, 43535, 27363, 64350, 59362, 57761 and 21289. Here there are $72 = 9 \times 8$ differences to calculate and then check whether they are in the table. Lietwiler refers to working out these differences one by one until the desired one turns up. Without any special knowledge we list the differences systematically as set out below:

---

[5]One could speculate on the extent to which the construction of the 24,000 cards was automated. Presumably Lietwiler considered (almost) halving the length of the table by using minor differences only to be a side issue not worth mentioning.

By July 1943 the preparation of tables of differences had been automated to a considerable extent. The *GYP-1 Bible* mentions on page 579 that this could be done using an NC3 and an NC4. These were specially adapted IBM equipment manufactured under secret contracts between IBM and the Navy Department. A description of their functions would be out of the scope of this book. However Canberra NAA file A425 C1947/514 1947–1949 entitled *Hollerith tabulating equipment taken over by RAN from USN* records that Frumel had its NC4.

Edward Simpson's *Bayes at Bletchley Park* in *Significance* 7(2) 76–80 (May 2010) throws some extra light on Notes 4 and 5.

[6]Section 10.9 below explains how reliably found correct decryptions of greater depths (15 or more) may be used to test for correctness a potential decryption method for small (9 or less) depths.

$$32815 - 30763 = 02152 \qquad 30763 - 32815 = 08958$$
$$32815 - 35252 = 07663 \qquad 35252 - 32815 = 03447$$
$$32815 - 43535 = 99380 \qquad 43535 - 32815 = 11720$$

$$\cdots \qquad\qquad\qquad \cdots$$

As 24 % of the 100,000 groups arise as common differences, we would expect about 24 % of these 72 to have differences given in the table.

Let us suppose that the observed difference $07663 = 32815 - 35252$ occurs in the table of differences. (So its reciprocal 03447 must also occur there.) The table would give an expression for 07663 as the difference of two frequently occurring scanning groups, say $07663 = 11685 - 14022$. We now set $\mathbf{a} = 32815$, $\mathbf{b} = 35252$, $\mathbf{c} = 11685$ and $\mathbf{d} = 14022$, so that

$$\mathbf{a} - \mathbf{b} = 32815 - 35252 = 07663 = 11685 - 14022 = \mathbf{c} - \mathbf{d}.$$

We now copy Lietwiler's words 'By reference to the table, the groups which produce these differences are found and tried in their proper spots'. Thus to 'try 11685 in its proper spot' it is necessary to write the original nine groups in the column given on the left below and then 11685 to the right of the 32815. Likewise 14022 is written to the right of the 35815. This yields the display:

32815 A		*11685* C
30763		
35252 B		*14022* D
43535		
27363	$\rightarrow$	
64350		
59362		
57761		
21289		

It may then be calculated that: $21230 = 32815 - 11685 = 35252 - 14022$, that is, $21230 = \mathbf{a} - \mathbf{c} = \mathbf{b} - \mathbf{d}$ and so it is worth while trying subtracting 21230 (equivalently, adding 89870) to each of the groups in the left hand column. This yields a candidate for the correct stripping of the nine groups in depth.

32815 A		*11685* C
30763		19533 $\checkmark$
35252 B		*14022* D
43535		22305 $\checkmark$
27363	$\rightarrow$	06133 $\times$
64350		43120 $\times$
59362		38132 $\times$
57761		36531 $\checkmark$
21289		00089 $\times$

This is where the JN-25B book groups being multiples of three comes in. As not all of the groups obtained for the right hand column are multiples of 3, the difference 07663 is indeed a common difference but this is purely a matter of chance unrelated to the stripping of the nine groups in depth.

The next common difference noted might be $25956 = 21195 - 06249$, observed to arise as the difference $57761 - 32815$ of two of the GATs in depth. Another common difference might be $19584 = 66126 - 57642$, which arises as the difference $30763 - 21289$ of two GATs in depth. The calculations are:

32815 B		*06249* D			32815		68278 ×
30763		04197 √			*30763* A		*66126* C
35252		09686 ×			35252		61615 ×
43535		17969 ×			43535		79998 √
27363	→	91797 √	and		27363	→	53726 ×
64350		38784 √			64350		90713 ×
59362		23796 √			59362		85725 √
57761 A		*21195* C			57761		83124 √
21289		95613 √			*21289* B		*57642* D

Such calculations would be carried out with each common difference arising until one obtains a right hand column in which all groups are multiples of three. For example $88591 = 95898 - 17307$ may be the difference of two common groups. It is the difference $35252 - 57761$ of two of the nine groups. The subtractor here is $35252 - 95898 = 40464$.

32815		92451 √
30763		90309 √
35252 A		*95898* C
43535		03171 √
27363	→	87909 √
64350		24996 √
59362		19908 √
57761 B		*17307* D
21289		81825 √

The seven √ signs show that the nine groups are all multiples of three. So this is probably (but see Appendix 2) the correct stripping of the original nine given groups in depth. The required additive is $92451 - 32815 = 40464$.

10.3   Differences on a Subtractor Machine

The example of Sect. 9.10 is repeated. Here there is a systematic search for groups **a** and **b** in the given depth or a displacement of it such that the difference **a** − **b** is in the list. Then the two common groups in the difference are tried 'in their proper spots'. There is again no need to return to the original nine groups (60621, etc)—the differences are the same!

60621 ✓	*60621* A	*73479* C	*73479* A	*51459* C
31945 ×	31945	44793 ✓	44793	22773 ✓
16464 ✓	16464	29212 ×	29212	07292 ×
54959 ×	*54959* B	*67707* D	67707	45787 ×
98825 × →	98825 →	01673 × →	01673 →	89653 ×
63959 ×	63959	76707 ✓	76707	54787 ×
53626 ×	53626	66474 ✓	*66474* B	*44454* D
43861 ×	43861	56619 ✓	56619	34699 ×
03930 ✓	03930	16788 ✓	16788	94768 ×
	51459	29999 ×	29999	14605 ×
	22773 A	*90213* C	*90213* A	*85929* C
	07292 B	*75732* D	75732	60448 ×
	45787	13227 ✓	13227	08933 ×
→	89653 →	57193 × →	57193 →	42809 ×
	54787	22227 ✓	22227	17933 ×
	44454	12994 ×	12994	07600 ×
	34699	02139 ✓	*02139* B	*97845* D
	94768	62208 ✓	62208	57914 ×
	14605	00807 ✓	00807	38674 ×
	85929	71121 ✓	71121	09998 ×
	60448 A	*56640* C	*56640* A	*84417* C
	08933	94135 ×	94135	22902 ✓
→	42809 →	38001 ✓ →	38001 →	66878 ×
	17933 B	*03135* D	03135	31902 ✓
	07600	93802 ×	93802	21679 ×
	97845	83047 ×	*83047* B	*11814* D
	57914	43116 ✓	43116	71983 ×

38674		07965 ✓		07965		20859 ✓
09998		78289 ×		78289		91173 ✓
84417		53708 ×		53708		76692 ✓
22902 A		91293 C		91293 A		14187 C
→ 66878 B	→	35169 D	→	35169	→	58053 ✓
31902		00293 ×		00293		23187 ✓
21679		90960 ✓		90960 B		13854 D
11814		80105 ×		80105		03099 ✓
71983		40274 ×		40274		63168 ✓

10.4   The Method Used by Yeoman Myers

Suppose in the previous example Yeoman Myers had worked out from the strippings of adjacent columns that the seventh GAT 53626 was likely to be a relatively common group. He would then call it the 'master' GAT and proceed as follows:

60621 ✓		60621 B		51459 D		51459	20859 ✓
31945 ×		31945		22773 ✓		22773	91173 ✓
16464 ✓		16464		07292 ×		07292	76692 ✓
54959 ×		54959		45787 ×		45787 B	14187 D
98825 ×	→	98825	→	89653 ×	→	89653	→	58053 ✓
63959 ×		63959		54787 ×		54787	23187 ✓
53626 ×		53626 A		44454 C		44454 A	13854 C
43861 ×		43861		34699 ×		34699 ×	03099 ✓
03930 ✓		03930		94768 ×		94768 ×	63168 ✓

Here the master GAT is marked **A** and the 'columnar' GAT marked **B**. With the aid of a bit of good fortune, this shortens the calculations.[7]

Phase 2 and Phase 3 methods depend upon the 100–150 most common book groups occurring sufficiently frequently.

---

[7]Edward Simpson confirms Lietwiler's comment on pages 134–135 of his account in *The Bletchley Park Codebreakers*, the augmented second edition (2011) of *Action This Day*, edited by Ralph Erskine and Michael Smith.

'Horizontal decryption' or 'horizontal stripping' would also be useful when it was known that a particular message was likely to contain GATs which were encrypted from a short list of book groups.

10.5 The GYP-1 Bible on Additive Recovery

This remarkable report, some 600 pages in length, describes in detail the procedures that Op-20-GY had developed and were applied to JN-25 systems by mid-1943. It begins with a careful account of IJN 'rules for use' for preparing a JN-25 message for transmission and then on how it is dealt with upon receipt. These were changed on certain occasions. Chapter VII, headed *Additive Recovery*, confirms that the Phase 2 and Phase 3 methods were useful. A lengthy chapter is devoted to the commonly occurring problem of detecting and (hopefully) correcting garbled indicator GATs in a message. Op-20-G wanted to decrypt as many JN-25 signals as possible.

A chapter on the problem of indicator (key) recovery contains (pages 191–192) the following important statement:

> Although tailing is responsible for most of our key recoveries, the Japanese apparently believe that tailing contributes to security by insuring that the same text additives are not used repeatedly by one sender. As a matter of fact, copies of Japanese code books in our possession [?JN-25C from Guadalcanal?] reveal that Japanese senders are *instructed to tail*, beginning their initial message at the S/P (starting point), which is the same as the CODE group identifying the sender.

At the very least tailing results in the encryption of a sequence of related indicators, leading to redundant encryption overall. This unsound practice is discussed at length in Appendix 2 of Chap. 14.

10.6 The Dayton Machine on Display in Pensacola

Pages 230–233 of the *GYP-1 Bible* are quoted extensively below. They deal with the use[8] of the Pensacola 'subtractor machine' seen by Turing.

> Differencing is usually effected by manual means, but can also be carried out by machine. From time to time, various types of machines, some power-operated and some hand operated, have been in use in the additive recovery rooms. It is undesirable to describe here in detail the mechanisms of all of these.
>
> The machines in use at present are electrically driven, manufactured by the National Cash Register Company, and embody virtually all of the essential principles of previous machines devised. Each digit of the five-digit group is registered separately by rotating a short section of a right circular cylinder on whose periferal surface the digits from 0 to 9 are marked. Groups of five of the 'wheels' can then be made to rotate together and, further, groups of these groups of 'wheels' can be caused to turn simultaneously. The 'face' of the machine thus exhibits sets of five-digit numbers which may be made to alter according to a

---

[8]See Erskine and Smith (cited in Note 6 above), Appendix VII for 'At one stage when we were struggling to keep up with the increasing quantity of incoming traffic Washington sent us a dozen or so calculating machines made by the National Cash Register Company'.

variety of synchronous transformation, governed by banks of keys similar to those of most computing machines.

a. **To clear the machine**: If any numbers appear on the face of the machine, look to see whether any of the '1, 2, 3, ... Keys' or 'A, B, D ... Keys's are depressed. If they are, strike the Minus-Bar before striking the Clear-Bar. If none are depressed, simply strike the Clear-Bar.

b. **To register numbers on the face of the machine**: The machine having been cleared, register the desired five-digit number by the *brown* numbers on the '1, 2, 3, ... Keys', leaving the keys untouched where a zero is desired. Thus 43713 would be registered by depressing the 4 key in the first column, the 3 key in the second column, depressing the 7 key in the third column, etc. After this depress the 'A, B, D ... Key' corresponding to the space in which it is desired to enter the number. Strike the Plus-Bar. This last operation automatically registers the group and restores all keys to their original position. (After five more such steps,) the completed column is then:

A	43713	O	00000
B	74751	P	00000
D	89854	R	00000
E	78330	S	00000
H	98366	T	00000
J	66055	V	00000
K	00000	W	00000
L	00000	X	00000
M	00000	Y	00000
N	00000	Z	00000

c. **To 'zeroize' any desired group**: Depress the 'A, B, D ... Key' corresponding to the number to be 'zeroized' and strike the Minus-Bar. This automatically alters all values on the face of the machine to a set of differences—i.e. the differences between the values first registered and the zeroized group. Zeroizing the first value in the column:

A	00000	O	67397
B	31048	P	67397
D	46141	R	67397
E	35627	S	67397
H	55653	T	67397
J	13442	V	67397
K	67397	W	67397
L	67397	X	67397
M	67397	Y	67397
N	67397	Z	67397

The values opposite B, D, E, H and J are the differences to be looked up in the table. When this has been completed and it is desired that the original groups be restored on the face, the Minus-Bar is again struck. The zeroization of the code opposite B is achieved by depressing the B key and the Minus-Bar. The column then reads:

A	79062	O	36359
B	00000	P	36359
D	15103	R	36359
E	04689	S	36359
H	24615	T	36359
J	82304	V	36359
K	36359	W	36359
L	36359	X	36359
M	36359	Y	36359
N	36359	Z	36359

The value opposite A may be ignored (as explained above) and when 15103 is looked up in the table, the code value, 89172, is found.

d. **To add any number to groups already registered**: Depress the '1, 2, 3, ... Keys' according to the number desired, but in this case using the *red* numbers on the keys. Strike the Minus-Bar. The, in order to 'try' the additive, 15421, first restore the original enciphered groups using the Minus-Bar. Then, using the red numbers on the '1, 2, 3, ... Keys', register 15421, and strike the Minus-Bar. Note that in this case no 'A, B, D ... Key' was struck. The face then reads:

A	58134	O	15421
B	89172	P	15421
D	94275	R	15421
E	83751	S	15421
H	03787	T	15421
J	61476	V	15421
K	15421	W	15421
L	15421	X	15421
M	15421	Y	15421
N	15421	Z	15421

This column is, of course, the desired one. The scanning values are at once recognised by the trained machine worker since the individual digits are coloured in this fashion:

The 'trained machine worker' should have confirmed divisibility by three of the groups in the A, B, D, E, H and J positions by checking the patterns of colours visible for each group. See Sect. 9.6. Sadly, the example given in the 1943 text is wrong: 03787 is not a multiple of three!

10.7 Flags on the Machine?

The section *Differences on a Subtractor Machine* shows incidentally that a capacity to mark where one is up to by 'A' and 'B' flags would be useful. And the Pensacola machine seems to have such a capacity.

10.8   The Real Mistake

A central error[9] made by the IJN in December 1941 was not replacing the JN-25B code book. Thus *the table of common differences constructed earlier by the team on Corregidor was still good*. Ensign Ralph Cook, a communications officer with experience with IBM, who was co-opted into the Cast team on 9 December 1941, recalled being shown immediately the method of attacking JN-25B8. Curiously, the Combat Intelligence Unit at Pearl Harbor appears not to have been sent a back-up copy of the table but it did receive a basically correct list of most of the 100–150 most common book groups in approximate order of frequency. This would be just as good. The greatly increased traffic and the extra staff, made available by telling the band of the *USS California* to learn new skills quickly, would have started Rochefort's team on the road to Midway. Lietwiler's letter makes it clear that between them Corregidor and FECB had 'recovered' (worked out meanings for) at least some common book groups and had likely meanings for some others.

The error made by the IJN may be expressed in an alternative form: *when enemy cryptologists attacking a cipher of JN-25 type have reached Phase 3, only a minimal enhancement of security is achieved by change of only the table of additives*.

Jasper Holmes,[10] an intelligence officer stationed in Hawaii throughout the Pacific War, commented on the practice of using only multiples of three as book groups in the JN-25 code books:

> This gave the Japanese cipher clerks a simple test for garbles and mistakes. It also gave the cryptologists a handy tool to work with, but that was unintentional.

Holmes would not have been able to see the use of the scanning distribution (Chap. 15) in 1942. If he had, he may have used a more emphatic phrase than 'handy tool'.

10.9   Finding the Correct Additive

The *GYP-1 Bible* asserts that:

> It is obvious, however, that (usually) several possible additives exist which will produce the mere property of 'scanning' in a large proportion of groups in a given column. In order to decide which of these additives is the correct one, additive recovery must depend on two other factors: (a) the proportion of well-recognised groups which a certain additive produces and (b) the context.
>
> Necessarily, the number of groups in a column (depth of the traffic) is a limiting factor in respect of (a). If the traffic is very shallow, the proportion of high-frequency groups may be high throughout without the additive being correct. Also, accidental garbles may exist in the

---

[9]A list of the principal errors made with JN-25A and JN-25B is given in Chap. 9. This error is the seventh in the list of seven given there.

[10]See page 56 of Holmes' *Double-Edged Secrets*, Naval Institute Press, 1978.

column produced by the application of the correct additive, whereas an incorrect additive may bring out a column entirely free of garbles. Generally speaking, a depth of about eight lines may be considered a reasonable minimum for judging correctness by internal characteristics alone though occasionally four very high frequencies in four lines of traffic will occur—in which case, the additive is quite probably correct.

The matter of context presents a more difficult problem. In the first place, the highly compressed, telegraphic style of the text renders the sequence of groups more difficult to predict than in conventional or literary Japanese. In the second case, the study of Japanese language is, for security reasons, discouraged. A translator is provided for each recovery section who assists in deciding whether the context is satisfactory in the case of questionable additives. Often, however, it is not possible to decide whether an additive is really the correct one. Only additives of which one can be reasonably certain are recorded in the 'Additive Book'.

One very powerful factor, in respect of the evaluation and recovery of additives, is the existence of certain stereotyped forms of expression. These standardized formulas are called 'patterns'. These patterns aid greatly in the evaluation of additives and in their meaning.

The following section details the main additive recovery methods. It stresses that:

All types of collateral information are useful—particularly those obtained from other messages, from other portions of the same message or even from newspapers.

Whatever the recovery method, the conclusion is that:

In general, the surest check on the correctness of an additive is context, but ... it is necessary in practice to judge an additive's correctness by both 'horizontal' and 'vertical' characteristics.

Phase 2 and Phase 3 methods yield possible decryptions with at least one 'high-frequency' book group. Indeed Phase 3 yields possible decryptions with at least two. But with reasonable depth most correct decryptions would have rather more. The occurrence of common book groups is, in the above sense, a 'vertical characteristic'. Rules of thumb (called *scoring systems*) were needed to enable semi-skilled decrypting clerks to decide whether to accept a potential decryption.

This design of such rules of thumb is discussed in Appendix 2. The general issue is of supreme importance in WW2 Allied cryptology, and not just that of the Pacific War.

In a much earlier chapter, the following pertinent comment appears:

One other minor function of the decryption room is the reporting to the additive recovery room of additives which are suggested by the text of a current message. These suggestions are sometimes of great value in additive recovery, since an additive which is obvious from the total context of a message may be much less obvious to a person who is working with a small portion of that message on a few additive recovery worksheets.

Op-20-GY extended the use of common book groups and their differences by compiling tables of these restricted to data obtained from intercepts of specific types of message, such as weather reports, geographical locations, number-rich reports, those containing Roma-ji, etc.

There was an evident method of testing decryption methods. A depth of (say) 15 signals could be found and the correct decryption determined without too much difficulty. The top (say) six could then be covered up and any proposed method for depths of (say) 9 tested on the remainder. The question of how to construct efficient scoring systems in the above sense is left to Appendix 2.

10.10   JN-25 at Bletchley Park 1943–1945

Edward Simpson[11] has recently given a detailed account *Bayes at Bletchley Park* of the work of the JN-25 team at Bletchley Park from October 1943 onwards. He includes comment on the extremely challenging alterations made by the IJN for several versions of JN-25 between December 1943 and August 1944 which sometimes completely frustrated both Washington and Bletchley Park. He confirms that successful additive recovery depended on a successful interplay between vertical and horizontal stripping. He wrote:

> The most important asset of a first-rate stripper is a 'faculty of association', a kind of photographic memory which not only stores each incident but equips it with 'hooks and eyes' by which the whole can be brought to the surface again, consciously or unconsciously, by the recurrence of some part of it. This faculty builds up for the stripper such a store of experience that eventually almost every group in the code book calls up some image of a phrase previously seen which included it.

The word 'stripper' was jargon for a cryptological clerk seeking the additives (or, equivalently, subtractors) for given depths of JN-25 intercepts.

The main thrust of *Bayes at Bletchley Park* is that appropriate use of Bayesian methods—now generally called sequential analysis—enables JN-25 to be decrypted with greater reliability (depths of 8 or more) and often to be decrypted at depths of 6 or 7.

In order to make available more easily usable information derived from the list of commonly occurring book groups and their frequencies, a 'scoring system' was derived from the list. It assigned a 2-digit score to each group in the list. This reflected the various frequencies of occurrence. Each listed group found in the column of scanning groups obtained from a speculative additive applied to a column was given its score and the sum of the resulting scores was calculated. If this sum exceeded a previously determined threshold, then that additive was deemed to be the correct entry in the additive table. This procedure was simple to apply. The scores were recalculated each time the list was updated in order to reflect the most recent information on frequencies of the common book groups. (All this is an over-simplification of that given by Simpson—see Note 7.) The scoring system was derived using Bayesian inference applied to the frequencies in the list. More on this is given in Appendix 2.

---

[11] See Erskine and Smith (cited in Notes 6 and 7 above).

Appendix 1 Minor Differences

Although this Appendix is particularly relevant to the interpretation of Lietwiler's letter, it is generally applicable to additive cipher systems with groups of any fixed size. Indeed, the examples are 4-digit groups! No special role for scanning groups comes in. Section 12.2 examines the use of differences in attacking additive cipher systems in general.

There is a simple device that almost halves the number of differences to be stored. A group $\mathbf{x}$ is said to be *minor* if $\mathbf{x} \leq -\mathbf{x}$. With a few exceptions, for groups $\mathbf{p}$ and $\mathbf{q}$, exactly one of $\mathbf{p} - \mathbf{q}$ and its negative $\mathbf{q} - \mathbf{p}$ will be minor. As these differences occur with the same frequency, in counting frequencies it is sufficient to work with the minor differences. So for the groups:

0011	0120	0177	0310	0321	0378	1286	1297	1502
4832	4909	4918	5105	5106	5194	5514	5541	5553
0099	*0980*	*0933*	*0790*	*0789*	*0732*	*9824*	*9813*	*9508*
6278	*6101*	*6192*	*5905*	*5904*	*5916*	*5596*	*5569*	*5557*

the first 18 (ordinary type) are minor while their negatives, the next 18 (slanted type), are not. The groups made up of the digits 0 and 5 only are their own negatives and would all be taken as minor.

Appendix 2 Bayesian Inference

In this appendix 'JN-25X' denotes an additive system whose book groups are all 5-digit and scannable. The Allied cryptanalysts are assumed to have some statistics on the frequency of the 200 (say) most common book groups. A method is needed to determine whether a proposed decryption of a depth of intercepted GATs should be accepted. As Friedman wrote (quoted in Sect. 4.4):

> Like the experimental scientist [the Cryptologist] is observing phenomena or occurrences to determine whether they are random or systematic.

So the cryptologist needed to use statistically sound methods that did not require computation other than the simplest arithmetic. Ideally the system being used would be based on assigning a 'score' to each of the 200 (say) most common book groups. Initially it is assumed that no 'horizontal' information is available and so the cryptologist should add up the scores in the decrypted depth and then check whether the total exceeds some preset threshold.

COUNTING			CRUDE			ACCURATE		
20859 ✓	→	1	20859 ✓	→	1	20859 ✓	→	6
91173 ✓	→	0	91173 ✓	→	0	91173 ✓	→	0
76692 ✓	→	0	76692 ✓	→	0	76692 ✓	→	3
14187 ✓	→	1	14187 ✓	→	3	14187 ✓	→	23
58053 ✓	→	0	58053 ✓	→	0	58053 ✓	→	0
23187 ✓	→	1	23187 ✓	→	1	23187 ✓	→	9
13854 ✓	→	1	13854 ✓	→	2	13854 ✓	→	14
03099 ✓	→	0	03099 ✓	→	0	03099 ✓	→	5
63168 ✓	→	0	63168 ✓	→	0	63168 ✓	→	0
TOTAL	=	4	TOTAL	=	7	TOTAL	=	60

Three examples (out of many) of scoring systems are shown above. Each would have its own threshold score for accepting the proposed decryption. This would vary with the number of GATs in the original depth. The 'counting' system simply allocates one point for each GAT in the most frequent hundred. This is not particularly natural: nothing is special about 100. But it was used at least initially by the Rochefort team at Pearl Harbor in 1942. The 'crude' system is in fact less crude than the 'counting' one. Here 3 points are allocated for the ten most common, 2 points for the next 20 and 1 point for the next seventy. The 'accurate' system illustrates (with totally contrived scores) the Bayesian method developed by Alan Turing in 1940–1941.

The cryptologist using the 'accurate' system would have instructions of the type 'for a depth of 9 the total score has to be at least 47'. So the decryption shown above should be accepted. As already mentioned, the 'threshold' 47 is determined experimentally using depths of 14 or more for which determining the correct decryption is much easier.

Any scoring system working on vertical evidence of an alignment being correct may be adapted to allow for bonus points being awarded for observed horizontal evidence. For example, if the previous column has 12345 in its ninth place and it is known that 63168 often follows 12345, supplementary rules might award bonus points to this decryption. Apparently no documentation has survived on how a numerical scoring system for evaluating vertical evidence may have accommodated scores for non-numerical horizontal evidence.

It is quite possible that Turing became aware that the 'counting' method was in use at FECB and started thinking about the general mathematical problem of designing an optimal scoring system. It is also quite possible that he realised from other problems that the general issue of designing scoring systems was of the greatest importance. By around August 1941 he wrote a general report on probabilistic methods in cryptology which may be read in conjunction with a key paper by his assistant, Jack Good.

The method amounts to choosing a constant $K > 0$ and then assigning the weight $K \log_{10}(p_g)$ to the book group g. Here p_g is chosen so that the book group g occurs on average p_g times in every 33,334 groups of decrypted traffic. Thus a group that occurs 7.3 times in every 10,000 decrypted groups occurs 24.334 times in every

33,334 groups. Tuing and Good found it convenient to take $K = 20$ and then round to the nearest integer. Here $20 \log_{10}(24.334) = 20 \times 1.3862$ is rounded and becomes 28. The unit of weight of evidence implicit in this was called the 'halfdeciban' or just 'hdb'.

One evident merit of this system is that if $p_g = 1$, that is if g occurs with the average of the frequencies of the 33,334 scannable groups, then $K \log_{10}(p_g) = 0$. The observation of such a g gives no evidence whatsoever as to whether a proposed decryption is correct.

In that era it was standard to use rounded base 10 logarithms to carry out multiplications. Here in effect the base 10 logarithms are being rounded to the nearest multiple of .05.

An interesting example of this method arises from the frequencies of the 26 letters in ordinary language with British spelling: These are displayed on a 'per 10,000' basis immediately below the letters in the table below:

E	T	A	O	I	N	S	H	R	D	L	C	U
1270	906	817	751	697	675	633	609	599	425	402	278	276
10	7	7	6	5	5	4	4	4	1	0	−3	−3
M	W	F	G	Y	P	B	V	K	J	X	Q	Z
241	236	223	201	197	193	149	98	77	15	15	10	7
−4	−4	−5	−6	−6	−6	−8	−12	−14	−28	−28	−32	−35

The frequencies are then converted by multiplication by .0026 to a 'per 26' basis (not shown). The third line above gives the hdb scores obtained by multiplying the base 10 logarithms of the converted frequencies by 20 and rounding. These can be used to test, by counting the occurrences of the 26 letters in a string of possible text, whether it has the distribution that would arise from natural language. For example,

H E S T O P P E T H O N E O F T H R E E   scores
$4 + 10 + 4 + 7 + 6 - 6 - 6 + 10 + 7 + 4 + 6 + 5 + 10 + 6 - 5 + 7 + 4 + 4 + 10 + 10 = 97;$
I F T U P Q Q F U I P O F P G U I S F F   scores
$5 - 5 + 7 - 3 - 6 - 32 - 32 - 5 - 3 + 5 - 6 + 6 - 5 - 6 - 6 - 3 + 5 + 4 - 5 - 5 = -90.$

Appendix 3   Turing and Bayes

The following is taken from Hugh Alexander's *Cryptographic History of Work on the German Naval Enigma*[12] (GCCS, 1945):

> Also in November [1942] Turing left the section for a visit to America. Although this did not mark the official end of his connection with the section he never did any more work in

---

[12] Alexander's report is HW 25/1 in the British National Archives and is available on the web.

it and therefore it is a fitting place to recognize the great contribution that he made. There should be no question in anyone's mind that Turing's work was the biggest factor in Hut 8s success. In the early days he was the only cryptographer who thought the problem worth tackling and not only was he primarily responsible for the main theoretical work within the Hut (particularly the developing of a satisfactory scoring technique for dealing with Banburismus) but he also shared with Welchman and Keen the chief credit for the invention of the Bombe. It is always difficult to say that anyone is absolutely indispensable but if anyone was indispensable to Hut 8 it was Turing. The pioneer work always tends to be forgotten when experience and routine later make everything seem easy and many of us in Hut 8 felt that the magnitude of Turing's contribution was never fully realized by the outside world.

Alexander previously gave due credit to the Polish cryptologists for their earlier work on Enigma. This of course included the exploitation of redundant encryption.

Edward Simpson's paper *Bayes at Bletchley Park* explains the importance of Bayes in decrypting Enigma.

Turing then worked with the cryptanalytic research section of the US Naval Communications Intelligence Staff. A series of internal *Cryptanalytic Research Papers*[13] has survived. The introduction to that series states:

RIP 450 is concerned mainly with the techniques themselves, while this series considers the cryptanalytic or mathematical theories which underlie the techniques. On the other hand machine research (from an engineering point of view) is not covered in this series. Some of the papers in this series are expository but most represent original work. It must always be borne in mind that we owe to the British the basic solution of the Enigma, and many of the basic subsidiary techniques, together with the underlying mechanical and mathematical theories. Much of what we call 'original' is only a retracing of steps previously taken by the British, and the Editor has striven to point this out in the Index. But there is also a great deal that extends or improves British methods, and some that strikes out in new directions.

The most sensational of these 'new directions' is the use of *Hall weights* (Sect. 15.3).

The 1941 report by Alan Turing on such methods ultimately led to mechanized applications of it, such as in Colossus. This was the machine that attacked the 'Tunny' encrypted teleprinter. Some quotations from the 1945 GCCS *General Report on Tunny*, by Jack Good, Donald Michie and Geoffrey Timms (British Archives HW25/4, available on the Web) are:

The fact that Tunny can be broken at all depends upon the fact that P, χ, Ψ', K and D have marked statistical, periodic or linguistic characteristics which distinguish them from random sequences of letters.

The importance of exploiting statistical characteristics is noted in such sentences as:

First method, stage 1. Solution of $Z = \chi + D$.
Various χ-patterns (or settings) are tried mechanically and the correct one is distinguished by the statistical properties of ΔD.

---

[13]The reference is NARA RG38 Radio Intelligence Publication Collection boxes 169–172.

Recall that 'In order to break a machine cipher, two things are needed'.

The special case of Bayes' Theorem $O[H|E]/O[H] = P[E|H]/P[E|\bar{H}] = f$ was first used in Bletchley Park by A. M. Turing. The fact that it was a special case of Bayes' theorem was pointed out by I. J. Good. The great advance of Turing consisted of the invention and application of the deciban in Hut 8. Deciban is abbreviated to 'db'. This is defined simply as $10\log_{10}(f)$ where f is the factor defined above.

Turing's use of the 'needle in a haystack' phrase turns up in this text:

In cryptography one looks for needles in haystacks and the object chosen has to have a large factor in favour of being a needle in order to overcome its prior odds. It will be observed that one could take a long time to find the needle if one could not estimate the factor very quickly—hence the necessity of machines in such problems.

It is worth noting here a few lines from Abraham Wald's highly relevant paper *Sequential Tests of Statistical Hypotheses* published in the *Annals of Mathematical Statistics* 12 (2) (1945) pages 117–186 but written in 1943:

The National Defense Research Committee considered these developments sufficiently useful for the war effort to make it desirable to keep the results out of the reach of the enemy.

Appendix 4   Contrived Examples

List A immediately below contains 144 groups in 12 rows of 12:

```
00001  00010  00022  00100  00112  00121  00202  00211  00220  01000  01012  01021
01102  01111  01120  01201  01210  01222  02002  02011  02020  02101  02110  02122
02200  02212  02221  03001  03010  03022  03100  03112  03121  03202  03211  03220
10000  10012  10021  10102  10111  10120  10201  10210  10222  11002  11011  11020
11101  11110  11122  11200  11212  11221  12001  12010  12022  12100  12112  12121
12202  12211  12220  13000  13012  13021  13102  13111  13120  13201  13210  13222
20002  20011  20020  20101  20110  20122  20200  20212  20221  21001  21010  21022
21100  21112  21121  21202  21211  21220  22000  22012  22021  22102  22111  22120
22201  22210  22222  23002  23011  23020  23101  23110  23122  23200  23212  23221
30001  30010  30022  30100  30112  30121  30202  30211  30220  31000  31012  31021
31102  31111  31120  31201  31210  31222  32002  32011  32020  32101  32110  32122
32200  32212  32221  33001  33010  33022  33100  33112  33121  33202  33211  33220
```

If List A is considered as a depth of 144 GATs, an electronic calculation shows that there are 8363 decrypting groups. As an exercise, how may this be used to produce a depth of 8363 GATs with 144 decrypting groups?

Next, the first 9 rows (108 groups) of List A may be used as a new *List B* which has 9558 decrypting groups.

Alternatively, the last 9 rows (108 groups) of List A may be used as a new *List C* which has 9557.

The reader may wish to seek a depth of 96 with 9408 decryptions.

One may also consider *List D*, which consists of the 72 groups common to both Lists B and C. This, considered as a depth of 72, has 10,752 decrypting groups. *List E*, consisting of the 54 italicized groups in List A, has 12,288. Next, 18 more groups can be deleted from List E to yield *List F*:

11101 11110 11122 11200 11212 11221 *12100* *12112* *12121* 12202 *12211* *12220*
13102 *13111* *13120* *13201* *13210* *13222* 21100 21112 21121 21202 21211 21220
22102 *22111* *22120* *22201* *22210* *22222* *23101* *23110* *23122* *23200* *23212* *23221*

List F has 36 GATs and 13,824 decrypting groups. *List G*, consisting of the 24 italicized groups of List F, turns out to have 15,552.

It is possible to use any of the above contrived examples to construct a totally non-historical JN-25 that resists some of the usual methods of attack. The use of such a system in the modern era cannot be recommended!

Chapter 11
Recovery of a Code Book

The recovery of the underlying code book forming part of an additive system depends upon the successful decryption of many intercepted messages, thus exposing the *bare code* equivalents. The skills required for successful decoding of a bare code message are more akin to the linguistic challenge of determining the nature and meaning of an unknown written language than to those needed for elucidating the operation of a cipher machine or the form of a superencipherment. This is particularly so when the underlying plain text is in a foreign language or in the jargon peculiar to a restricted class of users. Two historical examples are given.

11.1 The Rosetta Stone

One of the best-known historical incidents[1] in archaeology is the recovery of the language of ancient Egypt. Numerous inscriptions in what became known as *hieroglyphs* were available, but they were not easy to understand. However an ill-fated expedition by Napoleon into Egypt had acquired various antiquities, including one which has become known as the Rosetta Stone. In 1801 the British were able to force the French to hand over both the territory gained and the antiquities. The original stone was taken to the British Museum, where it is still on display. However a plaster copy made its way back to Paris.

The Stone had three texts, one of which was in easily readable ancient Greek, one in Demotic (derived from the Ancient Egyptian hieratic) and the third in Hieroglyphs. It was assumed that all three texts had identical meanings. Thus, the

---

[1] The well-known book *Gods, Graves and Scholars* by C. W. Ceram (Kurt Marek) gives much more information on the Rosetta Stone. A copy of it is displayed prominently in the National Cryptologic Museum in Fort George Meade, Maryland. Michael Coe's *Breaking the Maya Code*, Thames and Hudson, 1992, is also of interest here.

© Springer International Publishing Switzerland 2014
P. Donovan, J. Mack, *Code Breaking in the Pacific*,
DOI 10.1007/978-3-319-08278-3_11

Stone provided an obvious crib for breaking the hieroglyphic 'code' and so enabling the decipherment of all inscriptions in hieroglyphs. Many linguists and scholars devoted considerable effort to this task but without success. A common assumption about the structure of this 'code' was that, because its symbols appeared to be pictorial, they had to represent objects related to the pictures.

Thomas Young, a British polymath, after studying some distinctive sets of hieroglyphs encased in loops (*cartouches*), reasoned that these might stand for royal personages mentioned in the Greek text. Since these people were not Egyptian, he suggested that the enclosed symbols might correspond to the sounds in their names. That is, these symbols were phonetic rather than pictorial in their meaning.

A extremely gifted natural linguist, Jean-François Champollion (1790–1832), who had since childhood been fascinated by Hieroglyphics, learnt of Young's work, and applied it to other ancient inscriptions containing royal names. Using the phonetic interpretation, and his knowledge that the Coptic word for 'sun' was 'ra', he deduced that one royal name containing a sun symbol was Ramses. From this, with help from the text on the Stone and his wide knowledge of existing hieroglyphic inscriptions, Champollion 'broke the code' and demonstrated how to apply this to the deciphering of the ancient language.[2]

The story of the Rosetta Stone shows that great care must be applied in analysing a new code before making a final decision as to its structure and use. This will normally require a large amount of intense work testing possibilities and reflecting upon the linguistic and technical expertise available to the creators of the code.

On the positive side, this story is an early example of the use of cribs[3] in attacking encoded and/or encrypted material. Use of cribs became prominent in later attacks on sophisticated codes utilising very complex methods of encryption. In particular, the availability of a *plain language* version of just one intercepted message may be of great value.

---

[2] Another well-known example of the reading of ancient texts is the post-war work on *Linear B* using certain written material found in Crete. The work of Michael Ventris (1922–1956) and John Chadwick (1920–1998) initially attracted some controversy. Chadwick had worked on Japanese Naval Attaché material (not covered in this book) at Bletchley Park in 1944–1945.

A. P. Treweek was a classicist and mathematician who worked on Japanese material at Frumel. He published a paper *Chain Reaction or House of Cards: An Examination of the Ventris Decipherment* in the *Bulletin of the Institute of Classical Studies* 4 (1) December 1957, pages 10–26. He concluded that the Ventris work was basically sound and set an interesting related puzzle for the reader. This gave the names of various geographical locations in the Pacific in kana syllables. The challenge was to work out the meanings of the kana syllables used.

You begin by noting that ♠-♡-◇-◇ probably refers to HO-NO-LU-LU and then use the identifications ♠ = HO, ♡ = NO, ◇ = LU to work on the rest of the names in the list. The work on Linear B had made similar use of reference to geographical locations.

[3] The word 'crib' has one meaning in the context of code books and a quite different one in the saga of attacking the Enigma machine.

11.2 Major George Scovell and Military Intelligence

An early example of the breaking of a code book was achieved by Major George Scovell[4] of the British Army around 1811. Wellington's success in driving the French armies from Portugal and Spain in the *Peninsular War* (1808–1814) was an important precursor to Napoleon's eventual defeat at Waterloo. The real significance of Scovell's work, both as Wellington's superintendent of communications and as his main breaker of important French military codes from 1811 onwards, is clearly and fully developed in Mark Urban's recent biography of him.

Wellington's forces were always relatively close together. They operated in the sympathetic company of Portuguese and Spanish forces and guerrilla groups who were absolutely hostile to the French. The various French armies were often quite separated. In addition, their commands were subject to direction from Napoleon or his brother Joseph (installed in Madrid as king). The only means of communication among the French armies was by messenger, using either Army personnel or locals, carrying letters secreted on themselves or on their horses or companions. This was an extremely dangerous activity because the guerrillas and their supporters were well-dispersed and managed to intercept many carriers.

Scovell showed an aptitude in several areas. He efficiently organised the channels of communication needed for Wellington's purposes. He assisted in the important task of liaising with guerrilla groups and obtained from them information on

---

[4]The basic source of information on Scovell's code breaking work is Appendix XV *The Scovell Ciphers* in Vol. V of Charles Oman's *History of the Peninsular War*, published from 1902 onwards. Oman had access to papers of Scovell, now held by the British National Archives (TNA) as file WO 37. Mark Urban's stimulating biography of Scovell—*The Man Who Broke Napoleon's Codes*— appeared in 2001.

The French language would have been widely understood among the better educated officers of the British Army at the time of the Napoleonic wars. The Japanese language was understood by only a few members of the Allied armed forces in December 1941. Despite the introduction of special courses in that language, the decryption process was necessarily more difficult. The capture of the documentation for JN-25C in August 1943 taken together with the thousands of archived JN-25C intercepts must have provided invaluable insight into IJN telegraphic language.

Likewise Scovell would have been able to access a limited supply of intercepts without any assistance. In the case of JN-25, with thousands of book groups and thousands of intercepts, some mechanisation of the task of finding all occurrences of a given book group would have been most useful. This was one place where the IBM punch cards and associated machinery (see Sect. 1.15) came in very handy.

The *CBTR* include a somewhat jocular *Part L—Bibliography* put together by various students of Latin and ancient Greek in 1945. Thus one Aeneas the Tactician (360 BC) is mentioned but not Sun Tzu (Chap. 20) or Scovell. The authors of the *CBTR* also missed quoting Virgil's *Aeneid* (ii 390): Dolus an virtus, quis in hoste requirat? This translates to 'Trickery or courage: which is needed in conflict?'.

A fuller account of early codes could mention the French optical telegraph (semaphore line) dating from 1794. It signalled information by varying the position of movable wooden arms on towers. This could be read from the next tower in the chain and the signal re-transmitted. At one stage in the Napoleonic Wars the RN benefited from reading the signals transmitted from a tower near the coast.

enemy movements, as well as intercepted messages. From mid-1811 he was also responsible in general for 'codes and ciphers'. This occurred just as Marmont, the new commander of the French Army of Portugal, conscious of the very insecure nature of his communications, decided to develop a code book of 150 entries for use with his divisional commanders. Marmont also felt confident that the code book could be safely changed regularly because only a few copies needed to be issued on each occasion and these could be escorted to their respective destinations. The first intercepts written in this new code could not be deciphered using any of the standard methods based on letter frequency, and it was suspected that some code words stood for whole words, while some common letters, like 'e', might correspond to several code words.

Scovell was encouraged by the fact that the messages combined ordinary French words with code words and so suggested that the vocabulary of the code book was limited. Also some other intercepted messages were written completely in plain language. What he needed most was a sufficient number of intercepted messages to enable him to see contextual similarities or differences that would lead him to correctly guess the meaning of a code word. His main resource for this attack, apart from his own intelligent and sharp mind, was his exceptional familiarity with the French language and with its formal modes of address. Using the meanings suggested by the plain language portions to guess whether code words stood for persons, places, directions and the like, he was able, with the assistance of colleagues, to solve the new code. Information from it came in time to assist Wellington in deciding to withdraw his forces besieging Ciudad Rodrigo rather than face a much larger attacking French force.

Marmont changed his code book in October, soon after Wellington's withdrawal. (The code words remained as the numbers 1–150, but they were given new meanings.) By the end of the year, Scovell had managed to break this new version. This activity was in fact excellent training for his coming great challenge. King Joseph had persuaded Napoleon to use a version of the *Great Paris Cipher* (GPC) for highest level communications with himself and with Army commanders and select other officials. The GPC consisted of a code book with some 1,200 entries. The code book was based on another first used by Louis XIV a century earlier. It had been progressively refined for diplomatic use during the eighteenth century. The book randomly assigned a number from 1 to 1,200 to letters, syllables, words, etc, arranged in lexicographic order, while a reverse code book, arranged in numerical order, was used for decoding messages upon receipt.

The GPC represented a considerable increase in difficulty to a potential decoder. Two features of the version used in Spain and Portugal gave assistance and hope. Once again, messages were often found to combine plain language words, phrases or even sentences with sets of code words of varying length. Thus valuable contextual information was provided. Also, their original use as diplomatic ciphers meant that their code books contained a number of place names and titles that were quite irrelevant to those dealing with the Peninsular War. Some 200 names of places and people were therefore added to the original, not by replacement, but by adding another 200 numbers from 1,201 onwards! Once this was realised, it greatly reduced the number of possibilities of meaning for any code word in that range.

Once again, Scovell needed access to a number of messages, preferably with similar or overlapping content. And this took time. He was able to make some progress with early messages intercepted in April-May of 1812. As their volume increased, and as he grew more familiar with the code, more and more important partial decrypts were made available to Wellington. By the end of 1812, Scovell had, by patient analysis and testing, confidently assigned a meaning to some 900 code words—a remarkable feat and one which was completely unexpected by his enemy.

Scovell's work demonstrates some obvious points relevant to any attack on an encoding system based on such a book, namely that:

- the more intercepted messages one can collect, with a knowledge of their senders and intended recipients, the more likely is one to succeed;
- a good understanding of the structure of the underlying language and of the forms that may be used in military communication is essential;
- a willingness to work hard and long, using all relevant intelligence that may provide context, is necessary.

As mentioned briefly in Sect. 1.5, the political section of Room 40 in WW1 decoded a large number of German diplomatic messages and employed punched card machinery to expedite the process. The decoding team found that, on average, three guesses were needed in order to find the correct meaning of a code word used in this context.[5]

11.3   Recovery in JN-25 and Similar Systems

Stripping of the second-stage encryption from the GATs, even if possible, may not always be successful. For example, certain entries in an additive table may have been used so rarely that they cannot be determined. Sporadic garbled GATs may also obstruct the finding of entries in the table. Thus, it is not always possible to derive the full sequence of book groups contained in a message—there may be gaps and these will hinder the decoding process. Context plays a critical role in guessing the correct plain text meaning of a book group.

The task of *book building* (reconstructing or *recovering* the meanings of book groups) involves both analysis of messages in order to build up lists of commonly occurring groups or sets of groups, and the collection and application of a great deal of collateral information. These assist in identifying message contexts and guessing message content. Here, DF, TA, call sign identification and other ancillary information derivable from messages, would assist in forming a background data

---

[5]Gannon *Room 40*, page 169.

base.[6] With older messages, knowledge of contemporaneous events and the contexts of the sending and receiving locations might be available and help in guessing the context or the content. There must be a constant interplay between those who understand the context behind the messages (experienced naval personnel in the case of naval codes) and language officers who understand the underlying language and, if possible, likely jargon and message formats that the code users may be expected to use. Book building is a laborious task, prone to error and requiring constant testing of a claimed meaning for a book group against new message contexts.

The following quote from the *GYP-1 Bible* is repeated from Sect. 9.21:

> The actual text of the clear messages in JN-25 is terse and entirely innocent of the elaborate circumlocutions of polite Japanese. Full inflexion of verbs and adjectives is practically never observed. Honorific expressions are absent. It frequently happens that even after the full text of a message has been recovered only an expert in naval text can render it . . . into plain English text.

Difficulty in obtaining personnel with expertise in Japanese, with some experience of Japanese language forms, remained a problem for the Allies throughout the Pacific War and sometimes seriously inhibited the rapid derivation of intelligence from intercepted messages.

Cribs—plain texts giving the meanings of intercepted messages—are of immense value in getting the recovery process started. The simplest crib would be that obtained from the same text being transmitted in both an old and partly broken code and also in a new code. However operators known to send *stereotyped* messages (those having recurring format and content) or to use a limited vocabulary could be detected, particularly after the capture of JN-25C documentation. Monitoring these would provide entry into a new system. Any knowledge of some book groups is likely to indicate the meaning of others.

It would have been commonplace for only parts[7] of a message to have been read and yet the gist of the full text to be available for intelligence staff. Past experience might indicate what the 20 or so most commonly used groups were likely to mean, particularly in restricted contexts such as weather reports.

Lightly used code books were particularly difficult to recover. Conversely a developing major conflict would generate heavy use and facilitate recovery. This was to be crucial for work on JN-25B in the first 5 months of 1942.

---

[6]A letter from Lieutenant Fabian to Commander Safford dated 30 August 1941 (in RG38 of NARA but also available on pages 143–144 of Timothy Wilford's thesis via ProQuest Dissertations) makes it clear that the long USN experience in using TA and DF on IJN signals was instrumental in getting started on the JN-25B code book.

[7](This note repeats much of Sect. 7.1.) A bare code may be insecure without it being capable of being fully solved. Thus in NARA RG457 SRH-349 there is the following general comment: 'Reconstruction of a code book is a long laborious process. Each code group needs to be identified singly. The larger the code, the longer is the time needed.' Thus a decrypted (but incompletely decoded) message may look like 'xxxxx xxxxx submarine xxxxx attack xxxxx xxxxx merchant ship xxxxx xxxxx'. This state of knowledge must be what Alastair Denniston meant by the phrase *skeleton code books* in Sect. 3.2.

Best of all, the code book could be captured, as happened in August 1942 to a JN-25C code book.[8] The available files of about 80 days of intercepts in JN-25C must have provided valuable background knowledge. Both GCCS and Op-20-G had long experience with Japanese material but would not have had a capture like this before, which must have greatly assisted in understanding the style and jargon of the JN-25 systems.

Unfortunately for latter-day historians nothing like a complete record of captures of cryptological material or of the use of cribs has survived and made its way to the archives.

Even if the *values* (meanings) of only some of the commoner book groups were available, some intelligence might be obtainable. Thus one of the highlights of the work at Frumel in April and May 1942 was the decrypting (using a known piece of additive table) of part of an intercepted message. This particular message had been given lower priority but Yeoman William Tremblay had completed the material given higher priority and so turned to the next heap. One of the book groups obtained by stripping the intercept he knew to represent the Japanese word for 'attack'. And so the count-down to Midway started in earnest.

11.4   The GYP-1 Bible

Chapter IX of the *GYP-1 Bible* consists of some 30 pages describing techniques for recovering book group meanings. The general principle is to classify book groups into one of 11 classes:

1. Indicator groups of various sorts.
2. Numerals and Fractions.
3. Punctuation marks.
4. Dates.
5. Roman letters and digraphs.
6. Kana syllables, digraphs and trigraphs.
7. Proper names.
8. Individual Kanji.
9. Japanese words.
10. Word groups, sometimes of considerable length.
11. Dummy groups.

---

[8]The code book was saved from a fire by a Marine early in the Guadalcanal fighting. Possession of a complete code book in fact assists in finding entries in its replacement code book, provided data is available on the frequency of occurrence of commonly used book groups in both systems. One may reasonably assume that the meanings of, say, the 40 most commonly occurring book groups are much the same and use this to reduce considerably the work needed to identify individual book groups in the new system. Comparison of signals from an active base that regularly uses the same commonly used book groups could help accelerate the process.

It then describes methods in assisting in the correct classification of book groups, followed by guides to help arrive at the correct decoding of each group.

Surprisingly the ordering of the text groups in a typical message is here (page 305) first described in detail:

> In any transmitted message, the code groups are to be regarded as illustrating not merely serial order, but cyclic order—that is, reading from left to right across the sheet, in the order in which the groups have been intercepted, the last text group of the transmitted message is also the immediate antecedent of the initial text group of the transmitted message. This complete cycle is broken in *two* places: (1) at the point where the transmitted message is started and (2) at the point at which the proper 'reading' of the message is begun. In most text in JN-25 these points are distinct. This means that the Japanese introduce into almost all messages, an indicator which is usually termed 'begin message here' or, more simply, 'M/S', for 'Message Starts'. In short messages the 'M/S' is occasionally omitted ... but in over 99 % of the messages a 'M/S' indicator group is present.

Put simply, this states that, at the stage of encoding a message text, it is split at a point chosen by the operator, the latter part is then placed before the former and at the junction of the two a book group called 'Message Starts' is inserted. In the IJA Water Transport Code 'M/S' also appeared with book group 6666: this is mentioned in three other places in this book.

11.5　IBM Machine Assistance

The process for using IBM cards described below is a modern reconstruction. It may well differ in fine details from that used by the Combat Intelligence Unit[9] (Sect. 5.3) at Pearl Harbor in 1942.

Frequently the meaning of a book group would have to be discovered by examining numerous uses of it. It is here that IBM cards would have come in handy. Suppose that a message of 27 groups in a 5-digit decrypted code system was transmitted at 18.29 Tokyo time on 10 January 1942 apparently from source 24350 to an unidentified recipient:

---

[9]Joe Rochefort did give an oral history interview which is preserved in the United States Naval Academy. Key extracts may be found on the Naval Cryptological Veterans' Association website by asking Google to search for NCVA Cryptolog. In view of the insecure encryption of JN-25B messages from December 1941 to May 1942, his Combat Intelligence Unit *was in effect working on an unenciphered code.* (Even if this last comment is an exaggeration, the decryption had become a simple, standard chore.) He stated there that he needed and had a very good memory of previous messages to work out the likely meaning of key intercepts in which some but not all the book groups used were 'evaluated'. He described the use of 'two or three million' cards in a month. He states that the practice in his CIU was to use twice as many cards as there were groups in the message. This seems obscure.

In June 1943, Rudolph Fabian wrote 'A careful survey of the requirements of this unit, set against the possible space available, indicates that the total IBM machinery allowance should be two (2) each of the following: Type 405 Tabulators; Type 513 Summary (?) Producers; Type 075 Sorters; Type 077 Collators; Type NC4 machines; plus six (6) Type 035 Alpha punches.' See NARA RG457, SRH-275.

```
10713   87177   94098   50301   70653   60177   53703   00708   47646
40143   26466   76539   20742   37104   38976   36279   40134   17127
68616   06414   34545   19101   54504   15810   04119   04128   31502
```

27 cards would be prepared, each beginning 42011 01829 24350 (that is, year (42), month (01), day (10), transmission time, source). They would have this identification data on the left. Each of the 27 groups would be typed out on columns 46–50 of one of the cards, with the preceding five groups to the left and the following five groups to the right as shown below. Columns 76–80 would be left for other use if needed.

```
              1     1 2    2     3     3     4     4     5     5     6     6     7
1     6     1     6 1    6     1     6     1     6     1     6     1     6     1
42011 01829 24350                                     10713 87177 94098 50301 70653 60177
42011 01829 24350                               10713 87177 94098 50301 70653 60177 53703
42011 01829 24350                         10713 87177 94098 50301 70653 60177 53703 00708
42011 01829 24350                   10713 87177 94098 50301 70653 60177 53703 00708 47646
42011 01829 24350             10713 87177 94098 50301 70653 60177 53703 00708 47646 40143
42011 01829 24350       10713 87177 94098 50301 70653 60177 53703 00708 47646 40143 26466
42011 01829 24350       87177 94098 50301 70653 60177 53703 00708 47646 40143 26466 76539
```

The preparation of punched cards in this style would be greatly assisted by having a standard card-punch machine modified to carry out offset gangpunching. The first card would be punched digit by digit. On following cards the groups other than the last one would be mechanically copied from the previous card. Once punched, the cards could be mechanically sorted by columns 46–50. This would enable printouts to be made of all occurrences of a particular book group such as 87177, with surrounding context. It would also provide updated statistics of the frequency of occurrence of the more common book groups or pairs of consecutive groups.

This in turn would allow experts in telegraphic Japanese—in short supply in 1942—to work out gradually the meanings of group after group. Even with electro-mechanical assistance, this would be a slow process. In particular, with JN-25B after the Pearl Harbor raid, there was no great difficulty with stripping off the additive and so lots of decrypted signals were available, including some from the JN-25B7 era. So fragments of messages could be read in February 1942 with more and more becoming available in March, April and May.

11.6 Machine Decryption

Remarkably, the US Army moved a long way towards mechanising the process of decrypting, decoding and translating additive systems in the last 2 years of WW2. It was still most desirable to have a linguist check the printouts, but much could be done with IBM cards. The mechanisation of the process was found to be worthwhile, at least by the US Army. Indeed, George Aspden[10] remembered seeing it carried out in the machine section at Central Bureau (in the old Ascot Fire Station, which still survives).

A somewhat speculative account of the process is as follows.

Step 1. The intercept most likely arrived in the form of paper tape. If so, some machine would have been used to prepare a deck of cards, one for each GAT, with the message identification details in one fixed place, possibly the last ten columns, and one GAT per card in another fixed place, probably the first four columns. The IBM company did introduce a paper-tape-to-card machine for the commercial market soon after WW2.

Step 2. The additive table would be stored on IBM cards, probably with one additive per card on the left and with identification details on the right. These cards would be kept in boxes on a nearby trestle table. The operator would use the (assumed known) indicator system to find the appropriate additive cards and take them to a machine which would input one GAT card and one additive card at a time and output on new cards the GAT, the corresponding additive, the (false) difference (the *resultant*) and the identification details from the message cards. These message cards would then be replaced by the new cards, which may as well be called the *resultant cards*.

Step 3. The cards would then need to be serially numbered, presumably immediately before the identification details. The *NC1* machine developed for Op-20-G had this purpose and may well have been easy to produce. Most likely the resultant cards would need to be replaced by these *numbered cards* produced by the numbering device. It is possible that the numbering was carried out in the same process that produced the resultant cards, in which case steps 2 and 3 merge.

Step 4. The numbered cards would then be sorted to place the resultants in natural (that is, numerical) order. This would be done by a standard IBM sorting process on the sorting machine, and needing the involvement of the operator.

Step 5. The entries in the code book would be available on cards, with the book groups on the left, with the corresponding Japanese phrase in Romanised Kana or Roma-ji, that is the standard way of expressing Japanese in the Roman alphabet, next and the approximate English equivalent on after that. These might well use up columns 2–5, 16–40 and 41–65 respectively. A collator would then read in the sorted number cards and the sorted book cards and punch out *collator*

---

[10]This description was made possible by the memories of George Aspden, who was at CBB in 1945, and of Ian Watson, who worked with IBM cards in the early 1960s. It is confirmed on page 11 of Part G of the *CBTR*. Laurie Robertson's *Arlington Hall Station: The US Army's World War II Cryptanalytic Center*, in the *IEEE Annals of the History of Computing*, 2004, has also been of use.

The NAA Canberra file A6923 SI/5, being a technical report on CBB procedures as they were in April 1945, refers to automated decoding in the words 'The IBM stage decodes the message and produces a new work copy for the translators'.

cards each of which had one GAT, one additive group, one resultant, one Japanese phrase, one corresponding English phrase, the serial number and the identification details. (Usually knowledge of the code book would be incomplete. So some book groups would have something like 'RRRRRR' for the Roma-ji and 'TTTTTT' for the translation.) It might be sensible to run cards from a batch of messages through the collator together.

Step 6. These collator cards would then be sorted back into the order in which the original GATs were read. This would need the sorter again, which would now work on the serial numbers.

Step 7. The collator cards would be read by a tabulator which would print out the GAT, the additive, the resultant, the Roma-ji and the translation entries line by line.

Step 8. The printout would then need to be examined by a translator to check whether it was likely to be of interest. If it was so judged, the translator would write out an improved translation and add comments where appropriate.

11.7 Language Officers

The problem of finding experts in the Japanese language who might be suitable for intelligence work was very serious[11] until late in WW2.

Eric Nave (see Sect. 3.3) shared a house in Japan with a British Naval officer, Harry Shaw, and Australian Army officers George Capes and Ray Broadbent. Like Nave, Shaw was to play a key role in the GCCS work on IJN codes in the 1930s. Capes had been seriously injured at Gallipoli and so returned to Australia in 1915. He was involved in getting the Navy interested in some study of the Japanese language. He continued on as an intelligence officer but died around 1940. Broadbent was also a Gallipoli veteran and became an instructor at the Royal Military College, Duntroon in Administration and Japanese for a while but quit the Army in 1926. He later rejoined the Australian army serving in a senior capacity, but not in intelligence, in New Guinea and elsewhere.

Although very few Australians were sent to Japan by the services for language experience between the wars, significant numbers of British and American officers undertook prolonged language training there. Some of these, including Patrick Marr-Johnson of the British Army, later went into Intelligence work. Marr-Johnson was skilled in cryptanalysis and for much of WW2 occupied a senior position at the Wireless Experimental Centre (WEC), near Delhi. This was an important interception site that made significant contributions to the breaking of Japanese codes.

---

[11]NAA Canberra item A5799 5/1940 is entitled *Japanese Language in the Services* and contains a draft report abandoned on 20 January 1940 due to the war in Europe. It contains the rather dismal sentence 'At the present time the services have no first class interpreters of the Japanese language.' See also Note 12 of Chap. 7.

Sidney F. Mashbir of the US Army served as a language officer in Japan in the early 1920s. In WW2 he headed the Allied Translator and Interpreter Service (ATIS) attached to General MacArthur's Brisbane HQ. It grew in size and importance as more and more Japanese service personnel and documents were captured. Early in 1944 it translated major IJN battle plans, brought to Brisbane by submarine after having been found in the Philippines following the crash offshore of planes carrying, it seems now, Yamamoto's successor Admiral Koga and some of his senior staff.

One USN language officer highly relevant to WW2 cryptology was Arthur McCollum. He is now notorious as the author of a secret discussion paper, written in October 1940, about whether it would be to the advantage of the United States to engineer an earlier start to an apparently inevitable war with Japan. Later, in Brisbane, he was USN liaison officer with General MacArthur's SWPA headquarters and so was able to ensure that Water Transport Code decrypts reached the submarines. Conversely he was able to assist SWPA and to a lesser extent Central Bureau (Chap. 17) with USN Sigint.

Others skilled in the Japanese language were Tex Biard and Tom Mackie. The former had been a radio intelligence officer aboard the ill-fated aircraft carrier USS *Lexington*. Early in 1944 they were both working as translators for Frumel in Melbourne but were seconded briefly to Brisbane to assist Central Bureau exploit a magnificent haul of IJA cryptological material picked up by the Australian Army at Sio in New Guinea. Although this coup was not pure cryptanalysis, it enabled MacArthur to target weakly defended locations along the north coast of New Guinea and get much closer to the Philippines quickly and inexpensively. (See Chap. 23.)

Other language officers obtained their expertise from family connections with Japan. For example, Hugh Erskine, later to become the senior linguist at Central Bureau in Brisbane, was born to American missionary parents in Japan. Walter Abraham (1923–2006) was born in Japan to a father working in the family business there. His mother was German and his maternal grandmother was Japanese. He migrated to Australia with his parents in 1941 and later joined the RAAF. His expertise assisted in understanding the technology of crashed Japanese planes. Perhaps Abraham's ancestry excluded him from the extremely secret Sigint work.

Once the Pacific War had begun both Britain and the USA developed special courses in the Japanese language, with emphasis on its military telegraphic form. Young people were carefully selected for them. The 6-month courses set up by the GCCS at Bedford have been mentioned in Sect. 3.5.

Chapter 12
Breaking Additive Systems

This chapter discusses methods of attacking additive systems whose book groups are randomly chosen, rather than all being multiples of three. The whole process is much harder. Captured documents often helped in the later years of the Pacific War.

12.1 Breaking the Indicator System

The indicators of an additive cipher system are its first line of defence. A modest amount of creativity can produce quite new indicator encryption, encoding and/or concealment methods.[1] One reason why the Sio capture[2] (January 1944) of

---

[1] The science (or art?) of going from the intercepted messages in a given additive cipher to the original book groups is recorded in at least four places. One of these is *Cryptanalysis of an Enciphered Code Problem* which uses British training material from about 1941. This was edited by Wayne Barker and published by Aegean Park Press. The process is called stripping the messages and Central Bureau found it convenient to put together its own instructional booklet called *Stripping 2468*. This has survived in NARA RG457. The third reference is the Budiansky *Battle of Wits* book already mentioned. The keen reader eventually learns what *flags* were and various other tricks. The fourth is NARA RG457 Box 831 which gives a report on the breaking of the Water Transport Code in May 1943. By then the indicator systems presented no problem (see Chap. 14), the additive table was well on the way to full recovery but much work remained to be done.

[2] NARA RG38 Box 185, 5750/119. This was written a few years after the event and contains a comment that memories of captures of cryptological material were already fading. The AWM Research Centre file AWM124 4/269 contains the ATIS standing operating procedure for the SWPA including: 'All captured enemy codes, ciphers and cryptographic material of any kind will be delivered as promptly as possible to the representative of the Chief Signal Officer designated for the purpose with the Task Force capturing the material.' Captures did happen and their exploitation was given priority.

© Springer International Publishing Switzerland 2014
P. Donovan, J. Mack, *Code Breaking in the Pacific*,
DOI 10.1007/978-3-319-08278-3_12

cryptographic materials was especially valuable was that it included information about indicator systems used by the IJA.

12.2 Stripping by Differences

In this section we assume that the indicators of the additive system under attack can be read, or that at least some intercepts can be placed *in depth*. As before, this means the assembly of intercepts on forms as below:

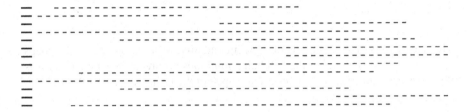

Here the longer and heavier dashes represent details of the message (date, time, source and the like). Each of the other dashes represents a GAT, with GATs constructed with the same additive in vertical alignment.

The task is the recovery of the corresponding portion of the table of additives.

Although it is assumed throughout that a 4-digit additive system is in use, analogous comments apply to 5-digit systems.

The first step is to examine the differences of a large sample of GATs in depth. For this process depth of two is quite sufficient. If the book groups used were $\mathbf{x}$ and $\mathbf{y}$ with the additive being $\mathbf{a}$, the GATs are $\mathbf{x} + \mathbf{a}$ and $\mathbf{y} + \mathbf{a}$. Then the algebraic identity:

$$\big((\mathbf{x} + \mathbf{a}) - (\mathbf{y} + \mathbf{a})\big) = (\mathbf{x} - \mathbf{y})$$

remains valid with false addition of groups. Thus *the difference of two GATs in any column is the difference of the underlying book groups*. As some book groups are going to be used much more frequently than others, some differences are going to occur more frequently than others.

Part F of the *CBTR* remarks that much cryptological material was captured in 1945. This is scarcely surprising: an army in retreat is likely to leave code books and associated material in places where they can be captured or found.

No systematic record of captures of either IJN or IJA material appears to have survived. Likewise there is no record of cribs, that is signals sent both in a known cipher system and a new system.

We assume, that from a reasonably large sample of intercepts placed in depth, about 200 of the most commonly occurring differences have been tabulated in order of decreasing frequency of occurrence. We use the fact that if $\mathbf{u}$, $\mathbf{v}$ and $\mathbf{w}$ are (unknown) commonly occurring book groups and if $\mathbf{p} = \mathbf{u} - \mathbf{v}$, $\mathbf{q} = \mathbf{v} - \mathbf{w}$ and $\mathbf{r} = \mathbf{w} - \mathbf{u}$, then $\mathbf{p}$, $\mathbf{q}$ and $\mathbf{r}$ are commonly occurring differences and $\mathbf{p} + \mathbf{q} + \mathbf{r} = 0000$. The cryptanalyst should then seek from the above list the *most* common triples $\mathbf{p}$, $\mathbf{q}$, $\mathbf{r}$ of differences satisfying $\mathbf{p} + \mathbf{q} + \mathbf{r} = 0000$. For example, 5225, 0980 and 5905 may all be in the list of 200 observed common differences and so the cryptanalyst could take $\mathbf{p} = 5225$, $\mathbf{q} = 0980$ and $\mathbf{r} = 5905$. The next task is to recover $\mathbf{u}$, $\mathbf{v}$ and $\mathbf{w}$ from $\mathbf{p}$, $\mathbf{q}$ and $\mathbf{r}$. One possible solution, probably non-primary, is:

$$\mathbf{u} = 0000, \qquad \mathbf{v} = -\mathbf{p} = -5225 = 5885, \qquad \mathbf{w} = \mathbf{r} = 5905$$

The negatives of the three differences 5225, 0980 and 5905 are 5885, 0120 and 5105 and would appear to yield another three groups. In fact the freedom to add any fixed group to all three and that of being able to choose which is $\mathbf{u}$, which is $\mathbf{v}$ and which is $\mathbf{w}$ may be used to show that the apparently different solution is the same. The differences are now as set out in the table:

		u	**v**	**w**
$\mathbf{u} = 0000$	**u**		5225	5105
$\mathbf{v} = 5885$	**v**	5885		0980
$\mathbf{w} = 5905$	**w**	5905	0120	

This would be written as a table of differences in the sense of Sect. 10.2.

$0120 = 5905 - 5885 \quad 0980 = 5885 - 5905 \quad 5105 = 0000 - 5885 \quad 5225 = 0000 - 5885$
$5885 = 5885 - 0000 \quad 5905 = 5905 - 0000$

Three[3] non-primary book groups cannot be enough. The next step is to seek further triples of common differences with sum 0000 and such that one of the three is 5225, 0980 or 5905. For example the new triple could be 0980, 7418, 3712. Then the fourth common non-primary group $\mathbf{x} = 8597$ is obtained by solving:

$$0980 = 5885 - 5905, \quad 7418 = \mathbf{w} - \mathbf{x} = 5905 - \mathbf{x}, \quad 3712 = \mathbf{x} - \mathbf{v} = \mathbf{x} - 5885.$$

Checking is needed around here. If indeed $\mathbf{u} = 0000$, $\mathbf{v} = 5885$, $\mathbf{w} = 5905$ and $\mathbf{x} = 8597$ are common non-primary groups, then $\mathbf{u} - \mathbf{x} = 2513$ is likely to be a common difference. If 2513 is not in the list of observed common differences, the cryptanalyst should be quite doubtful about the progress achieved. It was always possible that one of the three initial common differences was among the 200 by

---

[3]It is quite possible that the triple of common differences with sum the 0000 group does not come from three common book groups as hypothesised. The cryptologist who stumbled into such a situation would eventually recognise it, go back and start again.

chance. If this has happened it will become obvious later that the method is not
working and that it is necessary to go back to an earlier stage. The situation is
much more precarious than that prevalent in working on an additive system that
uses only multiples of three as book groups. The calculations are then more or less
self-checking.

Thus one obtains a 4 × 4 table of non-primary common book groups and
differences looking something like that immediately below.

	u	v	w	x
	u		5225	5105	2513
u = 0000
v = 5885	**v**	5885		0980	7398
w = 5905	**w**	5905	0120		7418
x = 8597	**x**	8597	3712	3692

This would be written as a table of differences in the sense of Sect. 10.2.

0120 = 5905 − 5885 0980 = 5885 − 5905 2513 = 0000 − 8597 3692 = 8597 − 5885
3712 = 8597 − 5905 5105 = 0000 − 5885 5225 = 0000 − 5885 5885 = 5885 − 0000
5905 = 5905 − 0000 7398 = 5885 − 8597 7418 = 5905 − 8597 8597 = 8597 − 0000

This method may be continued to produce a fifth non-primary book group **y**
and a 5 × 5 table as before. One would eventually get to a 10 × 10 table and
then start wondering when to stop. The aim is to be able to strip given depths to
the corresponding non-primary groups as set out in the two examples below. As in
Sect. 10.2, the markers A and B indicated a pair taken from the GATs in depth with
a common difference. The markers C and D indicate the common (non-primary)
book groups that give rise to that difference.

6062 A	*5885* B		4307	2999
3194	2917		*2468* A	*0000* B
1646	0469		7243 B	5885 D
9774 B	8597 D		3780	1322
9882	→8605		7177	→5719
6395	5118		4680	2222
5362	4185		3657	1299
4386	3109		2671	0213
0393	9116		*0955*	*8597*

The left-hand example uses only one common difference which could have been
produced by chance. Much time could be wasted working out book groups on the
(perhaps false) assumption that the correct decryption is shown. And allowing time
to be wasted conceded a minor victory to the other side. The right-hand example
has three common differences and so is much more likely to be correct.

Part G of the *CBTR* calls this method 'stripping by differences' and comments[4] that despite its theoretical advantages it needs a great deal of work.

If a partial knowledge of the code book was available, it would sometimes provide good grounds for guessing the book group from which a particular GAT had been obtained and thus produce a candidate for an entry in the additive table.

12.3   Primary Book Groups

As noted in Sect. 8.16, stripping by differences will determine at best non-primary book groups. These are obtained from the *primary* book groups by adding the same fixed group to each. Likewise the additives found will be only non-primary (relative) additives. This does not impede the process of recovering the values (meanings) of these book groups.

However in practice most Japanese code books did have patterns that eventually revealed the primary book groups. This is discussed by Professor Room in parts F and G of the *CBTR*. He went so far as to say that such patterns were a major factor in the total breakdown in security in Japanese Army communications in 1945.

An interesting example is the Water Transport Code, discriminant 2468. One frequently used book group stood for 'MESSAGE STARTS' and was initially taken to be 0000. This enabled various other commonly occurring words and phrases to be given relative book groups. The *CBTR* paragraphs marked K and L in Sect. 14.2 give more information on this. The primary book group of 'MESSAGE BEGINS' turned out to be 6666, and a high proportion of the commonly occurring words and phrases had primary book groups that were multiples of 11.

12.4   A Different Situation

At one stage Central Bureau had to handle a 3-digit additive system used for air-ground communication. The code book had been captured. However the additive table, which consisted of 1,000 3-digit groups, was replaced daily. In such circumstances knowledge of the code book would sometimes lead to the cryptanalysts working out what a particular GAT had to strip to and so work out an additive.

---

[4]The relevant part of the *CBTR* is quoted as 'Paragraph J' in Sect. 14.2.

12.5 Baby Brute Force

A *Baby Brute Force* situation[5] arises in an attack upon an additive system where
the indicators are understood to the extent that the number of the page used for
enciphering messages can be worked out but not the co-ordinates of the starting
place. Ideally the method discussed here should not need to be taken too far: instead,
evidence can be accumulated about the row and column co-ordinates of enough
starting places and then the encipherment used by the other side should become
clearer. Once this has happened the method discussed in the previous section takes
over.

It should be borne in mind that, with most Japanese Army additive systems,
when the transmitting operator reached the end of a page of a book of additives
he continued from the start of that same page. The different pages were in effect
different additive cipher systems.

The first step was to divide the intercepted messages into heaps according to the
page of the additive used.

To get started, at least some intercepts needed to be put into depth (alignment).
This can be done by searching for *hits*, that is the same GATs occurring in different
messages. Hits are likely, but not certain, to have arisen from the same book group in
two different messages being enciphered by the same additive. They are thus some
evidence of depth. They are much easier to find if the messages have been divided
up by pages of the additive table already.

Double hits, that is two hits involving the same two messages with the same
number of GATs between them, were more convincing as evidence of alignment
than single hits. Hits and double hits are more likely to be found if the other side has
made excessive use of a relatively small number of pages of the additive table.

Special machines were introduced[6] to search for double hits.

Intercepts apparently in alignment may then be used to put together a list of
differences between corresponding GATs and so to produce a draft list of common
differences.

---

[5]In the very special context relevant to the JN-25 cipher systems, the method of Hall weights can
be used here. This is discussed in Sect. 15.3.

[6]See for example NARA RG457, A1 9032, Box 705, files 1741 and 1743 on *Tape scanner for
double repeats, Japanese JN-25 traffic, 1943–1945*. The *Copperhead I* device was introduced by
Op-20-G to search for double hits. It was made in the NCR factory at Dayton, Ohio, but will not
be discussed further in this book.

Sections 15.3 and 15.4 of Chap. 15 deals with hits for JN-25.

12.6   Example: JN-25A1 in 1939

In August 1939 John Tiltman and three others of the GCCS (Sect. 9.4, and Note 5 of Chap. 9), discovered that the table of additives of the system that would retrospectively be called JN-25A1 appeared to have 300 pages numbered 002, 005, 008, ..., 890, 893, 896, 899. (These numbers were decoded by adding 1 and then dividing by 3.) The indicator was a five-digit group with the first three denoting the page number. A table available to both transmitting and receiving operators provided an indicator encryption group for each day: this was added to the indicator for the message by the sender and subtracted by the receiver. A method by which the four GCCS officers probably worked this out has been published[7] by Peter Donovan in *Cryptologia*. It was reasonable to guess that the fourth and fifth digits of the indicator represented the page co-ordinates of the starting place chosen in the selected page.

It would have been natural to write out the available intercepted messages— about 40 per day for about 10 weeks—on separate pieces of paper. These would have been sorted by page number into 300 batches. As was almost inevitable in a newly-introduced system, the operators would have not been properly trained and would have used some pages, such as those in the middle of the table, much more than others. The team would have then taken the larger batches one by one and written out the GATs observed—perhaps 1,000 in a batch of 30—on a suitably large piece of paper to find the hits (repetitions). Some of these would have been causal, some accidental. However it would have be observed very quickly that when two intercepts were placed in apparent depth using such a hit, the displacement between the starting places of the two messages would usually be a multiple of 10 groups. It would then become apparent that this was a general phenomenon with JN-25A1 messages in depth and that hits for which this failed should be rejected as accidental.

It thus became clear that there were exactly 10 permissible starting places on each page, with fairly large numbers of intercepts coming from the common pages arranged in depth. The GCCS team would then have suspected that the fourth digit of the indicator specified the starting place and then checked that this seemed to work. There was every reason to suspect that there was a daily additive encipherment of the fourth digit of the indicator and this turned out to hold. The fifth digit was just a null, that is randomly selected padding.

12.7   Scrambled Row and Column Co-ordinates

The Water Transport Code and certain other Japanese Army systems used 10×10 pages of additives with the rows and columns not numbered in natural order. The transmitting and receiving operators would have code books with pages looking something like the following:

---

[7]The paper may be found in *Cryptologia* 30(3), July 2006, 212–235.

Book 1, page 23 of WTC = 2468 table of additives

	2	8	9	6	4	5	1	0	3	7
9	4884	2843	4217	7745	4322	4746	9152	7870	8724	4612
8	6859	0840	9240	3899	6300	4929	8070	3518	7237	9417
0	8621	6442	3729	8991	0965	1456	3991	8746	0288	3360
2	3813	9862	9677	9341	7502	3313	8108	7319	4758	4163
6	3839	3165	1507	0845	9639	5445	5001	1202	4021	5531
3	8256	2727	3504	8148	0067	6872	0832	6387	8363	0221
4	1907	5100	9671	7081	8582	9571	8480	0690	2411	0503
5	2931	5542	1699	3517	4816	7249	8091	1505	4469	2625
1	8205	4831	5602	5346	4786	4528	9948	7633	4686	2177
7	7347	8954	6055	9835	8859	6783	7362	1560	2411	9387

Even when the indicators were decrypted there would still be some considerable work in recovering the row and column co-ordinates. The Allied cryptanalysts would need to work out the co-ordinates for each of hundreds of pages by Baby Brute Force methods. This scrambling of co-ordinates was essentially an encoding of the co-ordinates intended to hold back the attack of the other side. And as there would be minimal extra strain thrown on the operators by scrambling, it was a quite reasonable thing to do.

12.8   Recovering Page Co-ordinates

Let us see how the cryptanalyst would recover information[8] about the page co-ordinates on the sample 'page 123' given earlier. Thus, in the situation of March 1943 in breaking into code 2468, the messages as transmitted would be written out in the form:

123 1 5   5772 1566 4901 5328 6060 6512 0900 8240 2431 0421 5933 5939 9235

This would indicate that the GATs 5772, 1566, etc, used additive that started on page 123, row 1, column 5. The encrypted message would usually be longer than the 12 GATs shown in the example. However the page co-ordinates would be erratic and so row 1 could be any one of the ten rows of the table. The first step would be to sort the messages by page number. They could then be sorted by rows. The messages with the same page and row would then be considered together. They might start:

123 1 5   5772 1566 4901 5328 6060 6512 0900 8240 2431 0421 5933 5939 9235
123 1 3   9880 5767 2348 8486 7726 7776 6467 0936 9470 6329 1746 7495 1463
123 1 6   1447 2498 0708 2480 9605 0401 4489 8805 7672 3488 4867 7267 5933

---

[8] A good reference for this is in the *CBTR* paragraph marked 'J' in Sect. 14.2.

One now searches for hits, such as 5933 in the example. These most likely arise from the same commonly-used code group being added to the same additive. The probable alignment of the first and third messages is:

123 1 5   5772 1566 4901 5328 6060 6512 0900 8240 2431 0421 5933
123 1 6   1447 2498 0708 2480 9605 0401 4489 8805 7672 3488 4867 7267 5933

At this stage one may provisionally assume that column 5 is two columns to the right of column 6.

Some use may be made of common differences here. The code book may be known to have two frequently occurring book groups with difference 1524 or, if viewed in the other order, with difference 9586. Examination finds a GAT 7726 in row 3 above and a GAT 8240 in row 5. These have (false) difference 9586 and so are likely to come from an alignment:

123 1 3   9880 5767 2348 8486 7726 7776 6467 0936 9470 6329
123 1 5   5772 1566 4901 5328 6060 6512 0900 8240 2431 0421 5933 5939 9235

This time we may provisionally assume that column 3 is three columns to the right of column 5.

This process has to be continued, no doubt with occasional irritation being caused by the need to discard provisional assumptions. A similar process may be used to recover the order of the rows. Thus if there are messages starting at positions **123 5 6** and **123 1 8** such that the tenth GAT of the former is the same as the second GAT of the latter one may assume (provisionally, as always) that row 1 immediately follows row 5 and that column 8 is two columns ahead of column 6.

Intelligent use of hits and common differences may thus enable the page co-ordinates to be inferred. This is done on a page-by-page basis. If the transmitting clerks make excessive use of certain pages, those will tend to have co-ordinates established earlier. It is evident that the best strategy for the transmitters is to use the various pages—and there might be 300 of these—with approximately equal frequency. As the page usage should also be unpredictable, it is best to choose the starting point in the additive table by lot each time. (Admittedly this will not produce perfect equal usage, but it will make life hard enough for the other side.) It is evident that the basic weakness in these cipher systems is that certain code words are used much more frequently than others.

As already noted, erratic numbering of rows and columns does in fact hold back the other side for a while and costs the transmitters next to nothing. Wasting the time of the enemy cryptanalysts enhances the overall security of any cipher system.

A careful consideration of the scrambled co-ordinates device reveals that the enemy cryptanalysts can determine only a non-primary order of the rows on any page. This matter will not be examined further here.

12.9 Brute Force

The ultimate task in stripping additive ciphers would be that when the indicator system is totally incomprehensible. One is then reduced to trying to put intercepted messages into *depth* (alignment) by any available means. This process was called *Brute Force*. The first thing to look for is hits or (much better) double hits.

The extract from the *CBTR* given in Sect. 8.18 suggests that the Brute Force method is likely to be extremely slow. The amount of searching needed to carry out the Baby Brute Force process in this greater generality must have been very tiresome in the pre-electronic era. And there was much greater potential for effort to be wasted investigating false hypotheses.

The Japanese practice of replacing old systems by new ones every few months made slow and tedious decryption methods very unattractive.

Appendix 1   Recovering the Primary

The motivation here is to explain the extract from the *CBTR* quoted in Sect. 8.18:

> It seems that at least one of four conditions is needed to solve one of these codes:
> . . .
> (3) The code book must be heavily patterned and these patterns are clear from the differences of groups in the same column when the messages have been set in depth [that is, are clear from a set of non-primary book groups].

In general the systematic allocation of book groups in a code book has a strong tendency to produce insecurity.

Initially we work with a contrived example of recovering the primary book groups slightly different from the 1940 work of Driscoll and Currier on JN-25A. (See Sect. 5.5.) It is assumed that decryption by differences of a 5-digit additive cipher has been carried out to some extent and that one common book group has been taken to be 00000. The following non-primary book groups have been 'evaluated', that is the meaning established.

1	60630	4	60936	10	61530
27	62246	50	65548	256	85154

Unevaluated common non-primary book groups are assumed to be:

40697	43425	52736	92477	77018	24423	60461	95132	83413	56951
53722	79043	18641	67858	69202	38293	95183	10030	71895	38920

No information is needed about the meaning ('value') of 00000 or any of these 20 other non-primary book groups.

There seems to be a pattern in the non-primary groups for 1, 4, 10, 27, 50 and 256. Indeed if 60500 is (false) subtracted from each of these groups a new set of non-primary book groups is obtained:

1	00130	4	00436	10	01030
27	02746	50	05048	256	25654

The new batch of non-primary groups looks much more promising. However the fourth and fifth digits for the numerals are still messy. The best chance of fixing this up is to try successively (false) subtracting 00000, 00001, 00002, . . ., 00099 from these six groups and examining the 100 line table that begins: One line of this table

(subtr.)	1	4	10	27	50	256
60500	00130	00436	01030	02746	05048	25654
60501	00139	00435	01039	02745	05047	25653
60502	00138	00434	01038	02744	05046	25652
60503	00137	00433	01037	02743	05045	25651
60504	00136	00432	01036	02742	05044	25650

is: It is quite possible to jump to this line after writing out only three or four other

(subtr.)	1	4	10	27	50	256
60528	00102	00408	01002	02718	05010	25626

lines in the table.

In this line the fourth and fifth digits of each group form a pattern: in each case they correspond to the (ordinary, not false) double of the sum of the first three digits. And so we may try (false) subtracting 60538 from each of the original non-primary book groups. This yields: and

1	00102	4	00408	10	01002
27	02718	50	05010	256	25626

80169 83997 92208 32949 17580 64995 00933 35604 23985 96423
93294 19515 58113 07320 09774 78765 35655 50502 11367 78492

The recovery of the code book would now advance rapidly. The book groups for lots of numerals would now become obvious. Other sequences, such as January, February, March, . . ., might well also have book groups allocated systematically and so be easily worked out.

This method makes no use of the values of any common non-primary book group other than the six displayed above.

The experience[9] of Driscoll and Currier was somewhat different. They used GATs in depth to calculate differences of JN-25A book groups and obtained non-primary book groups as before. It was then noticed that 13343, 13445, 13547, 13649 and certain other common non-primary book groups are in almost arithmetic progression. (This time the non-primary book groups are those that actually occurred.) In particular, the last digit of each is odd. Something like this had been observed before in an old additive cipher. So it seemed reasonable to look at the new system of (possibly) non-primary book groups obtained by subtracting 13241 from each. (No doubt several other possible new systems of non-primary groups had been tried first.) This yielded 00102, 00204, 00306, and 00408 as common (possibly) non-primary book groups. It would now be natural to investigate whether these were in fact *primary* book groups for the numerals 1, 2, 3 and 10. It would then be noticed that 05010, 10002 also occurred frequently in the new system of non-primary book groups and work out that these most likely corresponded to the numerals 50 and 100. Thus Driscoll and Currier could work out that the primary book groups were likely to all be obtainable from the set of non-primary book groups by subtracting 13241.

Driscoll and Currier would then have examined a list of common primary book groups something like the list of 20 in the contrived example. They noticed fairly quickly that all were multiples of three. There was another pattern that confirmed that the new set of probably primary book groups were in fact primary.

Appendix 2   Another Cipher

This book has put much emphasis on the principal additive cipher systems used by the IJN and IJA. There were numerous other systems. The following Op-20-G document of 24 January 1945 entitled *Equipment for Decryption of Japanese Strip Cipher* is included as a start to restoring the balance. It is to be found in the excellent *Dayton code breakers* web site and is further evidence of the creativity of the USN Computing Machine Laboratory and its technical director, Joseph Desch.

1. The Japanese strip system recently introduced makes use of 48 Kana strips. Thirty of these strips are selected from a stock of 100. Certain of these strips are omitted while others are turned over. One interesting variant is the use of a transparent plastic strip board which permits hourly change from obverse to reverse side of the board. Hand methods of decipherment of this system are extremely laborious.
2. A project was initiated on 16 December 1944 to build a machine which would accomplish the complicated process of decipherment directly from a keyboard to a page copy. This equipment involves over 1,000 relays, 50,000 soldered connections, stepping switches, controls, etc. Less than 30 days elapsed between the original inception of the project and the completion of the equipment which is now in operation in our Pacific

---

[9]This is described by Stephen Budiansky on page 216 of *Battle of Wits*. The NARA College Park reference is *History of GYP-1*, RG38, CNSG, Box 116, 5750/202. Comparison may be made with *CBTR* paragraphs L and M quoted in Sect. 14.2.

section. Original rough designs were prepared by our research group. The equipment was designed, fabricated, wired and assembled by naval personnel attached to the U. S. Naval Computing Machine Laboratory, Dayton, Ohio.

3. The personnel most intimately connected with the project put forward tremendous efforts in making the equipment quickly available. It is suggested that enclosure (A) be forwarded as a commendation to the personnel most intimately concerned. Enclosure (B) will be forwarded to the Bureau of Ships for inclusion in the service records of these personnel.

Chapter 13
Making Additive Systems Secure

This chapter is to some extent a continuation of Sect. 8.19, which gave some general maxims about cipher security in the WW2 era, but its focus is now on additive systems. After some general remarks on secure use of such systems, subsequent sections investigate techniques for improving the security of their various components.

13.1 Generalities on Additive Cipher Systems

The author of part F of the *CBTR*, probably Professor Room, stated that the IJA 'narrowly missed making its traffic completely unreadable'. Admiral Luigi Donini[1] described his experience in communications security in WW2 and confirmed that from July 1941 the Italian Navy was able to make its additive systems secure.

An important question is whether the use of additive systems on the scale that occurred in the Pacific War was prudent or efficient anyway.[2] The Japanese language

---

[1]The switch to additive ciphers happened after the Battle of Cape Matapan (March 1941). From the Italian viewpoint it was successful. See Admiral Donini's paper *The Cryptographic Services of the Royal (British) and Italian Navies* in Cryptologia 14(2), April 1990, 97–127.

Additive systems were used in Australia during WW2 but were being phased out towards the end. NAA Canberra file A11093 334/81L deals with the RAAF Meteorological Committee and includes evidence of a general disillusionment with additive systems around March 1944.

[2]The book *Big Machines: Cipher Machines of World War II* by Stephen Kelley (Aegean Park Press) is of general interest as to what could have happened if the Japanese had introduced something like the Sigaba as a basic secure communications device. It gives useful information on the Purple diplomatic code machine. The interview of Solomon Kullback preserved by the National Cryptological Museum (adjacent to the American National Security Agency) mentions the capture of a Japanese enciphering machine but confirms that such machines were not made fully operational.

© Springer International Publishing Switzerland 2014
P. Donovan, J. Mack, *Code Breaking in the Pacific*,
DOI 10.1007/978-3-319-08278-3_13

was itself a problem: most Japanese radio operators would not be happy using a western-style alphabet.

The battle between the Japanese cryptographers and the Allied cryptanalysts was definitely not a walk-over. This book shows that a modest number of major blunders made a great difference. A fair conclusion is that the IJA and IJN should have had a small talented team researching the advantages and disadvantages of various encipherment methods throughout the 1930s. In fact, the Japanese made some experimentation with encryption machines[3] late in WW2, but these were never made fully operational.

A professional team, such as that envisaged above, would have been able to calculate approximate upper limits to the number of times a code or an enciphering system could be used with impunity. As a change of additive table would be easier to achieve than a change of code book, such a team should have been asked for information as to how many changes of additives were safe and how frequently such changes should be made. As this would depend upon the length of these tables, some recommendations should have been made about that too.

Admiral Mountbatten wrote a foreword to the book *Very Special Intelligence* (Hamish Hamilton, London, 1977) by Patrick Beesly and recalls there how he pressed before 1939 for the Royal Navy to switch to enciphering machines. The British Army and the RAF had switched to machine codes by 1939. Mountbatten comments that 'one of the lessons of Beesly's book is the astonishing overconfidence displayed by the British and Germans (and in fact by all nations) in the security of their ciphering arrangements'. This comment may be somewhat unfair: Ralph Erskine has shown in a paper that there was a shortage of suitable machines at the time. No detailed comments on Japanese ciphers appear in the book.

One argument against the use of code books and superencipherment is that if and when they are captured the enemy is able to read old messages and so obtain very useful background knowledge. This happened on various occasions, and in particular with JN-25.

The Canberra NAA file A1196 48/501/92 has figures to the effect that by 1945 the RAAF had switched from use of additives or OTPs to the Typex enciphering machine for most, but not all, of its secret messages. The file A705 201/23/453 includes a lecture by Wing Commander J. P. Lees on *Cryptographic Security*. This recommends the use of code books with a one-time pad as being the (only) method of obtaining absolute security. Good machine ciphers would yield only long-term security while additive systems would in practice give security for at least a month. This last would be quite ample for certain purposes. Lees was speaking in April 1945 at the Second RAAF Cipher Conference and so may be taken as expressing experience of over three years of very active use of ciphers.

[3]This is in Kullback's interview in the National Cryptologic Museum. By that time the US Army had enough resources to set up a team waiting to take on IJA machine codes if and when they were activated. In fact they were not activated. The ill-fated IJN submarine I-52, sunk in the North Atlantic on 23 June 1944, was intended to carry some Enigma machines back to Japan. For this, see Carl Boyd's paper in the *Journal of Military History*, April 1999.

Kullback's general conclusion about IJA communications in the Pacific is that they collapsed under their own weight.

Before the war a decimal Enigma that encrypted digital information was available commercially. When combined with an appropriate code book and with a new set of rotors it would have provided a nightmare task to the cryptanalysts.

Such a team would have made a statistical analysis of military and naval messages, determining the relative frequency of common words and phrases. In fact, the American Army cryptanalysis unit made such a study of a large sample of American traffic before the war and subsequently used decrypts of Water Transport Code messages to make such a study of Japanese traffic.

Above all, a competent Japanese communications security unit would have worked out enough of the weaknesses in the JN-25 series of ciphers in advance and stopped the use of only multiples of three as book groups.

13.2 Disguising, Enciphering and Encoding the Indicator

The operator of an additive cipher system is required to choose a starting point in the additive table randomly from among all possible groups or perhaps only from a list of preferred groups, and then to inform the intended recipient of this choice in some secure form. The indicator would be transmitted in some hopefully secure digital form in some fixed position in the message. Thus in the Water Transport Code (see Chap. 14) used by the IJA the second and third message GATs were indicator groups. This is in itself a quite sensible idea. Quite possibly it would hold back the enemy assaults on the system. Alternatively it may be convenient to transmit the indicator twice—say in the first and last GATs—and have other features defend its security.

Dummy indicators could be and sometimes were used. These could take the form of the same randomly selected group being placed at both the end and the beginning of the encrypted message. The ANDUSYN Allied meteorological code had such a feature at some stages in its career.

The use of scrambled row and column co-ordinates was mentioned in the previous chapter. A further refinement is available here. Suppose that there are 500 pages in the additive table. It would be possible to allocate non-systematically two different page numbers to each page, so that 'Page 123' might also be 'Page 674'. The operator who chooses the page would have to choose (by lottery) one of the two numbers to use. This device should avoid any 'missing digit' phenomenon as happened in JN-25A1.

The 1942 British interservice code was an additive system. There was an interesting method of indicating the starting point in the additive table. A table[4]

---

[4]The RAAF used the British Typex machine which used groups of five letters as indicators. NAA Canberra file A705 201/28/325 is available in digital form on the Recordsearch system and includes a table of indicators. 00001 denoted ZLLPC, 00002 denoted GFIHR, etc. Sending the code number rather than the indicator itself must have made life harder for the other side.

From 1941 to 1945 Georg Gluender was a communications operator in the German Army. On page 150 of Hugh Skillen's *Enigma and its Achilles Heel* Gluender records the indicator method used for the encrypted teleprinter from around January 1943. A list of serially numbered 12-letter indicators was written out on paper and a copy taken by courier to the other teleprinter

of authorised starting points was printed at the beginning of the book of additives.
A typical line of this would look something like:

1221 123-2-6   1222 013-2-7   1223 066-7-5   1224 271-8-8   1225 186-1-3

Thus 'starting point 1222' would mean 'page 013, row 2, column 7'. The additive
book also had a table of a special additive chosen for each day in its planned career.
The first two lines would look something like:

| 01Jan 3183 | 02Jan 0988 | 03Jan 6183 | 04Jan 7906 | 05Jan 6715 |
| 06Jan 3796 | 07Jan 7526 | 08Jan 7450 | 09Jan 2872 | 10Jan 4068 |

A sender using the starting point page 013, row 2, column 7 on 6 January would
work out the false sum $1222 + 3796 = 4918$ and then send the group 4918 as
indicator. Better still, the reversed fifth message GAT could have been (false) added
to this sum. The recipient would undo this process. The combination of encoding
and encrypting would provide the enemy cryptanalysts with a lot of work and so
delay if not defeat them.

Furthermore the sender was required to mark the number of the starting point—
here 1222—with a pencil and never use it again. Just one operator always making
the same choice would seriously damage the security of the system. Once again,
this had the effect of allowing a supervisor to check that the choice of starting
point in the additive table was reasonably random knowing that this would make
it harder for the other side to obtain useful depth of intercepted messages. It would
not be appropriate for all operators to use starting position 0001 first, then go on to
0002, etc.

Cryptologists sought to use the indicator(s) to obtain depth (alignment). Any
system that allowed other methods to reconstruct alignment would be insecure. See
Chap. 15.

13.3   The Choice of Starting Point

The transmitting staff were supposed to make certain random choices, notably of
a starting point in a table of additives. They[5] appear not to have been instructed
how to do this. One obvious way of achieving randomness in practice is the lottery.
Ten small wooden balls each engraved with a single digit and such that each of

---

station with which contact would be made. The message would refer to the indicators by number.
Usage was strictly one-time: once a numbered set of 12 letters had been used, it would be crossed
off each copy of the list.

[5]Brigadier Tiltman in his *Reminiscences* (NARA RG457, box 1417, item 4632) expressed the view
that the operators should be relieved of any responsibility for security. Presumably this includes
making arbitrary choices.

0, 1, 2, 3, 4, 5, 6, 7, 8 and 9 occurs can be shaken in a drinking vessel and one drawn out. This can be repeated as required. If, for example, a clerk has to choose a page out of 300 in a table of additives, the first draw would be made using only the balls numbered 0, 1 and 2. If the 0 were drawn then one of the first 100 pages would be used. If the 1 were drawn, then one of the second hundred pages would be used. If the 2 were drawn, then the third hundred pages would be used. All ten balls could then be placed in the drinking vessel and rattled around before a second draw is made to determine the tens digit of the starting page. A third such draw would complete the choice of page and then two more would be needed to choose the row and the column. However, considerable reading of the archives has failed to find any reference to mini-lotteries.

An alternative to the use of lottery methods would have been to use dice. The conventional cubic dice are not suitable for generating randomly digits from 0 to 9 but the 10-faced pentagonal trapezohedron or the 20-faced regular icosahedron have been used as shapes for dice for such a purpose. Once again, there appears to be no mention of dice of any sort being used by Japanese radio operators.

There is a third method that requires only a table of additives and a pencil. The operator had both of these anyway and could have been ordered to choose random digits by shutting his eyes, opening the book of additives and bringing the blunt end of the pencil down on a page. Once again, no mention of this or any other such procedure to generate a few digits randomly has been found.

In fact, there was a tendency[6] among operators to use only a relatively small set of starting places in the additive table and so present much more depth to the other side. This must be what the *CBTR* means in the text quoted in Sect. 8.18 as *used very badly*. Even if use of a lottery procedure is not obligatory, it is possible for management to prevent excessively repetitive choosing. For example, it would be easy to require operators to mark the book of additives at the starting point each time it is used. The marked book could then be inspected regularly. Re-use of a marked starting place could be forbidden. Such a system was used in Australia in the later years of WW2. At one stage the IJA tried to stop a general preference for starting on the right-hand page of an opened book of additive by printing every alternate page upside down. This at least made every page a right-hand page.

---

[6]The NAA file A1196 40/510/344 entitled 'Secret Manual of Instructions for Cypher Officers' is relevant here. It must represent the experience gained in working on Japanese codes when it states that stereotyped messages must be avoided. It also says that there is a tendency among operators to choose the top left of the page in the table of additives as a starting place and this must be most vigorously discouraged. One may see further comment about the weakness caused by these two practices in Part D of the *CBTR*.

The previously cited 1942 Barham report states that the initial break into the additive system JN-25A was greatly facilitated[7] by overuse of a few pages of the table of additives. This helped set going the processes that led to the Battle of Midway.

Suppose that the IJN practice of taking the additive immediately following the last additive on a page to be the first on the next page is in use. Then the practice of *tailing* was sometimes used. This was taking the first additive of a new message to be that immediately after the last additive of the previous one. Sections 9.5 and 9.22 showed that there is quite a lot wrong with tailing.

13.4   Alternative Book Groups

The previous chapter emphasised the use of common book groups and the associated common differences in breaking into an additive system. Furthermore, once intercepts have been stripped back to bare code, the common book groups will constitute a weakness open to exploitation. From the viewpoint of those trying to keep communications secure, such systems are much better off without excessively common book groups. This can be achieved by allocating two or more book groups to each commonly used word or phrase. The allocation should be random (see below). For example, 'hundred men' might be a common phrase and it could be allocated three book groups, say 7110, 8310 and 9773. The operator should be

---

Page 235 of Władysław Kozaczuk's book *Enigma* is part of an interview (July 1978) with Marian Rejewski, formerly of the Polish Cipher Bureau. It notes that in late 1932 German operators of Enigma machines tended to use AAA or BBB or the like for the keys to messages. This helped Rejewski in working out the wiring of the machines. This bad practice was stamped out in 1933, but the damage to the security of the German communications system had been done. (Hugh Skillen's book *Enigma and its Achilles Heel* reprints the 1937 Enigma instruction manual in which indicators such as AAA, BBB, etc are explicitly forbidden. Thus random methods in their selection were not in place.) Ultimately the GCCS work on Enigma was rendered possible by the gift of the fruits of the work of the Polish Cipher Bureau.

As mentioned in Note 32 of Chap. 8, letters may be randomly selected from the 26 by use of a pack of 52 playing cards. Two of the cards, say the two red aces, are marked with an 'A' on the face side. Two more, say the two red twos, are marked with a 'B' and so on. The random letter is then chosen by shuffling the pack and exposing one card.

The RAND Corporation in 1955 published a book of *A Million Random Digits* for statistical work. This has long been rendered obsolete by the random number generating capacity of modern statistical computer packages. The introduction advises the user to choose a starting point in the table by the 'third method' described in the main text: shutting one's eyes, opening the book at random and then stabbing the text randomly with the blunt end of a pencil. The 5-digit group thus selected would then be used to choose a starting point and marked so that it would not be used for this purpose again.

[7]Section 9.9 investigates the process of getting started in the cryptanalysis of what appears to be a new system expected to be using multiples of three for book groups. It justifies the use of the phrase 'greatly facilitated' here.

instructed to choose, by lottery methods each time, which of these three to use. Adoption of this simple strategy for the hundred most commonly used words and phrases would greatly increase the difficulties faced by the other side. Common differences and double hits would be much less common and much harder to exploit. Even after the additive cipher had been stripped, recovery of the code book would be more difficult.

13.5 Allocating Book Groups

Around 1935 Rowlett, Sinkov and Kullback constructed[8] a code book by tossing cards in the air, with one word on each card. This was done repeatedly to achieve some considerable shuffling of the cards. As it takes a bit of effort to shuffle even a deck of 52 cards, their method must have been considered the best available at the time. There appears to be very limited information on how book groups were assigned by the Japanese in that era. At least if IBM tabulating equipment and a lottery barrel were available there would be a reasonably efficient way of preparing a code book.

The lottery barrel used by the NSW State Lottery in that era contained 100,000 wooden balls of equal size, with diameter about 10 mm. These were numbered from 1 to 100,000 but could equally well have been numbered from 00000 to 99999. The barrel was large enough to allow the balls to move around and thus get mixed up when a handle was turned.

The standard IBM card was described in Sect. 1.15. Suppose that the the code words and phrases were typed on columns 1–60. The book groups could then be read out by someone drawing them out of the lottery barrel one by one and then dropping them into a bucket to avoid repetition. The book groups would be punched into columns 66–70.

The important point is that the cards could be mechanically sorted, either into lexicographical order of the letters marked in columns 1–60, or the groups marked on columns 66–70. Once sorted, the deck of cards could be used to control a line printer to produce the code book or its reverse. As with modern word processing, errors could be corrected by replacing a small part of the whole file.

Once one code book had been punched out on cards, gangpunching could be used to produce cards with the plain language meanings but with the book groups deleted and thus a replacement code book could be prepared at minimal extra human effort.

The process of preparing a 4-digit or a 3-digit code would be analogous. A lottery barrel for 10,000 balls and another for 1,000 balls would be needed. A further economy of effort would be available if another version of the code were needed:

---

[8]This incident is described in the oral history interview with Solomon Kullback conducted by the National Cryptologic Museum.

columns 1–60 of each card could be mechanically transferred to new cards and the lottery barrel used again.

There was very little IBM equipment in Japan before 1942 and not much more was captured in 1942. Furthermore, although the Japanese words could have been represented by the western alphabet using the Roma-ji system, this would have been somewhat awkward. The sorting process provided by IBM machinery could handle at most two holes in any column. Provision could be made for at most 66 letters being represented by punching two holes in a column with 12 sites. Yet the kana alphabet has more than 66 characters.

The fact remains that a lottery barrel could and should have been used to produce book groups even if the groups had to be manually written on cards as they were drawn. Any method of systematically allocating book groups to some or all of the words and phrases would be at least undesirable[9] and potentially quite pernicious.

13.6 Generating Tables of Additives

Overall, the generation of one-time pads and tables of additives appears to have been secure enough. However Michael Smith, on page 28 of *Station X*, mentions a case in the early 1920s when things went wrong: the method of producing the random digits was simply not good enough. A much more remarkable exception from 1944–1945 is described just before Appendix 1. One sound[10] way of doing it is to use the lottery barrel method already mentioned. Marbles would be drawn as before and dropped into a bucket after the number had been taken down. When the bucket was reasonably full its contents would be poured into the lottery barrel which would then be rotated several times to mix up the contents.

Some further information on the British production of OTPs is available. The GCCS had a *production section* that made cipher material and codes. Part of this was done at Mansfield College in Oxford, apparently using manual throwing of some form of counters. There was also a unit at Drayton Parslow (near Bletchley Park) that used tabulators. Its method appears to have survived only in the Canberra

---

[9]The relevant maxim in Sect. 8.19 is *Any easy method of complicating the task of the enemy while not throwing much extra work on one's own staff should be adopted*.

Although part F of the *CBTR* is hard to follow without access to various associated documents, it is evident that sophisticated patterning in various IJA code books from August 1944 onwards materially helped the Allied cryptanalysts.

[10]The account of the JN-25B7 system in NARA, RG38, states that no group occurred twice in any one of its 500 pages. The probability that this happened by chance with the groups being chosen independently and randomly is 1 in 56,539,923,500. To some extent this justifies speculation that the groups were chosen by taking marbles from a lottery barrel, typesetting that number onto a page but not replacing it in the barrel until at least that page was finished.

NAA, which lists several items, including A1196 12/501/133, under 'reciphering' in Recordsearch.

Nigel West's book on *Venona* mentions that the Soviet defector Vladimir Petrov told the Australian Security and Intelligence Organisation (ASIO) about some device used rather later that appears to have generated random numbers electrically, perhaps using output from a Geiger[11] counter.

In practice, production of OTPs was usually good enough to yield safe, albeit rather clumsy, encryption processes.

An interesting case of the recycling of tables of additives by interchanging the roles of the rows and the columns is mentioned in the *CBTR* (part G, page 18). Such a practice is dangerous.

13.7   Facilities for Destruction

An important practical issue is how to dispose of code and cipher material in a position about to be over-run by the other side. The RAAF material[12] on the Typex machine states that a heavy hammer should be kept near each machine to enable the rotors (wheels) to be damaged so much that the wiring could not be reconstructed. It also states that kerosene and matches should be kept near each such machine to facilitate rapid destruction of paper used for cryptanalytic work. There does not appear to have been any provision for rapid destruction available to the IJA operators at Sio, NG, in January 1944. (See Sect. 23.5.)

Section B-2 at Arlington Hall Station (AHS) dealt with IJA codes and was headed by the very significant pioneer Solomon Kullback. In his oral history interview,[13] Kullback stated very clearly that in the IJA those who lost cryptanalytical material or allowed it to be captured were subject to severe punishment. This encouraged those

---

[11]John Walker's paper *HotBits: Genuine Random Numbers Generated by Radioactive Decay*, which refers to the decay of Krypton-85, is of interest here. The process is available on-line: http://www.fourmilab.ch/hotbits. It produces on request fresh strings of hexadecimal (base 16 rather than base 10) digits.

[12]The key point is made in a document issued by the RAAF on secure communications (NAA Canberra A1196 48/501/171). 'Books which have been destroyed can be replaced, but those which have been captured may compromise thousands of copies throughout the world and may lead to large scale disasters before such compromise becomes known and action taken to overcome same.'

The design of the Typex machine and particularly of its rotors would have been of considerable interest to the Axis cryptographers. So RAAF file A705 109/3/1647 refers to keeping a thermite bomb around rather than a mere hammer to ensure total destruction. As already noted in Note 12 of Chap. 8, the German Army captured a Typex machine without the rotors in France in 1940 but decided that there was not enough information available to research it properly. Barry Fox pointed out on page 70 of the *New Scientist* of 24 December 2005 that the early commercial Enigma machine had been patented in Britain and elsewhere and so by 1930 full drawings were (and still are) available at the British Patent Office.

[13]This is also in the Oral History preserved in the National Cryptological Museum.

who had done so to concoct false accounts about what happened. Thus captured cryptographic material would continue being used for quite a while. This is at the heart of the find of so much material at Sio being so significant.

Inevitably the series of defeats suffered by the Japanese in the last 18 months of the war resulted in the Allies capturing code books, additive tables and other cryptographic material. Facilities were set up to enable this to be exploited promptly. The resultant mixture of Allied opportunism and cryptanalysis is covered minimally in this book.

13.8   Minor Field Codes

A sensible precaution was used in Melbourne around 1943–1944.[14] The Australian Army needed various simple field codes for short-term use. These needed to be changed fairly frequently. Lieutenant Colonel A. P. Treweek had been seconded to the Navy since 1941 and played no part in designing such codes. He was used to test them. The point is clear: any proposed cryptographic system, including any additive system, needed to be given adequate independent checking.

13.9   Code Changes

The description in Chap. 12 of the basic deciphering process shows that any knowledge of the code book, additive or indicator encryption system helps. Thus once it has been decided to use three, say, tables of additives with a given code book, it becomes apparent that a new table of additives should be introduced whenever a new code book comes in. This was not always done with JN-11[15] and JN-25.

---

[14]The extra function served by Treweek is recorded in NAA Melbourne file B5554.

This book scarcely mentions *field codes*, that is easily used systems designed to keep messages secure for several hours. The best-known of these was the use of a native American language totally unknown in Europe or Asia. See Sect. 22.3.

[15]The Jamieson report—Melbourne NAA B5554—recounts on page 59 of the digital version of how at one stage distribution difficulties with the cipher JN-11 resulted in various instances of code book and additive table not being changed simultaneously. In fact one problem with these systems is distribution: if things are going wrong difficulties must occur.

No explanation seems to have survived as to why the switch from the JN-25A code book to the JN-25B version was not accompanied by a simultaneous change of additive. Likewise it is not clear why the change in additive for JN-25 just before the raid on Pearl Harbor was not accompanied by a change in code book.

In Canberra NAA file A6923 37/401/425, page 182 of the digital version, is a letter (about 1 November 1942) from Henry Archer of GCCS to the Deputy Director of Military Intelligence giving reasons why the Diplomatic codes section of Nave's Special Intelligence Bureau should be continued even if the USN does not want to have it. When the minor diplomatic codes were being

There was a corresponding blunder with the rotors of the military Enigma machine in the 1930s. Additional rotors were introduced whereas total replacement of all previous rotors was called for. The Polish Cipher Bureau was able to recover[16] the new rotors on each occasion using knowledge of the old ones.

13.10 Repeated Encipherment

The Enigma machine was sometimes used for *Offizier*, that is officer-only, messages. The message would be encrypted once and then encrypted again. The operator who received it would decrypt once and then call for an officer to carry out the second decryption. If double encryption had been used throughout, the task of the GCCS would have been made much harder. There is an analogous way to make the super-enciphering process appreciably more secure for additive systems.

It is to construct the GATs by (false) adding, to the original book groups, groups taken from two different places in the additive table. This amounts to super-enciphering twice. Thus if the book groups in a message begin:

4096 8333 3564 8008 9409 2442 0849 0798 4190 2696

and if the additives from the table starting at page 123 row 7 column 3 are:

6028 7630 1866 5401 3673 8973 1056 3213 5186 3238

and the additives from the table starting at page 052 row 4 column 1 are:

2765 1326 3201 5349 7754 6555 4608 9233 8743 1519

the transmitted message would begin with the GATs:

2779 6289 7521 8748 9726 6860 5493 2134 7917 6333

One may use combinations of the methods already mentioned to find some way to transmit the indicator 1237305241 in some secure form. This would appear to give

---

replaced instructions for the new code would be sent out in the old code! Thus there would be much to be lost if the continuity of intercepting and processing these messages were broken.

On the night of 29–30 January 1943 two NZ corvettes, *Kiwi* and *Moa*, rammed the IJN submarine I-1 and forced it to beach at Kamimbo, Guadalcanal. It had been carrying new code books and additives in vast numbers. The resulting chaos did the Japanese cause no good.

[16]Stephen Harper's book *Capturing Enigma* describes how two members of the crew of HMS *Petard* entered the sinking *U-559* on 30 October 1942 and recovered some material used with the upgraded submarine Enigma introduced earlier that year. They were drowned when they went back inside for more. This material was more than welcome at Bletchley Park.

a method of transmitting messages with not much more work than from a single use of additive and making the task of the cryptanalysts on the other side very much harder.

Let us see why this has good prospects of working. The standard method of attacking these ciphers is to write messages *in depth*, that is with GATs produced by the same additive in columns. The double encipherment process produces for, say, 2,000 starting points in a table of (say) 10,000 random groups 1,999,000 possible methods of double encryption. Thus enormous numbers of messages would need to be received before any useful depth would be achieved.

Rather sadly, the only example of use of double encipherment known to the authors was a fiasco. This was the German diplomatic code known as Floradora. Solomon Kullback in an oral history interview describes how it had an additive table of 1,000 groups and so had in principle around 500,000 different starting configurations. Singularly bad use of the additive by one operator did ruinous damage to its security.

The Floradora code book was thus recovered by the team at AHS. It was later used for some messages being transmitted by double encipherment to some addressees and by one-time pads for others. It was possible to first recover the book groups of messages and then use these to recover portions of some one-time pads. Quite remarkably[17] it was possible to work out how the one-time pads were being produced and then infer the layout of some of the pads being used for messages without cribs. This made it possible for AHS to read some German diplomatic messages to and from Tokyo in 1945. This nicely complemented the Magic intercepts of Japanese diplomatic material.

Even if repeated encipherment using two different additives was considered too much of a strain on the operators it could have been used to encrypt indicators. A special table of 1,000 additives would have been needed. Two of these would have been added to the unencrypted indicator. Secondary indicators would have been needed to inform the recipient as to how to decipher the encrypted indicators.

Appendix 1   Codes Within Codes

It was customary at the time to use code words for operations and locations of current interest. Thus OVERLORD was the code word for the 1944 invasion of Normandy. The five invasion sites were GOLD, SWORD, JUNO, UTAH and OMAHA. The point here is that someone who breaks into the encrypted messages still has to work out just what GOLD refers to.

---

[17]See the work of Peter Filby in *Intelligence and National Security*, July 1995 and Cecil Phillips in *Cryptologia* 24(4), October 2000, 289–308. Filby has another version of his study on pages 37–40 of *The Enigma Symposium 1995* edited by Hugh Skillen. In Chapter 2 of *Station X*, Michael Smith mentions a case in the 1920s when GCCS worked out a method used to generate OTPs and exploited it.

A variant of this was used in 1941 by the IJN to confirm that the Pearl Harbor raid was to go ahead. It was the instruction 'Climb Mount Niitaka'. It was, however, a one-time usage and so secure.

A much more complicated example of this was used by the German submarines to indicate positions in the Atlantic Ocean. It would be possible to encode something like 30° 18′ North, 21° 43′ West by writing

$$\text{CJGRADAHMINNBAGRADDCMINW}$$

where 'Grad' is German for degree, 'MinN' stands for 'minutes north', 'A' denotes 1, 'B' denotes 2, etc, but this is not going to delay the other side for very long. Instead a rather complicated system was used for encrypting latitude and longitude. This was successful in delaying the Allied exploitation of breaking into submarine Enigma. Although some aspects of the system had been recovered earlier from intercepted messages, it was finally mastered only when a submarine,[18] the U-505, was captured intact on 4 June 1944. Indeed, such a capture had been the aim of the USN task force involved. The capture was kept secret until the European war was over and rendered void the extra protection given to the German submarines by this system. The Germans would have done better by changing it every 3 months.

The bulky series MP1049/7 of files in the Melbourne branch of the NAA contains an inward message from the Admiralty dated December 1941. The security of the merchant navy code was considered somewhat dubious. The ACNB was explicitly warned against using geographical[19] names or latitude and longitude in such messages. Elsewhere may be found documentation that the shipping routes across the Indian Ocean had been divided into 35 regions numbered from 1 to 35. These were to be referred to not by these numbers but by a 2-letter code which was

---

[18]For information on the U-505 incident, one may consult the USN website. Much more detail is given by Jeffrey Bray in volume V of *Ultra in the Atlantic*, published by Aegean Park Press. This volume has 75 pages and is entitled *The German Naval Grid and its Ciphers*. It is based on the almost contemporary account of the cryptanalytical task written down by some of the participants and now available in the American National Archives. More information is available in various places in David Syrett's *The Battle of the Atlantic and Signals Intelligence*. NAA item A11093 370/2G notes that by May 1942 the 'British lettered co-ordinate system' had been adopted in the South Western Pacific Area, which included Australia.

The work on German Naval codes was assisted by certain other captures. Patrick Beesly's *Very Special Intelligence* mentions three others, including the capture of the U-110.

W. J. Holmes describes on page 74 of his book *Double-Edged Secrets* (Naval Institute Press) how there was a systematic latitude-longitude code used for the JN-25 series of ciphers. A captured map showing a few examples of this gave the whole system away.

[19]David Syrett on page 50 of his book mentioned in the previous note quotes a standing order for German submarine communications banning the use of geographical names in radio communications.

As is noted elsewhere systematic allocation of code names for locations in JN-25B helped in the identification of Midway as the target for the operation planned for early June 1942. Rafferty (now known in the phrase *Rafferty's rules*) would have found a patternless allocation of code words for locations.

changed weekly. No pattern was evident in the letters allocated for the following eight weeks.

The IJA Water Transport Code did not have such a protection for the locations of shipping. The penalty imposed for this bad practice was heavy.

An interesting file at the Australian War Memorial gives a list of words suitable as code names for geographical locations. The list appears to have been taken from a dictionary and contains words like APPLE, ORION, and SHOVEL. Some of these have been crossed out by pencil. The standard practice was that whenever a few such words were needed, they would be selected randomly from the list and then marked to prevent reuse. Thus ORION, if still available, could be used to denote Sio in NG.

Likewise the name CARTWHEEL was well chosen for the general Allied SW Pacific strategy in 1943. By itself it could mean anything. Use of the names FORTITUDE NORTH and FORTITUDE SOUTH for the deception operations associated with the invasion of Normandy in 1944 was less appropriate: the 'North' and 'South' parts did correspond with reality.

In 1943, 'Archie' Cameron (Sect. 19.9) noted[20] that certain code words used by the IJA were not allocated in a patternless way and did provide considerable assistance to Allied intelligence staff. He pointed out that if the Japanese discovered that their codes were extensively compromised, the inevitable review would recommend that 'Rafferty' would do the allocation of code words thereafter. He would have had no difficulty guessing which of the Normandy invasion beaches GOLD, JUNO, OMAHA, SWORD and UTAH were allocated to the Americans.

The same sort of usage was detected in JN-25A and JN-25B. In the former, code groups for 'first day of month', 'second day of month', etc, were allocated systematically. With 33,334 book groups available it would have been easy to choose 366 of them randomly to represent the days '1 January', '2 January', etc, ending with '31 December'. In attacking JN-25B the code word for Midway could be almost inferred from code words used for other locations once these had been worked out.

In Sect. 9.17 another such poor practice was described. A somewhat complicated system was used to encode dates in JN-25B. Towards the end of May 1942 skilled analysis of past usages of this system, some of which corresponded to dates by then known by Rochefort's unit at Pearl Harbor, obtained full information on the system. A random allocation of book groups for a good range of future and recent past dates would have been much more secure.

Another related poor usage was the systematic assignment of numbers to convoys of Japanese merchant ships. The point is clear: without any guarantee on security of communications, no information should have been given away unnecessarily to the other side.

---

[20]Cameron's views of December 1943 are preserved in AWM54 225/2/3.

Appendix 2   Conversion Squares

The following is taken from page 2 of part G of the *CBTR*.[21]

> Originally the additives were added to the code text but early in 1943 the practice was adopted of using a *conversion square* consisting of 10 rows of 10 digits, with no repeats in a row. The cipher, given by plain text a and key b, was the digit within the square given by column a, row b.

Here, 'plain text a' means a single digit in a book group and 'key b' means the digit in the additive table group that lies directly below b when set out in the usual form. The example given below explains the new encryption system via transition from the usual additive encryption.

Suppose we wish to encrypt the message shown below by the additive printed immediately underneath using the new method:

message	CABLE	DAMAGED	OPERATION	CANCELLED
book group	5786	1221	5346	8426
keys	8363	9221	1907	5100
GATs	3049	0442	6243	3526

The necessary false addition of digit b (from the *key* or additive) to a (part of the book group) can be found by looking at the entry in the following table underneath the boldface entry a and to the right of the boldface b.

	0	**1**	**2**	**3**	**4**	**5**	**6**	**7**	**8**	**9**
0	0	1	2	3	4	5	6	7	8	9
1	1	2	3	4	5	6	7	8	9	0
2	2	3	4	5	6	7	8	9	0	1
3	3	4	5	6	7	8	9	0	1	2
4	4	5	6	7	8	9	0	1	2	3
5	5	6	7	8	9	0	1	2	3	4
6	6	7	8	9	0	1	2	3	4	5
7	7	8	9	0	1	2	3	4	5	6
8	8	9	0	1	2	3	4	5	6	7
9	9	0	1	2	3	4	5	6	7	8

Square 1 (False Addition)

Thus the digit $a = 5$ is encrypted by the key $b = 8$ to be transmitted as a 3. The table set out above is a particularly simple example of a *conversion square*. The entry in the table in column a (in boldface text at the top of the table) and row b (in boldface text on the left) is the false sum $a + b$.

---

[21]The conversion square is the most general way to encrypt a single digit using a key of one digit. So it is a basic prototype of encryption. Thus in some sense the redundant encryption of system 2468 indicators discussed in Chap. 14 is the prototype of all redundant encryption.

The method generalises easily to give a more subtle encryption, using a less evident square, provided each of the ten digits occurs once in each row to the right of the key (in bold face type). An example is:

	0	**1**	**2**	**3**	**4**	**5**	**6**	**7**	**8**	**9**
0	4	5	6	7	8	0	1	2	3	9
1	7	9	3	6	1	4	2	0	8	5
2	9	2	5	0	3	1	8	7	6	4
3	1	4	8	2	0	7	6	5	9	3
4	3	7	1	8	6	5	4	9	0	2
5	6	0	7	5	4	3	9	8	2	1
6	8	6	4	3	2	9	7	1	5	0
7	5	3	2	1	9	6	0	4	7	8
8	2	1	0	9	5	8	3	6	4	7
9	0	8	9	4	7	2	5	3	1	6

Square 2 (Encrypting)

Here the key $b = 2$ encrypts the digit $a = 5$ as 1, the key $b = 7$ encrypts the digit $a = 7$ as 4, the key $b = 4$ encrypts the digit $a = 8$ as 0 and the key $b = 7$ encrypts the digit $a = 6$ as 0. So the group 2747 encrypts the book group 5786 as 1400. The full message encryption becomes:

message	CABLE	DAMAGED	OPERATION	CANCELLED
book group	5786	1221	5346	8426
keys	2747	0889	9103	5900
GATs	1400	5008	2686	2761

Cryptanalysts, not supplied with the square, would have to spend much time trying to work out what was going on. Encryption of this type cannot be called 'additive'. Indeed, the difference methods useful in breaking additive systems do not work against most conversion squares.

The message could be made even more secure by using seven different conversion squares, one for each day of the week. The extra effort and extra risk of error might be justified for particularly high-level messages. Appendix 4 of Chap. 14 looks at the use of such squares to encrypt just the indicators. In itself this was a reasonable compromise between security and practicality. As a great deal was at stake this little extra effort[22] would be more than justified.

---

[22]Two of the general principles listed in Sect. 8.19 are relevant here:

Indicator systems should be seen as being particularly important and no effort should be spared in concealing, encoding and encrypting them.

Any easy method of complicating the task of the other side while not throwing much extra work on one's own staff should be adopted.

It is not difficult to work out other simple devices that also make life harder for the cryptanalysts on the other side.

Given the encryption of digital information produced by a known square and known keys, the decryption can be carried out without too much trouble on a small scale. Otherwise it is easy enough to produce the inverse of the encrypting square. For example, with the above 'Square 2' the inverse, or *decrypting square*, is

		0	**1**	**2**	**3**	**4**	**5**	**6**	**7**	**8**	**9**
	0	5	6	7	8	0	1	2	3	4	9
	1	7	4	6	2	5	9	3	0	8	1
	2	3	5	1	4	9	2	8	7	6	0
	3	4	0	3	9	1	7	6	5	2	8
Square 3	**4**	8	2	9	0	6	5	4	1	3	7
(Decrypting)	**5**	1	9	8	5	4	3	0	2	7	6
	6	9	7	4	3	2	8	1	6	0	5
	7	6	3	2	1	7	0	5	8	9	4
	8	2	1	0	6	8	4	7	9	5	3
	9	0	8	5	7	3	6	9	4	1	2

For example, with Square 2 and key 4, 9 is encrypted to 2. With Square 3 and this key, 2 is decrypted to 9.

Appendix 3   An Anachronism

A modern mathematician might ask if it was feasible to allocate book groups to the more commonly used words in a new code book in a way that actively slowed down cryptanalysis. Apparently searching for such methods did not happen before or during WW2. Yet there is a fairly evident method of resisting attack by differences. In a 4-digit code, the 16 groups 0000, 0005, 0050, 0055, 0500, 0505, 0550, 0555, 5000, 5005, 5050, 5055, 5500, 5505, 5550, 5555, have the nice property that only 15 groups arise as differences of pairs taken from the 16. If an additive cipher system uses these 16 groups for its most commonly used book groups and has alternative book groups for any other commonly used words, each of the 15 common differences will be the difference of 16 different pairs of book groups and so the stripping by differences method (Sect. 12.2) will produce confusion for the enemy cryptanalysts.

Appendix 4   Reducing Alignment

In the previous chapter an example was given in which the GATs of various messages were written out in depth (alignment). Thus the second GAT of message 1, the eighth of message 2, the fifth of message 3 and the fourth of message 4 were in depth. With a standard additive system one can deduce that the third GAT of message 1, the ninth of message 2, the sixth of message 3 and the fourth of message 4 had to be in depth too. Thus the cryptanalyst could use hits or (better) double hits to put entire messages into alignment.

There is a simple way to make this harder. The version presented here is closely related to the *grille* method[23] used with additive systems in the later stages[24] of WW2, but appears not to have been practised in this form. Viewed objectively, it is a fifth component to an additive system. Each table of additives could easily have included a table of *gaps for the day*, which would have looked something like:

GAPS FOR THE DAY

1 August 1942 00020-32212-03100-02101-01020-10010-01220-01020-31200-3

2 August 1942 01000-12010-01000-20020-00221-322

3 August 1942 12010-23220-10002-00211-23202-20000-32000-021

with one line as shown above for each day in which the table was to be used. Note that the numbers of gaps vary from day to day. The idea is that in using the additive table of the previous chapter on, say, 3 August starting at row 5, column 8 on page 123 one does not simply use the additives:

5									8363	0221
6	1907	5100	9671	7081	8582	9571	8480	0690	2411	0503
7	2931	5542	1699	3517	4816	7249	8091	1505	4469	2625
8	8205	4831	5602	5346	4786	4528	9948	7633	4686	2177
9	7347	8954	6055	9835	8859	6783	7362	1560	2411	9387

---

[23]NAA item A705 201/116/35 is a British report dated 1948 on the security of a stencil cipher system. It confirms that stencil ciphers were used by the British military in 1948. It concludes that use of such systems might become insecure if traffic had to be increased in an emergency. It gave instructions that effective 1 January 1949 additive systems were to be used with two different stencils giving two different additives, both of which were used. The GCHQ has released some contemporary reports on the security of additive ciphers.

By that stage the encryption procedure was that given a book group x to be encrypted by the 'additive' a, the GAT used was the false difference $a - x$. This has the advantage of making the decryption process the same as the encrypting one. However it does not change anything in the general theory of additive ciphers.

Confusingly the word 'stencil' was also used to describe the transposition system used in certain other non-additive ciphers. This is explained, with examples, by Hanyok and Mowry in *West Wind Clear*. The second usage is not needed in this book. However the Mamba process of Sect. 15.4 did involve IBM cards being used as stencils in the conventional sense.

[24]Perhaps the word 'grille' should be clarified: it referred to a piece of paper in which holes had been cut. When this paper was placed over a page of an additive table it would obscure some of the groups but not others. The operator would use only visible groups. Several alternative grilles would be provided and the indicators would need to inform the intended recipient which one was in use. As usual, the choice of grille should have been made with dice or other such lottery method.

Edward Simpson's contribution to Ralph Erskine and Michael Smith *Bletchley Park Codebreakers* (Biteback, London, 2011), being the second edition of *Action this Day* (Bantam, 2001) explains how in 1944 a version of JN-25 that used grilles could not be broken. Later versions did not have this feature and were insecure.

but instead leaves gaps as directed by the sequence of gaps indicated for this date: '12010' meaning 'after the first selected additive, leave one gap; after the second, leave a gap of two; after the next, leave no gap; after the next, leave a gap of one, and after the next again, leave no gap. After implementation, the whole sequence produces the following additives for use:

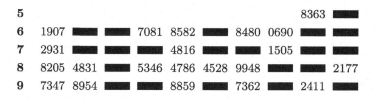

```
5                                                   8363  ■
6   1907  ■   ■   7081 8582  ■   8480 0690  ■   ■
7   2931  ■   ■   ■   4816  ■   ■   1505  ■   ■
8   8205 4831  ■   5346 4786 4528 9948  ■   ■   2177
9   7347 8954  ■   ■   8859  ■   7362  ■   2411  ■
```

The grilles in use were stencils that blotted out some of the additives. The effect was the same.

Thus a message consisting of the book groups:

1415	9265	3589	7932	3846	2643	3832	7950	2884	1971
6939	9375	1058	2097	4944	5923	0781	6406	2862	0899

would be encrypted by working out the false sums as set out below:

1415	9265	3589	7932	3846	2643	3832	7950	2884	1971
8363	1907	7081	8582	8480	0690	2931	4816	1505	8205
9778	0262	0560	5414	1226	2233	5763	1766	2289	9176
6939	9375	1058	2097	4944	5923	0781	6406	2862	0899
4831	5346	4786	4528	9948	2177	7347	8954	8859	7362
0760	4611	5734	6515	3882	7090	7028	4350	0611	7151

Any complication, like this gap system, delays or prevents the enemy cryptanalysts from getting serious quantities of GATs in depth even after they have broken into the indicator system. This in turn enhances security.

Chapter 14
Redundant Encryption

This chapter is quite technical. Readers without a sound basis in the mathematical or physical sciences may prefer to skim it and just accept that the Water Transport Code (WTC or 2468) could be and was broken. A full analysis requires the theory of permutations in the sense of modern algebra. The root cause of the insecurity was the practice of repeating the row and column co-ordinates of the first additive in the indicator before encryption. The method of exploiting this is a significant example of WW2 Allied cryptanalysis at its best. Remarkably, this error was analogous to the misuse of the Enigma machine that Marian Rejewski of the Polish Cipher Bureau exploited (Appendix 2) from 1933 onwards. Rejewski's work also involved permutations. Equally remarkably, a variant, known as 'tailing', of this same practice assisted Op-20-G in breaking JN-25 indicator encryption systems in 1942 and 1943. These examples show that redundant encryption in any form is potentially insecure.

14.1 The Task of the Cryptologists

At this stage an account of what cryptologists (cryptanalysts) actually did becomes very relevant. So the following quotation from Sect. 4.3 is repeated. It was written by William Friedman and his wife Elizebeth in a post-war analysis[1] of the rationality, if any, in published claims that Francis Bacon wrote the material generally attributed to William Shakespeare.

> The Cryptologist must discipline himself to follow certain procedures and to submit to certain checks. Like the experimental scientist he is observing phenomena or occurrences to determine whether they are random or systematic, and if systematic how they work—

---

[1] The quoted fragment is from page 286 of *The Shakespearean Ciphers Examined*, Cambridge University Press, New York, 1957.

© Springer International Publishing Switzerland 2014
P. Donovan, J. Mack, *Code Breaking in the Pacific*,
DOI 10.1007/978-3-319-08278-3_14

what principle can be detected in them. . . . He is trying to formulate an exact statement
about the phenomenon before him. . . . Cryptology is an application of scientific method.

The method by which the IJA constructed and encrypted the indicators of the
WTC is well documented. What the Friedmans called 'systematic phenomena'
were produced by this undoubtedly insecure method and exploited by the Allied
cryptanalysts.

14.2   The CBTR on the WTC

The following extracts, marked from **A** to **M** for reference, are taken from pages 17
and 18 of Part G of the *CBTR*.

   A The effort applied towards reading the Water Transport System, a higher echelon army
shipping system, was well repaid. Resulting intelligence gained concerning shipping
movements was extremely valuable and the cryptanalytic experience and cross-reference
of data aided in the solution of other systems. A large part of the original work on the
Water Transport problem was contributed by CBB, and it was a principal interest for
about eighteen months, beginning early in 1943.

   B The first traffic identified in the Water Transport system was intercepted on 24 December
1942, at a time when no Japanese Army codes were being read. The system used the
undisguised discriminant 2468. From a break into a Japanese system in 1939, from
captured message forms in other systems in 1942, and from an intensive[2] study of
the traffic, it was found that the Japanese were using code enciphered by an additive
system. Investigations proceeded on the assumption that the Water Transport system was
also enciphered by some such method. To recover additives, a depth of messages must
be available for each set of additives used and the messages superimposed in proper
relative order, as determined by the co-ordinates of the additive set. The co-ordinates,
known collectively as the indicators, were disguised, and their recovery became one of
the most difficult problems in the solution. From the captured messages it was observed
that they were not enciphered in the same manner as the text, which was enciphered by
non-carrying addition of additive and code groups.

The use of Comint in finding targets for submarines is discussed in Chap. 16. The
statement that no IJA codes were being read in December 1942 shows the great
importance of the weaknesses of JN-25 systems in the first 18 months of the Pacific
War. The *CBTR* makes the significance of the WTC extremely clear.

Presumably the 'break into a Japanese system in 1939' refers to the work of John
Tiltman of GCCS in Hong Kong and later London in 1938. Tiltman had only a
limited knowledge of oriental languages and so would not have been involved in
the final stages of working on that system. The FECB reported on 20 June 1939
that there were new difficulties with the Army cipher. Doubts were expressed about
whether work on it would be productive.

---

[2]The 'intensive study' most likely revealed that double hits were more common in 2468 traffic than
would have occurred with randomly generated intercepts.

C   In an examination of traffic in March 1943 two peculiarities were noted:

   (i) that the first digit of the third GAT was always one of nine numbers rather than ten;
   (ii) that for each number occurring in the first digit of the second GAT, there were only
        three numbers that would appear in the first digit of the third GAT.

Both (i) and (ii) draw attention to the first digit of the third GAT. Part (i) is both confusing and inconsequential. It is analysed in Appendix 1. Part (ii) is the basis of the section *Decoding the Ten Letters* later in this chapter.

The *CBTR* counted the discriminant 2468 as the first GAT in an intercepted message, ignoring the preliminary parts. It turned out that the second and fifth GATs were respectively the first and second enciphered groups of the actual message, while the third and fourth GATs were the encrypted indicators. All this helps in understanding the extracts given here.

D   These two facts were beyond coincidence and stimulated research. The next discovery
    was that if the first and second digits of the fifth GAT were the same, the first and second
    digits of the fourth GAT were also the same. The same relationship held between the
    third and fourth digits of the fourth and fifth GATs.

Joe Richard, an American assigned to CBB, has recorded his part in the saga in various[3] places. This was a collective effort: some progress[4] (observing peculiarities (i) and (ii)) had been made—perhaps earlier than March 1943—at the Wireless Experimental Centre (WEC) in Delhi and then Richard made the 'next discovery'. Richard's accounts state that he started in ignorance of the Delhi advance. Abe Sinkov then showed him a letter from the WEC and allocated more resources to attacking the indicator. Arlington Hall Station (AHS) joined the project with information being exchanged regularly.

E   Further examination revealed that the relationship of these doubles in the first and second
    digits of the fifth and fourth GATs and of the third and fourth digits of the same GATs
    existed only in one direction. That is, if the first and second digits of the fifth GAT
    were similar, so were the first and second digits of the fourth GAT, but they were not

---

[3] Joe Richard, by then the Grand Old Man of the National Security Agency, died on 8 April 2005. He had assisted the authors of various books and had published his account of the breaking of 2468. His later memories are fully compatible with what was written for the *CBTR* in 1945.

[4] Michael Smith in *The Emperor's Codes*, page 175, notes that Wilfrid Noyce and Maurice Allen, two classicists at WEC, made the key observation that the first digit of the third GAT was not random. Smith's claim is quite compatible with what is worked out later in this chapter. In a sense there were only three initial observations to make: Richard made one, the second would have followed immediately from that of Richard and the Noyce-Allen contribution is the third. Richard should also be credited with having the punching of cards arranged so that sorting on date was possible. This enabled the cryptanalysts to separate out batches corresponding to different squares.

The *CBTR* paragraph marked 'A' in the main text contains the interesting statement 'A large part of the original work on the Water Transport problem was contributed by CBB'. After the observation made by Joe Richard had been fully investigated a serious mathematical problem remained. The two people present with the necessary background and ability were Sinkov and Room. There is a strong possibility that Room made a serious contribution here. See also Note 7. A letter sent by Room in 1951 (NAA item A663 O130/2/1309 and available on line) rather supports this viewpoint.

necessarily similar in the fifth GAT if they were in the fourth. An explanation of this might be that the fifth GAT was used to encipher the fourth, a fact later proven.

F Considering together the fact observed above of the relationship of the first digit of the second GAT and the first digit of the third GAT, and the probability that the fifth and fourth GATs were used together for at least part of the indicator, it seemed quite plausible that the disguising of the indicators was performed by some use of a substitution table.

G One of the first natural assumptions for a substitution table would be a ten-by-ten [conversion] square. The recovery of the first square was a major accomplishment that determined a method to be used in all later solutions. It took much time and concentrated effort until it was found that by using the relationship between two digits of similar plain text, dissimilar cipher text, it was possible to recover a non-primary square.

The last sentence of paragraph G is crucial: redundant encryption of only two pairs of 'similar' digits in each of a reasonably large batch ended up betraying the indicator system.

The jargon 'non-primary' (related to that of Sect. 8.16) is explained later.

H For this method of recovering squares, see the mimeographed instructions 'Japanese Additive Systems'.

An account of the square recovery process, including a contrived example, and a (slightly speculative) account of how it was discovered, is given in Peter Donovan's paper[5] in Cryptologia 30(3), July 2006, 212–235. The explanation given later in this chapter uses the concept of *permutations* which was rather slurred over in the paper. The Seahorse Enigma[6] saga (not covered in this book) was another example of redundant encryption.

Paragraphs G and H confirm that the conversion squares were replaced from time to time. After the Allied cryptanalysts had worked out what was going on, a change of square was remarkably ineffective in enhancing security. Experiments with randomly concocted data indicate that about 130–150 intercepts would reveal a new square. The WTC was very heavily used.

I Deciphering this fourth GAT for all the traffic revealed in each case an *aabb* pattern which was assumed to be the row and column co-ordinates repeated. The relationship of the first digit of the second and third GATs having been observed, the second GAT was used to decypher the third, as the fifth had been used with the fourth. Since the square was non-primary, letters were used as column co-ordinates of the square. The first digit of the third GAT deciphered to one of only three letters in each case, bearing

---

[5]The paper *The Indicators of Japanese Ciphers 2468, 7890 and JN-25A1* by Peter Donovan appeared in *Cryptologia* 30(3) July 2006 212–235. This book contains a greatly up-graded version of it. Another example of square recovery is to be found there. The cited report *Japanese Additive Systems* may survive in NARA and/or the GCCS documents, but has not been seen. The NAA Canberra file A705 201/28/325 entitled *Typex Indicators* (March 1945) states once again the importance of indicator encryption.

[6]The reference is the paper by Ralph Erskine and Philip Marks, *Seahorse and other Kriegsmarine Cipher Blunders*, *Cryptologia* 28(3), July 2004, 211–241. Section 8.19 above listed maxims about good cipher practice for the era. Avoiding redundant encryption and taking extra care with indicators were two of them.

out the observed relation of the second and third GATs. The assumption followed that this was the book or hundreds digit of the page used. If this were so, the next two digits would be the ten's and unit's digits, and the fourth might be a summing digit, as a summing digit was commonly used as a check. From this assumed summing property the square was made primary, i.e. the actual values of column co-ordinates were discovered. No obvious denials appeared, so that premise was apparently correct. The pattern then became book-page, tens-page, units-page, sum (of Book and Page digits), Row, Row, Column, Column.

Here, as elsewhere in the WW2 cryptanalysis saga, knowledge of past enemy practice[7] was extremely useful.

J   All available traffic was then de-paged, sorted by page and page co-ordinates, and an IBM index of cypher text made. When identical cypher text [hits] appeared in two messages, it was assumed that identical plain text had been enciphered by the same additive. When a page was set up, the additives were recovered column by column by the established method of differences until as much as 75 % of the page was stripped. Stripping by differences is theoretically ideal, but in practice is very difficult, and it was found that 150,000 differences had to be taken before satisfactory results were obtained. For a treatment of stripping by the method of differences, see the mimeographed instructions for stripping 2468.

This incidentally identifies the significance of using only multiples of three for JN-25 book groups. Stripping by differences of a code with so many book groups would have been 'very difficult'. The decryption devices and the tables of common differences made this process much easier, particularly if the indicator system was readable.

K   Progress to this point had recovered the substitution square and deciphered the indicators of the traffic, the traffic had been placed in proper relative position by page, the co-ordinates of the page determined, and the additives for these pages had been stripped, revealing plain code text. But so far the plain code values had not been defined. A frequency chart was made of the several pages of text available, and, with observations of the text sequences, relative code group values could be assigned for many of the frequent groups. For instance, relative values at first set BEGIN TEXT as 0000 rather than 6666; CTC was 8778 rather than 2112. From previous observation, the Japs used a group for BEGIN TEXT and which occurred in every message. It could then be assumed that certain of the frequent groups preceding it were numbers and part numbers (i.e. numbers of the part of the message).

Once again the value of Tiltman's work on the 1938 system emerges—the 'BEGIN TEXT' group was anticipated. Likewise the use of CTC—see Sect. 1.10—had been detected in 1938–1939. In retrospect it seems to have been a fundamentally unsound practice. Here '8778 rather than 2112' should be '2112 rather than 8778'.

L   All the tentative code groups were then reduced to primary values when a study of the most frequent number and part number groups indicated that the relative values should

---

[7] Apparently Abe Sinkov suggested that the otherwise useless fourth digit might be the sum of the first three and so a check for error. In general, military reports—and the *CBTR* is such—avoid mentioning individuals by name. Once the practice of using the fourth digit as a check in this way had been detected in the *CBTR*, cryptanalysts would be looking for it in other systems.

be increased by six in each digit. In explanation: of the values assumed to be numbers, the most frequent would be the first ten, in ascending order, and it was observed that the second digit of each of these groups was four. When part numbers were used, parts one and two would necessarily be used equally and be by far the most frequent. The relative values for the most frequent groups considered—parts one and two—were 1551 and 7667. By the addition of six to each digit, code groups with plain text values in the second and third digits were recovered for all the numbers and all but Part Five of the parts. The change was made in all the tentatively identified groups.

The usage made here of *primary* was discussed in Sects. 8.16 and Appendix 1 of Chap. 12. Long messages were sent divided into parts. Evidently a message would have a 'part 1' only if it had a 'part 2'. The parts would usually be sent consecutively. Thus the groups for 'part 1' and 'part 2' could be identified. Part F of the *CBTR* commented that any patterning in a code book was potentially a weakness.

M When the code groups were made primary, it was necessary to change all the additives accordingly, which was but a simple arithmetic change. All the number groups were found to fall into a ten-by-ten garble chart[8] square. With these primary code groups, plain text values were rapidly built up by the use of numbers and their checks, brackets and the practice of checking basic kana by CTC, principally the latter. It was soon discovered that all basic kana and all functional groups were patterned $ABCD$, so that A plus C equalled B plus D, except that when either pair summed to more than ten it was one greater, when reduced by ten, than the other pair. This summing property speeded the recovery of many of the basic groups.

So some important book groups (including 1551 and 6666) were multiples[9] of 11, rather than the multiples of 3 used in the JN-25 series!! It would appear that the discovery of the primary code groups was achieved in much the same way as Mrs Driscoll (Appendix 1 of Chap. 12) went through to obtain the primary code values for JN-25A.

Part G of the *CBTR* sets out about ten phases in the evolution of 2468. These will not be considered in this somewhat simplified account.

---

[8] A discussion of *characteristics* is given in Sect. 15.2. As remarked in Note 4 of Chap. 4, some of the jargon used for WW2 cryptology is rather obscure. 'Garble chart' is an example. The web page www.codesandciphers.org.uk/documents/cryptdict gives access to the Bletchley Park *Cryptographic Dictionary* that survives in NARA RG457, Box 1413, item 4559. It reveals that a 'garble table' was 'a table designed to facilitate the construction or checking of kana or other groups that have a characteristic sum'. Here the reference must be to what is called a *conversion square* in this book.

[9] A 4-digit number $abcd$ is a multiple of 11 if and only if $a - b + c - d$ is a multiple of 11. The *CBTR* is rather obscure here. Thus as $9 - 1 + 7 - 4 = 11$ so is 9174. So in the 2468 code book all basic kana and all functional groups were multiples of 11. NARA file RG457, Box 926, item 2649 (vintage 1943) analyses the frequency of the more common book groups in the Water Transport Code. It confirms the *CBTR* remark.

Appendix 3 of Chap. 15 discusses a possible 'Mamba-11' process.

According to Part F of the *CBTR*, at least some early commercial codes had the feature that the more common words and phrases were encoded by numbers that were multiples of 11. This has not been verified and may be an error. If so, it is possible that the use of multiples of 3 in the JN-25 code books can be traced back to this practice.

14.3   The Dependence

The *CBTR* does not state explicitly that the second digit (w) of the fourth GAT is determined by its first digit (t) and the first two digits (u and v) of the fifth GAT. Likewise the fourth digit ($\bar{w}$) of the fourth GAT is determined by its third digit ($\bar{t}$) and the second two digits ($\bar{u}$ and $\bar{v}$) of the fifth GAT. However this is indeed the case and can be inferred from the given extracts of the *CBTR*. This matter will be examined somewhat later using the mathematical concept of *permutations*.

14.4   The Indicators of Cipher 2468

The next six sections set out at some length a conceptually simple example of redundant encryption of indicators. This is intended to explain the CBTR text just quoted. It may be looked on as being the prototype of exploiting redundant encryption.

Code 2468 used 4-digit groups. The additive table consisted of three booklets with the pages of each being numbered from 00 to 99. As the books were numbered from 1 to 3, the pages were in effect numbered from 100 to 399. There were 100 additive groups on each page set out in a 10×10 display. Three digits were needed to convey the page number (from 100 to 399) of the starting place in the additive table. Two more were needed for the row and column co-ordinates of the selected additive group. This is a total of five. For example, if the proposed starting place was the italicized group in the sample table below, the indicator would be 14785.

The column and line co-ordinates were erratically numbered by single digits as in the following modern reproduction:

Book 1, page 47 of WTC = 2468 table of additives

	2	8	9	6	4	5	1	0	3	7
9	4884	2843	4217	7745	4322	4746	9152	7870	8724	4612
8	6859	0840	9240	3899	6300	*4929*	8070	3518	7237	9417
0	8621	6442	3729	8991	0965	1456	3991	8746	0288	3360
2	3813	9862	9677	9341	7502	3313	8108	7319	4758	4163
6	3839	3165	1507	0845	9639	5445	5001	1202	4021	5531
3	8256	2727	3504	8148	0067	6872	0832	6387	8363	0221
4	1907	5100	9671	7081	8582	9571	8480	0690	2411	0503
5	2931	5542	1699	3517	4816	7249	8091	1505	4469	2625
1	8205	4831	5602	5346	4786	4528	9948	7633	4686	2177
7	7347	8954	6055	9835	8859	6783	7362	1560	2411	9387

Here the indicator information would need to be spread over two groups of four. As $8 - 5 = 3$ the designer of the system had three digits to use for checking purposes or otherwise. If the book number was p, the page number was qr, the row co-ordinate was m and the column co-ordinate was n, the unencrypted indicator groups used were $pqrs$ and $mmnn$, where s is the false sum $p + q + r$. These groups were encrypted[10] before transmission by means of a conversion square using the first two GATs of the message text as keys.

Thus suppose that the decision had been made to start on page 47 of book 1, in the second row and the sixth group of that row. (This position is italicized above.) The second row has co-ordinate 8 while the sixth column has co-ordinate 5. The indicator for these choices would consist of two 4-figure groups. The first would begin with the book and page number (here $pqr = 147$) followed by the false sum of these three (here $s = 2$). The second group would consist of the row co-ordinate repeated (here $mm = 88$) followed by the column co-ordinate repeated (here $nn = 55$). Thus the indicator groups are 1472 and 8855.

The additive chosen for use begins:

    4929   8070   3518   7237   9417   8621   6442   3729   8991

Suppose that the message is that of Sect. 8.10:

    CABLE   DAMAGED   OPERATION   CANCELLED

with corresponding book groups as before. The operator would calculate:

CABLE	DAMAGED	OPERATION	CANCELLED
5786	1221	5346	8426
4929	8070	3518	7237
1867	3251	2838	1299

The first group of the encrypted message, 1867 in this example, was then used to encrypt the first group of the indicator, here 1472. The second group of the encrypted message, here 3251, was used to encrypt the second indicator group, here 8855. This was not done simply by false addition but by means of a conversion square such as 'Square 2' of Appendix 2 of Chap. 13:

The 1 (the bold-face number beginning the second row) encrypts 1 to give 9. The 8 (in the ninth row) encrypts 4 to give 5. The 6 (in the seventh row) encrypts the 7 to give 1. The 7 encrypts the 2 to give 2. Thus the GAT 1867 encrypts 1472 as 9512.

Likewise the first digit 3 of 3251 encrypts the 8 that begins the second indicator group 8855 as 9. The second digit of the former, 2, encrypts the second digit of the latter, 8, as 6; the third digit 5 encrypts 5 as 3; the fourth digit 1 encrypts 5 as 4. So 3251, the second GAT of the message, encrypts 8855 as 9634. Thus the transmitted message begins:

    2468   1867   **9512**   **9634**   3251   2838   1299

The encrypted indicators are printed in bold face to emphasise their special role.

---

[10]Enciphering all four of p, q, r and s is redundant. Enciphering both of m and n twice is even more redundant. There are various ways of avoiding such redundancy. For example, *nuls*, that is randomly chosen digits whose sole function is padding, could have been used. Chapters 8 and 13 make it clear that some such secure method should have been adopted.

		0	1	2	3	4	5	6	7	8	9
	0	4	5	6	7	8	0	1	2	3	9
	1	7	9	3	6	1	4	2	0	8	5
	2	9	2	5	0	3	1	8	7	6	4
	3	1	4	8	2	0	7	6	5	9	3
Square 2	**4**	3	7	1	8	6	5	4	9	0	2
(Encrypting)	**5**	6	0	7	5	4	3	9	8	2	1
	6	8	6	4	3	2	9	7	1	5	0
	7	5	3	2	1	9	6	0	4	7	8
	8	2	1	0	9	5	8	3	6	4	7
	9	0	8	9	4	7	2	5	3	1	6

14.5 Recovering a New Conversion Square

Suppose that the conversion square 'Square 2' was replaced by a new 'Square 4', not available to the Allied cryptanalysts. Instead, they would soon have available records of intercepts with indicators encrypted by the new square. The beginnings of these will look something like:

discr	GAT2	ENCR	IND	GAT5
2468	6989	1507	8216	2193
2468	5455	9279	2244	1100
2468	9388	8685	7579	8415
2468	6704	9793	9354	0117
2468	6315	1413	1734	8008
2468	5228	9308	5789	6771
2468	6844	1630	2947	9665
2468	5819	9692	4827	6882
2468	9675	8884	2708	4530
2468	4614	8324	5893	7301

The unencrypted indicators are definitely not transmitted. When Square 4 is recovered (later) it may be checked that 6989 encrypts 2248 to 1507, 2193 encrypts 4466 to 8216, etc.

The message beginnings given above are the first ten of 2,000 randomly generated examples. Usually the first 200 alone are enough to reveal the square and then the original indicators of all the intercepts.

14.6 Tabulating the Dependence

As already noted, the second digit (w) of the fourth GAT is determined by its first digit (t) and the first two digits (u and v) of the fifth GAT. This can be tabulated as below, using 100 columns for the $100 = 10 \times 10$ possibilities for the pair t, u and

10 rows for the 10 possibilities for v. The other[11] possible methods of tabulation turn out to be less useful. The same table may as well contain the information about how $\bar{w}$ depends upon $\bar{t}$, $\bar{u}$ and $\bar{v}$. It turns out[12] that with the 2,000 randomly generated messages in the sample only 981 of the possible 1,000 entries in the table are obtained.

For practical reasons the table is printed in two parts below.

```
t = 0 0 0 0 0 0 0 0 0 0 1 1 1 1 1 1 1 1 1 1 2 2 2 2 2 2 2 2 2 2 3 3 3 3 3 3 3 3 3 3 4 4 4 4 4 4 4 4 4 4
u = 0 1 2 3 4 5 6 7 8 9 0 1 2 3 4 5 6 7 8 9 0 1 2 3 4 5 6 7 8 9 0 1 2 3 4 5 6 7 8 9 0 1 2 3 4 5 6 7 8 9

0   0 2 5 8 9 3 4 7 1 6 1 6 0 9 4 5 8 2 7 3 2 1 9 5 6 0 3 8 4 7 3 9 6 7 1 4 0 5 8 2 4 7 8 . 5 1 6 0 3 9
1   5 0 8 7 3 9 6 4 2 1 2 1 5 3 6 8 . 0 4 9 0 2 3 8 1 5 9 7 6 4 9 3 1 4 2 6 5 8 7 0 6 4 7 0 8 2 1 5 9 3
2   1 9 0 4 2 5 7 6 8 3 8 3 1 2 7 0 4 9 6 5 9 8 2 . 3 1 5 4 7 6 5 2 3 6 8 7 1 0 4 9 7 6 4 9 0 8 3 1 5 2
3   7 4 2 0 1 6 8 3 9 5 9 5 7 1 8 2 0 4 3 6 4 9 1 2 5 7 6 . 8 3 6 1 5 3 9 8 7 2 0 4 8 3 0 4 2 9 5 7 6 .
4   9 5 4 6 0 7 1 8 3 2 3 2 9 0 1 4 6 5 8 7 5 . 0 4 2 9 7 6 1 8 7 0 2 8 3 1 9 4 6 5 1 8 6 5 4 . 2 9 7 0
5   . 6 1 9 8 0 3 5 4 7 4 7 2 8 3 1 9 6 5 0 6 4 8 1 7 2 0 9 3 5 0 8 7 5 4 3 2 1 9 6 3 5 9 6 1 4 7 2 0 8
6   . 8 6 1 7 2 0 9 5 4 5 4 3 7 0 6 1 8 9 2 8 5 7 6 4 3 2 1 . 9 2 7 4 9 5 0 3 6 1 8 0 9 1 . 6 5 4 3 2 7
7   4 1 3 2 6 8 5 0 7 9 7 . 4 6 5 3 2 1 0 8 1 7 6 3 9 4 8 2 5 0 8 6 9 0 7 5 4 3 2 1 5 0 2 1 3 7 9 4 8 6
8   6 7 9 3 5 4 2 1 0 8 0 8 6 5 2 9 3 7 1 4 7 0 5 9 8 6 4 3 2 1 4 5 8 1 0 2 6 9 3 7 2 1 3 7 9 0 8 6 4 5
9   8 3 7 5 4 1 9 2 6 0 6 0 8 4 9 7 5 3 2 1 3 6 4 7 0 8 1 5 9 2 1 4 0 2 6 9 8 7 5 3 9 2 5 3 7 6 0 8 1 4
```

```
t = 5 5 5 5 5 5 5 5 5 5 6 6 6 6 6 6 6 6 6 6 7 7 7 7 7 7 7 7 7 7 8 8 8 8 8 8 8 8 8 8 9 9 9 9 9 9 9 9 9 9
u = 0 1 2 3 4 5 6 7 8 9 0 1 2 3 4 5 6 7 8 9 0 1 2 3 4 5 6 7 8 9 0 1 2 3 4 5 6 7 8 9 0 1 2 3 4 5 6 7 8 9

0   5 0 3 6 2 7 1 4 9 8 6 4 7 3 8 2 5 . 0 1 7 8 4 0 3 6 9 1 2 5 8 5 1 4 7 9 2 3 6 0 9 3 2 1 0 8 7 6 5 4
1   . 5 9 1 0 4 2 6 3 7 1 6 4 9 7 0 8 3 5 2 4 7 6 5 9 1 3 2 0 8 7 8 2 6 4 3 0 9 1 5 3 9 0 2 5 . 4 1 8 6
2   0 1 5 3 9 6 8 7 2 4 3 7 6 5 4 9 0 2 1 8 6 4 7 1 5 3 2 8 9 0 4 0 8 7 6 2 9 5 3 1 2 5 9 8 1 4 6 3 0 7
3   2 7 6 5 4 3 9 8 1 0 5 8 3 6 0 4 2 1 7 9 3 0 8 7 6 5 1 9 4 2 0 2 9 8 3 1 4 6 5 7 1 6 4 9 7 0 3 5 2 8
4   4 9 7 2 5 8 3 1 0 6 2 1 8 7 6 5 4 0 9 3 8 6 1 9 7 2 0 3 5 4 6 4 3 1 8 0 5 7 2 9 0 7 . 3 9 6 8 2 4 1
5   1 2 0 7 6 5 4 3 8 9 7 3 5 0 9 6 1 8 2 4 5 9 3 2 0 7 8 4 6 1 9 1 4 3 5 8 6 0 7 2 8 0 6 4 2 9 5 7 1 3
6   6 3 2 4 8 9 5 0 7 1 4 0 9 2 1 8 6 7 3 5 9 1 0 3 2 4 7 5 8 6 1 6 5 0 9 7 8 2 . 3 7 2 8 5 3 1 9 4 6 0
7   3 4 8 9 1 0 7 5 6 2 9 5 0 8 2 1 3 6 4 7 0 2 5 4 8 9 6 7 1 3 2 3 7 5 0 6 1 8 9 4 6 8 1 7 4 2 0 9 3 5
8   9 6 4 8 7 1 0 2 5 3 8 2 1 4 3 7 9 5 6 0 1 3 2 6 4 . 5 0 7 9 3 9 0 2 1 5 7 4 8 6 5 4 7 0 6 3 1 8 9 2
9   7 8 1 0 3 2 6 9 4 5 0 9 2 1 5 3 7 4 8 6 2 5 9 8 1 0 4 6 3 7 5 7 6 9 2 4 3 1 0 8 4 . 3 6 8 5 2 0 7 9
```

---

[11] The table is given with all the 't' and 'u' values spread out on the top. If instead the 't' and 'v' values or the 'u' and 'v' values are arranged at the top the solution is much less obvious.

[12] This figure of 2,000 intercepts being available is quite realistic. Those in doubt may examine NARA RG457, Boxes 531–575, containing the 1943 decrypts from the Water Transport Code. The heavy use of this code is confirmed in Part G of the *CBTR*.

The computer facility MAPLE has the useful feature that its random number generator produces the same output on different runs unless the *seed* is changed. To make this experiment as objective as possible, the seed was not changed and for each of 2,000 cards there were nine invocations of this generator using 1+irem(rand(),3) (once), irem(rand(),10) (four times) and irem(rand(),10000) (four times).

A run with a different seed produced another example. Only 136 messages with both parts of the fourth GAT being used turn out to be just adequate for square recovery.

The opportunity is taken to state that all the other computer experiments needed for this book were carried out with MAPLE.

The remaining 19 blanks in this table may now be filled in so that each block of 10 columns in it contains, in some order, the 10 columns of 'Square 6' below. The unknown Square 4 will be some re-ordering, yet to be found, of the columns of Square 6.

		A	B	C	D	E	F	G	H	I	J
	0	0	2	5	8	9	3	4	7	1	6
	1	5	0	8	7	3	9	6	4	2	1
	2	1	9	0	4	2	5	7	6	8	3
	3	7	4	2	0	1	6	8	3	9	5
Square 6	**4**	9	5	4	6	0	7	1	8	3	2
'Non-primary'	**5**	2	6	1	9	8	0	3	5	4	7
	6	3	8	6	1	7	2	0	9	5	4
	7	4	1	3	2	6	8	5	0	7	9
	8	6	7	9	3	5	4	2	1	0	8
	9	8	3	7	5	4	1	9	2	6	0

14.7 Using a Smaller Sample of Intercepts

Now suppose that fewer intercepts are available. For example, if the first 200 of the randomly generated sample are used, the 10×100 table turns out to have 329 digits entered with 671 missing. It is:

```
t = 0 0 0 0 0 0 0 0 0 0 1 1 1 1 1 1 1 1 1 1 2 2 2 2 2 2 2 2 2 2 3 3 3 3 3 3 3 3 3 3 4 4 4 4 4 4 4 4 4 4
u = 0 1 2 3 4 5 6 7 8 9 0 1 2 3 4 5 6 7 8 9 0 1 2 3 4 5 6 7 8 9 0 1 2 3 4 5 6 7 8 9 0 1 2 3 4 5 6 7 8 9
0   0 . 5 8 . 3 . 7 1 6 . . . 9 . . . . 7 . . . . . . . . . . . . 7 . . 6 . . . . . . . . 4 . . . 5 . . . . 9
1   . . . . . . . . 2 1 . . 5 3 . . . . 9 0 2 . . . 9 7 6 . 9 . . . . . . 8 7 . 6 4 . . . . . . . 9 .
2   . . 0 4 . 5 . . . . . . . 0 . . . 5 . 8 . . . . 7 6 . . . 8 . 1 . . 9 . . . . . 8 . . 5 .
3   7 . . . 1 . . 3 . . 9 . . . . . 0 . 3 6 . . . 2 . . . . 3 6 . . . . . . . . . . . . 2 . . 7 . .
4   . 5 . 6 . 7 1 . . . . 2 . . 1 . . 5 . 7 . . . . . . 9 . . 1 8 . 0 . . . . . . 9 . . . . . . . . 7 .
5   . . 1 9 8 0 . . . . 4 . 2 . 3 . . . . . 6 4 8 . 7 2 . . . . . 8 . . 4 . . 1 9 . . 5 . . . . 7 . . .
6   . . . . . . . . 5 . 5 . . . 0 6 1 . 9 2 8 . . . . . 2 . . 9 2 7 . 9 . 0 . . 1 . 0 . . . . 5 . . . 7
7   . . . . 6 . . 0 . . . . 4 . . . . 0 . 1 . 6 3 . 4 8 . . . . 6 . . 7 . . . . . . . 2 . . . . . . 8 .
8   6 . . . . . 2 1 . . 0 8 . . . . . . . . . . . . . 6 . . . . . 4 . . . . . . 9 . . 2 . . . 9 . 8 6 4 .
9   8 3 . . . . . . . 0 . . . 4 . . . . 1 3 . . . . 8 . 5 . . . . . 0 . . 9 . . . 3 9 . 5 . . . . . . .

t = 5 5 5 5 5 5 5 5 5 5 6 6 6 6 6 6 6 6 6 6 7 7 7 7 7 7 7 7 7 7 8 8 8 8 8 8 8 8 8 8 9 9 9 9 9 9 9 9 9 9
u = 0 1 2 3 4 5 6 7 8 9 0 1 2 3 4 5 6 7 8 9 0 1 2 3 4 5 6 7 8 9 0 1 2 3 4 5 6 7 8 9 0 1 2 3 4 5 6 7 8 9
0   . . . . 2 . 1 . . . . . . 3 . . 5 . 0 . . 8 4 . . . . . . 2 . . . . 4 7 . 2 . . 0 9 3 2 . . 8 7 . . .
1   . 5 . 1 . . 6 3 . 1 . . 7 . 3 . 2 . 7 . . . . . 2 . . . 2 . . 3 . 9 1 . 3 . . . . . . . 1 . .
2   . 1 . 3 . . . 4 . . . 5 . . . 2 . . 6 4 . . . . . 0 4 . 8 . . . 9 . 3 . . . . . 1 . 6 . 0 .
3   2 . . . . . 9 8 1 0 5 8 3 6 . 4 2 . . . 3 0 . . 6 . . . . 4 . 0 . 9 8 . 1 . 6 . . 1 . . . 7 . . 5 . 8
4   . 9 . . . . 3 . . . . . . 5 4 . 9 3 8 . . . . 2 0 3 5 . 6 . 3 . 8 0 5 . . 9 . 7 . 3 . . . . .
5   1 . . 7 . . . . . . . 3 . 0 9 6 . . 2 . . 9 . 2 0 . . . 4 . 1 9 1 4 . 5 . 6 0 . 2 . . . . . 2 9 . . . 3
6   . . . 4 . 9 5 0 . 1 . 0 . . . 8 6 . 3 . . 1 . . . . 7 . . 6 1 . . 0 9 . 8 2 . 3 . 2 8 5 3 . . . . .
7   . 4 . . . . 7 . . 9 . . . 2 . 3 . 4 . 0 2 5 4 8 . . . . 1 3 . . 7 . 0 6 1 . 9 . 6 8 . 7 4 2 0 . . .
8   . . . . . . . . 2 . . . 2 1 . 3 7 . . . . 0 . 3 2 . . . . . 7 . 3 . . . . . . . . 6 . . . . . . .
9   . . . . . . 6 . 4 . . . . 1 5 3 7 . . . 2 5 9 . . . . 6 3 . . . . . 2 . . 1 0 8 . . . . . . . . . .
```

It is still possible to work out what most of the blank entries have to be. The trick is to consider first the columns with a zero in the top entry. These are the three columns on the left in the tabulation below. The next seven columns there are chosen to be those with an entry in common with those on the left. These may be collated to give the full column 'A' on the right.

```
t =   0  6  8  4  9  2  3  5  1  9        A
u =   0  8  9  7  4  5  6  1  1  4

0     0  0  0  .  .  .  .  .  .  .        0
1     .  .  .  .  .  .  5  .  .           5
2     .  .  .  .  1  .  1  1  .  1        1
3     7  .  .  7  7  .  .  .  .  7        7
4     .  9  9  .  .  9  9  9  .  .        9
5     .  2  2  .  2  2  .  .  2  2        2
6     .  3  3  .  3  .  .  .  .  3        3
7     .  4  .  .  4  4  .  4  4  4        4
8     6  .  6  6  .  6  .  .  .  .        6
9     8  .  8  .  .  8  .  .  .  .        8
```

Columns 'B', 'C', etc, can be worked out analogously. This method fills in the missing entries in the 10×100 table and so recovers Square 6.

14.8   Decrypting the Indicators

Square 6 may be viewed as specifying a system for using a single digit as a key to encrypt one of the letters **A, B, C, D, E, F, G, H, I, J**. For example, 3 encrypts **G** as 8. It is possible to consider Square 6 as an encrypting conversion square in its own right and calculate the corresponding decryption square. This turns out to be:

	0	1	2	3	4	5	6	7	8	9
0	A	I	B	F	G	C	J	H	D	E
1	B	J	I	E	H	A	G	D	C	F
2	C	A	E	J	D	F	H	G	I	B
3	D	E	C	H	B	C	F	A	G	I
4	E	C	F	G	C	B	D	F	H	A
5	F	C	A	G	I	H	B	J	E	D
6	G	D	F	A	J	I	C	E	B	H
7	H	B	D	C	A	G	E	I	F	J
8	I	H	G	D	F	E	A	B	J	C
9	J	F	H	B	E	D	I	C	A	G

Square 7
(Decrypting)

Square 7 may now be used to decrypt the sample of beginnings of intercepts given earlier. This yields the following:

discr	GAT2	ENCR	IND	GAT5
2468	6989	DDIC	IIFF	2193
2468	5455	DJJD	IIGG	1100
2468	9388	AFJE	BBDD	8415
2468	6704	HIEI	EEAA	0117
2468	6315	DBJG	HHFF	8008
2468	5228	DJCJ	IIFF	6771
2468	6844	DAIE	HHJJ	9665
2468	5819	DAFH	JJGG	6882
2468	9675	ABFI	JJDD	4530
2468	4614	HAIC	GGEE	7301

Here the ten capital letters from A to H are being used to encode the ten digits from 0 to 9. These have to be decoded to recover the indicators.

14.9   Decoding the Ten Letters

The decoding is remarkably easily[13] achieved by writing out the decrypted third groups of the first 200 intercepted messages and sorting them. In the randomly generated example this yields:

AAAE	AABH	AAFD	AAGA	AAHG	AAHG	AAJF	AAJF	ABBA	ABCF
ABDJ	ABFI	ABHE	ABHE	ABID	ABID	ACAI	ACAI	ACBF	ACBF
ACCE	ACDA	ACEJ	ACEJ	ACFG	ACGC	ACHD	ACIB	ACIB	ADBJ
ADCA	ADCA	AEAB	AEBG	AEBG	AECJ	AFBI	AFDH	AFEC	AFIA
AFJE	AGAA	AGCC	AGDD	AGII	AGJJ	AHAG	AHBE	AHCD	AHGH
AHGH	AHGH	AHGH	AHJI	AHJI	AIAJ	AICB	AIDE	AIGI	AIGI
AIIH	AJAF	AJAF	AJAF	AJAF	AJCH	AJCH	AJCH	AJED	AJFE
AJFE	AJIG	DAAC	DADG	DAEI	DAFH	DAGD	DAGD	DAHF	DAHF
DAIE	DAIE	DAIE	DBCB	DBEF	DBJG	DBJG	DCFF	DCGG	DCGG
DDAG	DDFJ	DDGH	DDIC	DEAI	DEFG	DEGC	DEJH	DFFI	DFHE
DFJC	DGAD	DGAD	DGBI	DGCG	DGEC	DGGF	DGHJ	DGHJ	DHBC
DHED	DHFE	DHFE	DHGJ	DHGJ	DHGJ	DHHI	DIAE	DIAE	DIBH
DICI	DIFD	DIFD	DIFD	DIHG	DIIJ	DIJF	DJAB	DJCJ	DJCJ
DJGE	DJHA	DJIF	DJJD	HAAG	HAAG	HABE	HACD	HACD	HADF
HADF	HAGH	HAHB	HAIC	HBAE	HBAE	HBBH	HBCI	HBDC	HBEB
HBEB	HBIJ	HBIJ	HCAD	HCDH	HCDH	HCDH	HCEC	HCFB	HCHJ
HDAF	HDED	HDGJ	HDHI	HDIG	HDIG	HDJA	HEAA	HEHH	HEII
HEJJ	HFAJ	HFBD	HFDE	HFDE	HGBA	HGCF	HGDJ	HGDJ	HGHE
HGID	HGJC	HGJC	HGJC	HHDI	HHDI	HHDI	HHGE	HHHA	HHJD
HICA	HIEI	HIFH	HIHF	HJBF	HJCE	HJGC	HJIB	HJIB	HJJH

[13]Here 'easily' is the operative word. Like the switch made from JN-25B7 to JN-25B8 in December 1941, a change of conversion square created false confidence that security had been enhanced.

Here A, D and H are code for 1, 2 and 3 but not necessarily in that order and the fourth letter represents the (false) sum of the digits represented by the previous three letters. One notes the false sum $H + D + A = F$ and deduces that $F = 6$. Next, since $E \neq 3$ and $A + A + A = E$, $A \neq 1$. Similarly, since $E \neq F = 6$, $A \neq 2$. Hence $A = 3$ and so $E = 9$. Next the false sum $A+A+J = F$, that is $3+3+J = 6$ yields $J = 0$. These values may be substituted in the first two lines of the sorted decrypted third groups to yield:

```
3339   33BH   336D   33G3   33HG   33HG   3306   3306   3BB3   3BC6
3BD0   3B6I   3BH9   3BH9   3BID   3BID   3C3I   3C3I   3CB6   3CB6
```

The reader is now invited to finish the decoding and then to substitute the appropriate digits for the letters at the top of Square 6. Square 4 may now be recovered by re-arranging the columns of Square 6. It turns out to be:

		0	1	2	3	4	5	6	7	8	9
	0	6	7	8	0	1	2	3	4	5	9
	1	1	4	7	5	2	0	9	6	8	3
	2	3	6	4	1	8	9	5	7	0	2
	3	5	3	0	7	9	4	6	8	2	1
Square 4	4	2	8	6	9	3	5	7	1	4	0
(New)	5	7	5	9	2	4	6	0	3	1	8
	6	4	9	1	3	5	8	2	0	6	7
	7	9	0	2	4	7	1	8	5	3	6
	8	8	1	3	6	0	7	4	2	9	5
	9	0	2	5	8	6	3	1	9	7	4

The inverse, 'Square 5', will not be printed here.

14.10   The Algebra of Permutations

This section requires some understanding of modern algebra. The 'natural assumption' of quoted CBTR paragraphs F and G is analysed to determine the method of recovering the non-primary square and then the primary square in use by the enemy.

Here *permutations*, denoted by the lower case Greek letters σ, τ, sometimes with subscripts, are always of the ten digits 0, 1, 2, 3, 4, 5, 6, 7, 8, 9. They act on the left and can be specified by writing the ten digits once each between square brackets. So if $\tau = [6\,7\,8\,0\,1\,2\,3\,4\,5\,9]$ then $\tau(0) = 6$, $\tau(1) = 7$, $\tau(2) = 8$, etc.

The data of a conversion square Σ (Appendix 2 of Chap. 13) is thus equivalent to the data of ten permutations σ_0, σ_1, $\ldots$, σ_9. Thus, in Square 4, σ_0 is the permutation called τ above, σ_1 is given by the second row in the square: $\sigma_1 = [1\,4\,7\,5\,2\,0\,9\,6\,8\,3]$, and so on. In general a conversion square Σ may be represented symbolically by a display such as that on the left:

	0–9		0–9
0	σ_0	0	$\sigma_0 \tau$
1	σ_1	1	$\sigma_1 \tau$
2	σ_2	2	$\sigma_2 \tau$
3	σ_3	3	$\sigma_3 \tau$
4	σ_4	4	$\sigma_4 \tau$
5	σ_5	5	$\sigma_5 \tau$
6	σ_6	6	$\sigma_6 \tau$
7	σ_7	7	$\sigma_7 \tau$
8	σ_8	8	$\sigma_8 \tau$
9	σ_9	9	$\sigma_9 \tau$

The display on the right, in which τ denotes any fixed permutation, represents another square, which will be called $\Sigma\tau$.

Suppose then that a conversion square Σ, as yet unknown, has been used to encrypt the indicator group $m\,m\,n\,n$ using as key the fifth GAT $uv\bar{u}\bar{v}$. The (encrypted indicator) fourth GAT is then $tw\bar{t}\bar{w}$, where:

$$t = \sigma_u(m) \qquad w = \sigma_v(m) \qquad \bar{t} = \sigma_{\bar{u}}(n) \qquad \bar{w} = \sigma_{\bar{v}}(n)$$

and as $\sigma_u(m) = t$ is equivalent to $m = \sigma_u^{-1}(t)$ the dependence of w on u, v and t can be expressed[14] as $w = (\sigma_v\sigma_u^{-1})(t) = \sigma_v\left(\sigma_u^{-1}(t)\right)$. Likewise there is a formula $\bar{w} = (\sigma_{\bar{v}}\sigma_{\bar{u}}^{-1})(\bar{t}) = \sigma_{\bar{v}}\left(\sigma_{\bar{u}}^{-1}(\bar{t})\right)$.

As $(\sigma_v\sigma_u^{-1})(t) = \left((\sigma_v\tau)(\sigma_u\tau)^{-1}\right)(t)$, the formula expressing w as being determined by u, v and t cannot be used to recover the *primary* square Σ. Indeed, it could just as well recover $\Sigma\tau$ instead. The non-primary[15] (*CBTR* jargon, see paragraph G of Sect. 14.2) square is the best that can be hoped for.

When the dependence of w is tabulated with ten rows corresponding to the ten values of v and 100 columns corresponding to the 100 possible values for u and v, the ten possibilities for $\sigma_u^{-1}(t)$ give rise to ten possibilities for the columns. These may be called A, B, C, etc. These may be tabulated as before.

---

[14]This is a non-abelian form of 'non-primary'! The matter is best ignored by those without a strong background in classical mathematics.

[15]More on this aspect may be found in Peter Donovan's paper in *Cryptologia* 30(3) July 2006 212–235. The *CBTR*, Part G, notes that in most other Japanese Army mainline codes the second indicator group consisted of the row co-ordinate of the first group of additive used, the column co-ordinate of the first group used, the row co-ordinate of the last group used and finally the column co-ordinate of the last group used.

14.11   Remark on Indicators

As noted earlier, the practice in using certain other IJA *mainline* additive systems was to give an enciphered version of the starting place in the additive table somewhere near the beginning and an enciphered version of the ending place somewhere near the end. As the enemy cryptanalysts were capable of counting the number of message groups in the intercept this practice may have served as a check but has a considerable element of having two separate encryptions of the plain text indicators in it. At best, such a practice was potentially[16] insecure. If it is desired to have the capacity to check the indicators of messages, the safest method is just to transmit each one twice without attempting further concealment. All that is needed is an indicator encoding and encryption system that is secure enough not to require concealment.

Appendix 1   Back to the CBTR Text

The mathematically minded may care to deduce from (i) in the quoted *CBTR* paragraph C that the conversion square in use was not a Latin square! A 10×10 Latin square in the conventional mathematical sense is a conversion square in which each digit occurs not only once in each row but also once in each column. The two basic sample squares used here, called 'Square 2' and 'Square 4', are both Latin squares. The reader may find their source by asking Google for Graeco-Latin Square + Parker.

In 'Square 8' below the digit 7 does not occur in the columns headed by **1, 2** and **3**. It may be checked that when this square is used to encrypt 1, 2 or 3 the encrypted digit is one of those italicized in the square and so cannot be 7. Use of such a non-Latin square runs the risk of producing a 'missing digit' phenomenon perhaps comparable with that emerging from the indicators of JN-25A1. As the Friedmans remarked in the quoted text, this sort of thing is what the cryptologists of the era were looking for.

---

[16]A careful analysis shows that exploitation of the use of a check summing digit in the indicators is very difficult with only the calculating capacity of the 1940s available.

	0	1	2	3	4	5	6	7	8	9
0	6	9	8	0	1	2	3	4	5	7
1	1	4	3	5	2	0	9	6	8	7
2	3	6	4	1	8	9	5	7	0	2
3	5	3	0	1	9	4	6	8	2	7
4	2	8	6	9	3	5	7	1	4	0
5	7	5	9	2	4	6	0	3	1	8
6	4	9	1	3	5	8	2	0	6	7
7	9	0	2	4	7	1	8	5	3	6
8	8	1	3	6	0	7	4	2	9	5
9	0	2	5	8	6	3	1	9	7	4

Square 8 (rows **4**–**5**)
(not Latin)

Parts (i) and (ii) of paragraph C suggested that the third GAT warranted diligent study. Only part (ii) helped with the decryption.

Appendix 2   Back to Poland 1932

Richard's observation of the connection between the first and second digits of the fifth GAT of a 2468 message and those same digits in the fourth GAT looks far removed from the early 1930s Polish Cipher Bureau work on the German Army Enigma. In fact, there is an important analogy linking his observation with the observed properties of the first six letters of an Enigma message of the time. A brief explanation of how recognition of the connection between the first and second triples in these six letter indicators was exploited by Rejewski[17] is now given but it is convenient to reproduce three salient paragraphs from Sect. 8.3:

> Prior to sending a message, the operator was also required to choose a new initial position for each of the rotors. This choice had to be conveyed to the message recipient by including an indicator in the message. Of crucial importance to those attacking the cipher in the above period was that the indicator was encrypted twice.

> For example, suppose WDV was specified as the initial rotor positions and the operator randomly chose MHT as the new initial message setting. Having set the rotors to WDV, the letters MHT MHT would be typed in obtaining, say, ZXCHTJ. The rotors were then set to MHT and the message text encrypted. The message transmitted by the radio operator included both encryptions of the indicators, here ZXC HTJ, and the encrypted text.

> Enigma was designed to be reversible—the recipient, with machine set to the same initial configuration, typed in ZXC HTJ, obtaining MHT MHT. Now, with the rotors reset to MHT, typing in the received message text would produce the original plain text of the message.

---

[17]Rejewski published in *Cryptologia* 6(1), January 1982, 1–18 an account of the mathematics involved in the solution (decrypting) of Enigma. The book by Hugh Sebag-Montefiori on *Enigma* is quite useful, as is Władyslaw Kozaczuk's *Enigma*.

The paragraphs quoted above follow a short explanation stating the common daily instructions given to Enigma operators prescribing the machine's starting configuration (rotor order, initial settings for each rotor and plugboard connections) for the day. These instructions play no role in the method developed by Rejewski for extracting valuable information from the first six letters of the intercepts of a given day.

It is the 'double encryption' or 'redundant encryption' of the message setting that provides the link with the Richard observation. Rejewski and his colleagues would almost certainly have failed to make much progress in attacking the encrypted message indicators. Specifically, Rejewski exploited the fact that the first and fourth letters both encrypted the first letter of the actual indicator, the second and fifth did this for its second letter and the third and sixth did so for its third letter. His procedure applied in the same way to each of the three pairs and it suffices to describe it once, say for the first pair. He found that in practice some 60–80 messages sent on the same day were needed for it to work. His message did not seek to reveal the message indicator. But it did reduce very substantially the number of possibilities to be tested for establishing the initial configuration of the machine for a given day.

His idea was to identify the closed cycles that make up the sequence of first and fourth letters found in the intercepts. Suppose he started with the letters NWQ ZCJ He then looked for a message with the first letter Z (the fourth letter here). There would usually be one or more candidates: we suppose he chose ZTV GSU He had now obtained a chain N,Z,G and then he would seek a message beginning with G, say GHB CNL. So he could add C to the chain which became N,Z,G,C. Eventually he would complete the loop or closed chain, say (NZGCQAXVL), that is a message beginning with L would have fourth letter N and the loop has been closed. He would now take another intercept with encrypted indicators (say) BHP BDY beginning with a new initial letter. This would produce a loop of length one. Continuing in this way, he would finally be able to write down the following sequence of loops:

$$(NZGCQAXVL) (B) (KH) (IR) (YOUDJEMPT) (F) (S) (W)$$

giving the lengths 1, 1, 1, 1, 2, 2, 9 and 9, which sum as they must do to 26.

This process repeated for the second and fifth letters of the intercepts might yield the chains

$$(ACDWKRUT) (BGELXQFN) (HSV) (IMO) (JP) (YZ)$$

giving the lengths 2, 2, 3, 3, 8 and 8. Again the sum has to be 26. The process has to be repeated for the third and sixth letters, which might yield lengths 13 and 13.

Knowledge of the structure and operation of the Enigma machines explains why the lengths of loops occur in pairs. Hence the information derived from this process applied to (say) the first and fourth letter is a method of expressing 13 as a sum of positive numbers, $1 + 1 + 2 + 9$ in the example. Such methods of expressing 13 are called partitions of 13. It turns out that there are 101 of these, including 13, $1 + 12$, $1 + 1 + 11, 2 + 11, 1 + 1 + 1 + 10, 1 + 2 + 10, 3 + 10, \ldots$.

The number of possible such triples of patterns (on for each of the three pairs used) is thus $101^3 = 1,030,301$.

Once the wiring of the Enigma was fully known, it was possible to build a machine that, for each of the $6 \times 26^3 = 105,456$ choices of rotor order and starting positions, worked out the corresponding triple of partitions. Although there was a surprising amount of duplication with only about 7,000 triples actually occurring, it was possible to tabulate this—in fact using one card for each of the 105,456 possibilities—and then read off what was usually a modest number of possible order and initial positions for the rotors.

C. H. O'D. Alexander (1909–1974) was head of Hut 8 (Naval Enigma) at Bletchley Park for part of WW2 and played significant roles elsewhere. He wrote a detailed report *Cryptographic History of Work on the German Naval Enigma*, now item HW 25/1 in the British National Archives and available on the Web. The following text is sections 30, 31 and 32 of the Introduction to that report.

> Beside this system an earlier and much less cryptographically secure method was used on a number of keys. Here there was no K book or Bigram Tables and the operator chose the message setting out of his head. Suppose this was ASD. Then setting up to the Grund he encoded ASD twice, i.e. he encoded ASD ASD giving QJV RTS. This was then sent off generally with dummy letters inserted (in Mediterranean traffic the 2 extra letters indicated what particular key was being used). So the groups XQJV ARTS might be sent over the air.
>
> It is easy to spot when this system is being used because if in *two six letter indicators (ignoring the dummy letters) the first, second or third letters agree so must the fourth, fifth or sixth respectively and vice versa.* In our example ASD ASD encoded as QJV RTS. Now suppose we have another indicator group beginning with Q—then this implies that the message setting must be A** and therefore A** A will be encoded at the Grund which must produce an R in the fourth place to agree with the R of RTS. This 'throw-on' effect from Q to R is what gives the system its name.
>
> The 'throw-on' effect produces very serious weaknesses enabling a system to be broken on quite a small number of messages and the enemy's use of it undoubtedly played a very large part in our success. From the user's point of view the system is a good one since every indicator is checked and corruptions are automatically shown up—for this reason, no doubt, it was originally recommended by the makers and adopted by the Germans.

The text beginning 'two six letter indicators' has to be compared with the paragraphs D and E quoted from the *CBTR* earlier. Quite evidently much of the exploitation of 'throw-on' runs parallel to that of exploiting the redundant encryption of the Water Transport indicators. Abe Sinkov had been at Bletchley Park in February 1941 and had been introduced to the Enigma. One may speculate that this background knowledge helped Central Bureau in its attack on the redundantly encrypted indicators of the Water Transport Code. In both Pacific and Atlantic, 'the enemy's use of it undoubtedly played a very large part in our success'.

The practice of 'tailing' by JN-25 operators has been mentioned earlier. It amounts to choosing as the first additive to be used in a new message the successor of the last additive used in the previous message. The use of indicators for the new message was then essentially redundant. This was a very dangerous practice: enemy cryptologists could count the number of groups used. Likewise the use of encrypted indicators for both the first and last additive used was redundant. The secure way to guard against corruption of the encrypted form QJV of the indicator ASD is to transmit it three times: QJV QJV QJV.

It is possible to use the concepts of information theory to interpret the weakness in using only scannable groups in JN-25 as another form of redundant encryption. Briefly, the IJN was using 5-digit groups as encrypted forms of 5-digit scannable groups. The former contain 5 decimal digits of information while the latter contain only $\log_{10} 33334 = 4.52$ digits of information. Thus there is .48 digits of information redundancy in every 5, and so the whole JN-25 encryption process is 9.6 % redundant.

Chapter 15
The Scanning Distribution

Increases in the complexity of the JN-25 cipher systems from 1939 to 1943 must have suggested that developing new techniques in anticipation of future challenges was desirable. A section of Op-20-G devoted to Mathematical Analysis and Machines and called Op-20-GM was established early in 1942 and must have had the JN-25 systems as a high priority. A peculiarity found in the analysis of the 33,334 book groups underlying these systems turned out to yield important new methods of searching for alignments and did not depend on the breaking of indicator systems. This chapter explains these developments.

15.1 By-Passing the Indicators

A key theme has been mentioned previously in Sects. 9.3, 9.4 and 13.2:

> If cryptanalysts are unable to work out the rules for the indicators, then the additive system may successfully defy repeated attack.

> Cryptologists sought to use the indicator(s) to obtain depth (alignment). Any system that allowed other methods to reconstruct alignment would be insecure.

Until August 1942 there had been only one authorised JN-25 system in use at any one time. Three code books and nine additive tables had sufficed. But after then the IJN used combinations of several more code books and numerous additive tables with three or more systems in concurrent use. The small message volumes intercepted from some of these systems were already making decryption difficult. Worse was to come in 1944!

© Springer International Publishing Switzerland 2014
P. Donovan, J. Mack, *Code Breaking in the Pacific*,
DOI 10.1007/978-3-319-08278-3_15

Edward Simpson's history[1] of work on the JN-25 series of ciphers is stating something rather sensational on page 13 in recording a visit made to his team at Bletchley Park in November 1943. The text is:

> [Howard Engstrom of Op-20-GM] gave us the first news we had heard of a method of testing the correctness of the relative setting of two messages using only the property of divisibility by three of the code groups. The method was known as Hall's weights and was a useful insurance policy just in case [the indicators of] JN-25 ever became more difficult. He promised to send us a write-up of it.

Thus the use of only scannable groups in the construction of JN-25 code books not only made decryption easier but also left internal evidence of alignment. This made the cryptanalysts less dependent upon breaking the indicator system. This 'insurance policy' became relevant in 1944 when the indicator systems became more secure. So Hall's weights (usually written just as Hall weights) may have had only limited historical importance but deserve to be remembered as one of the best pieces of WW2 cryptology.

The NARA file RG457 series A1 9032 SRH-275 contains non-technical correspondence from Frumel to Op-20-G Washington. Two remarkable lines in it dated February 1944 are:

> The new method of cryptanalysis invented and exploited by Frupac (Fleet Radio Unit Pacific) is beyond praise.

This would have been written by Syd Goodwin, by then the Commandant. In all probability this refers to something in this chapter. In fact Hall weights are presented here as the first of four[2] associated weaknesses in JN-25 caused by the limitation of using only scannable groups in code books.

Alan Turing's work in statistics comes in here as a key ingredient in WW2 cryptology. The following text from the 1945 GCCS *General Report on Tunny* (the encrypted teleprinter)[3] is at the heart of the matter:

> The fact that Tunny can be broken at all depends upon the fact that P, χ, Ψ', K and D have marked statistical, periodic or linguistic characteristics which distinguish them from random sequences of letters.

---

[1]The Simpson report is in the British Archives as item HW 8/149. It should not be confused with the technical report by J. W. S. Cassels and Simpson available as item HW 43/34. This latter was declassified only in 2013.

The link with Marshall Hall is confirmed by the record of a presentation on the scanning distribution given by Hall in the Dayton Navy Computation Laboratory on 18 December 1943.

[2]In summary, the use of only scanning groups in most JN-25 code books was a major blunder.

Those who like that sort of thing may wish to read David Wragg's *Snatching Defeat from the Jaws of Victory* for an account of 28 of the biggest military blunders of the twentieth century, starting with the 1905 Battle of Tsushima. Other such books are Saul David's *Military Blunders: The How and Why of Military Failure*, Robinson, London, 1988 and Kenneth Macksey's *Military Errors of WW2*, Cassell, 2005. These books do not mention JN-25.

[3]This has already been quoted in Appendix 3 of Chap. 10.

Marshall Hall, Jr. (1910–1990) emerges as a principal figure in WW2 American cryptography. His postwar work includes constructing the simple group (in the mathematical sense) of order 604,800. He was elected to membership of the American Academy of Arts and Sciences. In a few pages of autobiography[4] Hall noted that Howard Engstrom (1902–1962) gave him much help with his Ph.D. thesis at Yale in 1934–1936 and later urged him to work in Op-20-G in WW2.

> I was in a research division and got to see work in all areas, from the Japanese codes to the German Enigma machine which Alan Turing had begun to attack in England. I made significant results on both of these areas. During 1944 I spent 6 months at the British Headquarters in Bletchley. Here there was a galaxy of mathematical talent including Hugh Alexander the chess champion and Henry Whitehead the eminent topologist . . .

Hall was one of about ten core members of an Op-20-G team of about 30 not too far from being another galaxy of mathematical talent.

15.2 Characteristics

The *characteristic* $\chi(\mathbf{x})$ of a 5-digit group $abcde$ is defined as a false sum: $\chi(\mathbf{x}) = a + b + c + d + e$. (This is the same as the ordinary sum taken modulo 10.) The following general algebraic relationship holds:

$$\chi(\mathbf{x} + \mathbf{y}) = \chi(\mathbf{x}) + \chi(\mathbf{y}).$$

There are 10,000 of the 100,000 5-digit groups with a given characteristic. Thus the characteristic distribution is uniform on the digits $0, 1, \ldots, 9$ with each value having probability $10\,\% = 0.1$.

A quite different and most surprising outcome is found if one computes the characteristics of the 33,334 5-digit scannable groups:

characteristics 6 and 9 each occur for 925 (2.7 %) of these,
characteristics 2 and 3 each occur for 1780 (5.3 %) of these,
characteristics 0 and 5 each occur for 3247 (9.7 %) of these,
characteristics 7 and 8 each occur for 4840 (14.7 %) of these,
characteristics 1 and 4 each occur for 5875 (17.6 %) of these.

The occurrences of relative frequencies in pairs is a consequence of the fact that $\chi(99999 - \mathbf{x}) = 5 - \chi(\mathbf{x})$.

'Symmetry' and 'wave' properties are more easily seen if the distribution of the characteristics is written out as in the following table:

---

[4]Hall's autobiographical notes are on pages 367–374 of Peter Duran, Richard Askey and Uta Merzbach *A Century of Mathematics in America*, volume 1, American Mathematical Society, Providence, RI, 1989.

$\chi =$	9	2	5	8	1	4	7	0	3	6
Number	925	1780	3247	4840	5875	5875	4840	3247	1780	925
Percent	2.7	5.3	9.7	14.7	17.6	17.6	14.7	9.7	5.3	2.7

The set of probability values $p(9) = \frac{925}{33334} = .027$, $p(2)$, etc., set out in the third row is called the *scanning distribution*. Its application to the decryption of the JN-25 systems is developed below. What remains obscure is why Hall (or a colleague) had the idea of working out these numbers at all. Once it is noted and analysed, a significant distinguishing feature of the typical JN-25 code book has been exposed. The quotation from the *General Report on Tunny* given above is now applicable to JN-25 and methods to exploit this feature may be anticipated.

15.3   Weakness 1: Alignment and Hall Weights

The methods described previously for building up the table of additives in a JN-25 system required the exploitation of insecure indicator systems enabling the alignment of intercepts to produce depths.

Sometimes in 1944 the page part of the indicators for JN-25 could be read but the row and column co-ordinates for JN-25 were chosen randomly as in the example from Sect. 8.12:

$$\textbf{69} \quad \textbf{04} \quad \textbf{15} \quad \textbf{75} \quad \textbf{98} \quad \textbf{23} \quad \textbf{42} \quad \textbf{95} \quad \textbf{54} \quad \textbf{56} \quad \textbf{30} \quad \textbf{11}$$

The cryptologists needed a method[5] to align two intercepts known to have come from the same page of additives.

With 33,334 scanning 5-digit groups, there are 33,334×33,334 pairs $\mathbf{a}$, $\mathbf{b}$ of such groups. The numbers of such pairs $\mathbf{a}$ and $\mathbf{b}$ with given values of $\chi(\mathbf{a})$ and $\chi(\mathbf{b})$ are the 100 products 925×925, 925×1780, etc. The number of such pairs with a given value of the false difference $\chi(\mathbf{a}) - \chi(\mathbf{b})$ may then be calculated. The probability distribution[6] q for the 10 differences is:

$$i = \quad 0 \quad\quad 3,7 \quad\quad 4,6 \quad\quad 1,9 \quad\quad 2,8 \quad\quad 5$$
$$\text{prob. } q(i) = .13051 \ .12467 \ .10939 \ .09056 \ .07535 \ .06955$$

---

[5]There is a variant of the Mamba method for putting JN-25 GATs into depth when there are several active operators who are reliable tailers. It is assumed that each of them chose a starting point in a new additive table independently and worked on from there. The indicator system in use may well be totally incomprehensible. The practice of tailing would produce long stretches of GATs encrypted with consecutive additives from the unknown table and one could use this method on, say, 100–200 such GATs to find possible alignments with another long stretch. The challenge of mechanising the attack on such a system is not considered here at all.

[6]This is essentially the convolution square of the scanning distribution.

Thus if GATs $\mathbf{a} + \mathbf{x}$, $\mathbf{b} + \mathbf{x}$ are in depth, the probability that the (false) difference $\chi(\mathbf{a} + \mathbf{x}) - \chi(\mathbf{b} + \mathbf{x}) = \chi(\mathbf{a}) - \chi(\mathbf{b})$ is 0, 3, 4, 6 or 7 is about 59.9 %. This helps put JN-25 intercepts into alignment. For example, consider the following 24 adjacent depths of 2:

```
32435 17488 89429 03048 90928 18433 74515 76330 20116 00079 11976 81852
14943 17034 03445 51285 47487 89870 65481 09310 06715 67753 43942 94831

81328 26799 36834 38719 97799 67640 43506 91287 36843 50688 61640 69630
82173 31701 42374 71281 39429 12930 05676 96682 93553 94128 11455 32106
```

The differences of the characteristics are then:

5	5	8	7	7	6	4	2	6	9	7	0
0	3	6	2	3	7	0	2	4	4	6	3

and as 16 of these are 0, 3, 4, 6 or 7 while only eight are 1, 2, 5, 8 or 9 this is quite likely to be a real alignment.

A more accurate approach is to use the ten Bayes (logarithmic) terms:

$i =$	0	3, 7	4, 6	1, 9	2, 8	5
$10q(i) =$	1.3051	1.2467	1.0939	.9056	.7535	.6955
$\log(10q(i)) =$	.26628	.22050	.08975	−.09920	−.28300	−.36308

If these logarithmic Bayes terms are scaled by the factor 18.777 the following table (A) is obtained:

$i =$	0	3, 7	4, 6	1, 9	2, 8	5
scaled (A)	5.000	4.140	1.685	−1.862	−5.314	−6.818
rounded (A)	5	4	2	−2	−5	−7

The next two reasonable scalings and roundings, denoted here by (B) and (C), are obtained from factors 42.31 and 112.66 respectively.

$i =$	0	3, 7	4, 6	1, 9	2, 8	5
scaled (B)	11.000	9.108	3.707	−4.098	−11.691	−14.999
rounded (B)	11	9	4	−4	−12	−15
scaled (C)	30.000	24.842	10.112	−11.176	−31.884	−40.906
rounded (C)	30	25	10	−11	−32	−41

As already noted, those with only WW2 technology available would have much preferred calculating using only small integers. Scaling (C) above yields only small integers as the 'weights' with relatively little distortion (about 1.5 %) caused by the rounding.

Presumably the 'Shinn weights' and 'Hall weights' used by Op-20-G were two of the three systems constructed above.

The weights (C) can be tested in the case of two signals of N GATs each set in alignment. The difference of the characteristics may be calculated with 0 occurring n_0 times, 1 occurring n_1 times, etc. One then evaluates:

$$S = 30n_0 - 11n_1 - 32n_2 + 25n_3 + 10n_4$$
$$- 41n_5 + 10n_6 + 25n_7 - 32n_8 - 11n_9.$$

The appropriate threshold values seem to be -100 and 100. So we consider the reliability of stating that

- if $S \leq -100$ the two intercepts are not in alignment;
- if $S \geq 100$ the two intercepts are in alignment;
- and otherwise more evidence is needed.

Computer calculation of larger numbers of randomly generated samples shows that:

If $N = 50$, this is correct in 58 % of cases, wrong in 9 %;
If $N = 100$, this is correct in 75 % of cases, wrong in 7 %;
If $N = 200$, this is correct in 90 % of cases, wrong in 3.5 %;
If $N = 400$, this is correct in 97.5 % of cases, wrong in 1 %;
If $N = 800$, this is correct in 99.8 % of cases.

The information tabulated above does not of itself solve the problem of alignment. The difficulty is that the prior odds that two messages are in alignment is so low (around .00002) and therefore a very great weight of evidence is needed to show that there really is an alignment.

Suppose that a hit, that is the occurrence of one GAT in two separate messages, has been noted as in the following concocted example:

```
32435  17488  89429  03048  90928  18433  74515  76330  20116  00079  11976  69630
14943  17034  03445  51285  47487  89870  65481  09310  06715  67753  43942  94831

81328  26799  36834  38719  97799  67640  43506  91287  36843  50688  61640  69630
82173  31701  42374  71281  39429  12930  05676  96682  93553  94128  11455  32106
```

The set of differences of characteristics provides some considerable evidence that the hit is not a coincidence but a strong indication of alignment. A Bayesian analysis is needed to assess rationally whether this proposed alignment is (almost certainly) genuine.

Note 24 of Chap. 1 refers to the search for double hits,[7] that is two separate hits the same distance apart in two signals. Indeed a device was invented to

---

[7] The reference is NARA RG38 CNSG Library Box 15 item 3222/54. This was discovered by Ralph Erskine of Belfast, who is most heartily thanked for drawing it to our attention. The document is anonymous and carries no date but presumably was written in the second half of 1943. Another

search for double hits. Here the Bayesian approach has good chances of being successful. Likewise, if it is known from partly read indicators that two intercepts were encrypted from the same page, the Bayesian approach is potentially valuable. It may help work out the row and column co-ordinates on that page.

15.4   Weakness 2: The Mamba Process

In previous chapters the exploitation of insecure indicator systems to place various signals in depth (alignment) has been discussed. Here a method of aligning JN-25 signals with a known portion of an additive table without using the indicators is described. This was quite useful when the indicator systems for JN-25 were upgraded and led to the 'Mamba' method which is based on the scanning distribution.

This chapter follows *Op-20-G usage*,[8] the reverse of the convention used earlier: encryption is carried out by subtracting the additive, decryption by adding it. The following contrived example will be used:

---

document in RG457 Box 622, item 1682, dated November 1943 and entitled *Research Report on Alignment of JN-25 Traffic* is also of interest.

The counts of the numbers of multiples of 3 with the various characteristics may be found in NARA but have been checked using a modern computer.

NARA RG457 Box 705, item 1742 includes copies of a report by LtCdr E. W. Knepper and Lt (Jg) L. E. Shinn entitled *A Proposed Method for Placement of Partially Keyed Messages in Additive Systems when some Additives are Available*. This appears to have been written in March 1944 or perhaps a little earlier. The document appears to refer to situations in which some progress has been made by a baby brute force procedure coupled with the methods set out in Chap. 12 to recover a run of additives from a page.

Presumably Knepper was the principal inventor of *knepperizing*, a third method of decrypting JN-25 systems once the common book groups are known. This is described in the *GYP-1 Bible*. More information is given in *History of GYP-1*, NARA RG38, CNSG 5750/202.

Other interesting references to the technology of the era are in RG457 Box 705, items 1741 and 1743 on *Tape scanner for double repeats, Japanese JN-25 traffic, 1943–1945*.

[8] Although the IJN operators added the 'additives' to the book groups to obtain the GATs and subtracted it from the GATs to recover the book groups, Op-20-G found it convenient to use the opposite practice after it had reconstructed the encryption process. Its version of the additive table was the negative of the IJN original. Thus Op-20-G decryption was carried out by (false) addition. This 'Op-20-G usage' is used in this chapter here for the sole purpose of making the surviving Mamba cards in NARA RG457, Box 705, slightly more intelligible. See Note 19 of Chap. 8.

The section entitled *A Distinction in Additive Table Usage* in Sect. 8.14 explained that the IJA and IJN used different practices when at the end of a page of additive. The Army usage was to go back to the first additive on the same page, while the Navy went on to the first additive on the next page. In effect the IJA was using numerous short tables of additives, while the IJN used only one long table.

EXTRACT FROM AN ADDITIVE TABLE (not necessarily a complete page)

01029	57300	13805	34355	99977	66304	06505	57230	56459	28004
87993	22644	89059	38243	16879	43063	33936	41315	22758	68843
76308	28803	10556	06863	65153	28330	31478	96184	69999	88839
09715	01806	81655	61293	28286	53769	21623	46992	46088	44751
88893	40675	74572	71207	65280	23840	25525	40102	36856	08898
72765	07250	76365	14246	24721	99754	77021	60964	24945	69026
87549	64979	94619	22575	50422	65411	89660	31238	14735	22614
98473	30106	49089	42050	34044	71141	78279	13005	75846	78719
72190	20071	90190	31745	45215	34413	84869	03880	29222	94548
48942	92655	90883	35305	12911	82067	46736	79346	55633	21216

We suppose that an intercepted message is being processed without full understanding of the indicator system. This message could be:

SAMPLE INTERCEPTED MESSAGE

07677	69565	04427	91272	25010	84004	47022	82314	17614
90239	78794	10432	36404	31042	19441	14434	47607	12093
75155	56038	81552	59618	58646	54446	69522	00165	53151
29655	65827	46221	52930	31341	31681	82835	13325	30091
79561	03924	88729	97369	39960	78192	37938	45850	90549
56061	44736	85233	86722	12055	37065	04888	70457	01148

Without any knowledge of the current JN-25 code book, it is possible to consider
successively possible alignments of the intercept with segments of the additive table
and calculate the corresponding false sums:

TEST RUN NUMBER 1

GATs	07677	69565	04427	91272	25010	84004	47022	82314	17614	90239	.....
Table	01029	57300	13805	34355	99977	66304	06505	57230	56459	28004	.....
FalseSums	08696	16865	17222	25587	14987	40306	43527	39544	63063	18233	.....
SumDigits	29	26	14	27	29	13	21	25	18	17	.....
Scans?	×	×	×	✓	×	×	✓	×	✓	×	.....

Here the *GATs* line consists of the first ten groups in the intercepted message.
The *Table* line consists of the first ten entries in the additive table. The *FalseSums*
line gives the false sum of the two groups immediately above. The *SumDigits* line
gives the sums of the digits of the groups immediately above. It is used to work out
that most of these false sums are not scanning. So this is not the correct alignment.

So it is necessary to try again, this time starting from the second group in the
table of additives. This also fails the scanning test. The process is repeated and
eventually the cryptologist finds that starting from the tenth group in the extract
from the additive table yields a sequence of scanning groups:

TEST RUN NUMBER 10

GATs	07677	69565	04427	91272	25010	84004	47022	82314	17614	90239	.....
Table	28004	87993	22644	89059	38243	16879	43063	33936	41315	22758	.....
FalseSums	25671	46458	26061	70221	53253	90873	80085	15240	58929	12987	.....
SumDigits	21	27	15	12	18	27	21	12	33	27	.....
Scans?	✓	✓	✓	✓	✓	✓	✓	✓	✓	✓	.....

As here all ten false sums are a multiple of 3, it is extremely likely that this is the correct alignment. This can be checked by calculating the 44 false sums needed to finish 'stripping off' the additive and obtain:

THE ORIGINAL UNENCIPHERED ENCODED MESSAGE

25671	46458	26061	70221	53253	90873	80085	15240	58929
12987	36537	86730	54207	41598	15204	79587	65937	43461
61239	15927	69381	58323	59442	35091	20715	28341	06810
40278	01719	82209	96681	19134	71256	56307	84522	95271
92301	28449	28821	23115	37758	40857	34188	11115	04785
70782	33480	52254	46686	36990	96081	81327	34326	95757

This process might well enable enough intercepts to be put in alignment with either end of the known block so that the block could be extended. The process was used sufficiently frequently that considerable thought must have gone into its possible improvement. Of course in the modern era this sort of thing is easy to program and scarcely needs adaption.

In 1944, a decision was made to use a quite different method, the *Mamba Process*, in automating[9] this alignment method. (The special machine was named Mamba following the then current usage of naming machinery used for naval cryptology after poisonous snakes.)

Only 16.2 % of the scanning groups have characteristics 2, 3, 6 or 9—far fewer than the 40.0 % that would have been guessed. These four characteristics were called *minimals*. The characteristics 1, 4, 7 and 8 were called *maximals*. It turns out that 64.3 % have one of these as false sum. A first version of a useful but radical thought was that one could test a sequence of groups of characteristics as above for divisibility by 3 by checking whether the proportion of them with false sums 2, 3, 6 or 9 is more like 16.2 % than 40 %. The merit of this method is that, instead of working with the original table of additives over and over again, *it is good enough to work with the characteristics of each group*. Thus the table of 100 additives given above can be replaced by a list of the characteristics (*Tab-chars* below) of the digits in each group:

---

[9]As noted earlier, large-scale mental arithmetic is tiring and unreliable. Automation is most desirable for problems like this.

2570196794	6810164449	4173063826	2551604061	6257179783
7477647543	3591379705	1001543902	9090755970	7786436922

Likewise the 54 GATs forming the intercepted message may be replaced by their characteristics (called *GAT-chars* below). Unlike the GATs themselves, the single-digit sums can all be displayed on a single line:

717186   589350   709645   321993   425785   929643   884477   027841   531834

The relevant characteristic sums may now be calculated. This may be compared with the calculation in the 'TEST RUN NUMBER 1' table above.

MAMBA PROCESS RUN NUMBER 1

GAT-chars	717186	589350	709645	321993	425785	929643	884477	027841	531834
Tab-chars	257019	679468	101644	494173	063826	255160	406162	571797	837477
Char-sums	964195	158718	800289	715066	488501	174703	280539	598538	368201

Here the minimals 2, 3, 6 and 9 occur a total of 16 times in the *Char-sums* line. As 16 is about 30 % of 54 rather than about 16 % this is most unlikely to be a correct alignment and so is discarded. As it is possible to go back to the full (non-Mamba) calculations in any case where the figures look marginal, a detailed analysis of the underlying Bayesian inference[10] is not necessary.

MAMBA PROCESS RUN NUMBER 2

GAT-chars	717186	589350	709645	321993	425785	929643	884477	027841	531834
Tab-chars	570196	794681	016444	941730	638262	551604	061625	717978	374776
Char-sums	287272	273931	715089	262623	053947	470247	845092	734719	805500

Here the minimals 2, 3, 6 and 9 occur a total of 22 times in the *Char-sums* line, and 22 is more than 40 % of 54. So here again the 54 sum groups are likely to be random rather than all multiples of 3. Eventually we try:

MAMBA PROCESS RUN NUMBER 10

GAT-chars	717186	589350	709645	321993	425785	929643	884477	027841	531834
Tab-chars	468101	644494	173063	826255	160406	162571	797837	477647	543359
Char-sums	175287	123744	872608	147148	585181	081114	571204	494488	074183

---

[10]See Appendix 2 of Chap. 10.

Here the minimals 2, 3, 6 and 9 occur a total of 8 times, and 8 is near enough to being 16 % of 54. Although checking is still necessary, the correct alignment has been identified.[11]

There is another way of working out the alignment. In the list of 100 table characteristics given above (257019 ...) the digits 0, 1, 2, 3, 4, 5, 6, 7, 8 and 9 can be replaced by the corresponding columns in the left hand (minimal) part of the following table:

```
              MINIMAL                                     MAXIMAL

 0   0  █ █  0   0  █  0   0  █          0  █  0   0  █  0   0  █ █  0
 1  █ █  1   1  █  1   1  █  1           █  1   1  █  1   1  █ █  1   1
 █ █  2   2  █  2   2  █  2   2          2   2  █  2   2  █ █  2   2  █
 █  3   3  █  3   3  █  3   3  █         3  █  3   3  █ █  3   3  █  3
 4   4  █  4   4  █  4   4  █ █          █  4   4  █ █  4   4  █  4   4
 5  █  5   5  █  5   5  █ █  5           5   5  █ █  5   5  █  5   5  █
 █  6   6  █  6   6  █ █  6   6          6  █ █  6   6  █  6   6  █  6
 7   7  █  7   7  █ █  7   7  █          █ █  7   7  █  7   7  █  7   7
 8  █  8   8  █ █  8   8  █  8           █  8   8  █  8   8  █  8   8  █
 █  9   9  █ █  9   9  █  9   9          9   9  █  9   9  █  9   9  █ █
```

(This display is about 1.2 times the standard size. The 'X' and 'Y' rows at the top of the standard IBM card are disregarded in this chapter.)

Here in the column corresponding to 0 the holes that are punched correspond to the four minimals. In the column corresponding to 1, the holes that are punched are those one less than the four minimals, that is the digits than when false-added to 1 give minimals. Likewise in the column corresponding to 2, the holes are punched for $2 - 2 = 0$, $3 - 2 = 1$, $6 - 2 = 4$ and $9 - 2 = 7$.

---

[11]The choice of starting place that yields the next lowest number of these digits turns out to be the 870th, with the first group used turning out to be 27330. This produces 11 occurrences of 2, 3, 6 and 9. Various experiments with similarly produced random data all found the starting place, sometimes just as one of two or three possibilities. Thus Mamba seems to work in practice. It would be less reliable for shorter messages, say with around 30 GATs.

Then the above digits   2570196794681016444941730638262551604061625717
9783 are replaced by:

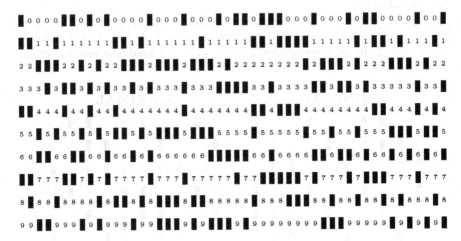

This should be considered as the first 50 of 100—or perhaps 1,000—such columns printed on standard[12] 80 column IBM cards, by a standard card-punch specially modified for the purpose. If, as here, there were more than 80 known additives, it would be necessary to use further cards as well. The left and right edges would then have to be cut off to enable the user to tape the cards together.

Likewise we can take the 54 false sums of the 54 GATs of the message:

717186   589350   709645   321993   425785   929643   884477   027841   531834

and punch these digits into a standard IBM card. Again, provided we disregard the 'X' and 'Y' rows at the top of the card and omit the last four digits—1834—to avoid exceeding the page width, the first 50 columns of the card will look like:

---

[12]The IBM card (see Sect. 1.15) was made of firm thin quality cardboard of standard size. Holes could be punched in at the intersection of any of its 12 rows and 80 columns. These holes were punched with considerable precision by a card-punch.

NARA RG457 Box 705, item 1742 includes an envelope marked MAMBA and dated 2 May 1944. The contents are described as 'Samples of maximal and minimal cards, punched on an I.B.M. printing punch modified by (illegible) for Op-20-GY-P.' The initials 'J. H. H.' are those of Lieutenant John H. Howard of Op-20-G who was heavily involved in this project. The 'X' (top) rectangles have all been punched out: this reflects the practice (NARA RG38 CNSG Library Box 15, item 3222/54 page 2) 'A measure of the overlap at any given alignment can be obtained by brushes reading the number of overlap holes in common between the additive card and message card'. The 'X' and 'Y' rows are not shown in the displays representing cards in this chapter.

These cards must be among the most remarkable relics of WW2 cryptology in NARA. They may be compared with the 'Banburismus' used at Bletchley Park against Enigma. It also was based on alignment of holes punched in pieces of paper.

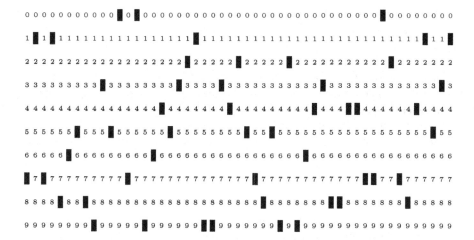

Now suppose that the first punched card is placed over a piece of blue cardboard. The point is that if the second punched card is now put on top of the first, *blue rectangles visible through the holes correspond to minimal differences.* Thus a prospective alignment may be detected by sliding the second card over the first and seeking places where the number *m* of visible blue rectangles is relatively low.

The process as described above is somewhat arbitrary in that it makes no use of any counting of the maximal digits that occur as false sums. It is quite possible to work with these instead. In the list of 100 false sums of additives given above:

$$
\begin{array}{ccccc}
8530914316 & 4290946661 & 6937047284 & 8559406049 & 4853931327 \\
3633463567 & 7519731305 & 9009567108 & 1010355130 & 3324674188
\end{array}
$$

the digits 0, 1, 2, 3, 4, 5, 6, 7, 8 and 9 can be replaced not as before but by the corresponding columns in the right hand (maximal) part of the previous table. Then the first 50 of these digits are replaced by

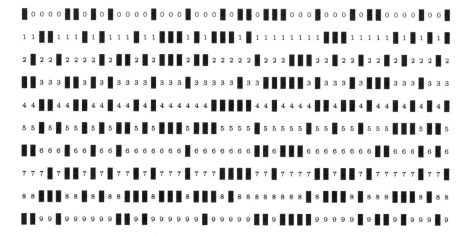

A third card punched as above can now be placed on top of a piece of red cardboard. If the second card is slid over this third card as before, a prospective alignment may be detected by seeking places where the number M of visible red rectangles is relatively high.

It is even possible to test for both maximals and minimals at once. The two special cards may be used as stencils[13] to paint red and blue rectangles on an appropriate sheet of white cardboard! Red rectangles are used for the maximal digits and blue rectangles for the minimal.

Op-20-GM gave the matter considerable attention in 1943–1945 and eventually decided that the best criterion for a possible alignment would be:

$$M - m \geq .3B + 4.$$

Here M denotes the number of occurrences of maximal digits as characteristics, m denotes the number of occurrences of minimal digits as characteristics and B denotes the number of differences calculated. The constant .3 was called the *slope* while 4 was the *intercept*. This version has the advantage of not artificially preferring the minimals over the maximals. One would expect $M - m$ to be small when there is no alignment and to be quite big when there was one. Apparently slightly different values for the slope and intercept were sometimes used but .3 and 4 are good enough here.

In the example $M = 38, m = 8, B = 54$ and

$$30 = 38 - 8 \geq (.3) \times 54 + 4 = 20.2$$

If the extended additive table with 1,000 groups is used instead of just the initial 100, two more displacements occur with $M - m \geq 20.2$. If one starts at the 870th group the value $M - m = 22$ occurs. If instead one starts at the 132nd group one finds $M - m = 21$.

In fact around April 1944 the JN-25 indicator system was not exploitable by Op-20-G but (presumably) double hits were detected in sufficient numbers to put GATs arising from some commonly used parts of the additive table into alignment. The additive was being changed every two to four weeks. It was noted that[14]:

---

[13]NARA RG38 CNSG Library Box 15 item 3222/54 also contains minutes of a meeting held at Op-20-GYP on 12 April 1944. Page 3 makes explicit mention of the use of stencils in the Mamba process.

The reader may wish to compare this process with the 'Banburismus' used at Bletchley Park against Enigma.

[14]NARA RG457 Box 705, item 1742. This indirectly reveals the magnitude of the dependence upon the use of multiples of three as book groups in JN-25. Note 2 is relevant here: the Mamba method would be much less useful against the IJA practice of using numerous short tables of additives.

At present there is very slow recovery of the complete keying system but early recovery of substantial blocks of key system. The present technique is to extend these initial blocks by aligning additional messages with the recovered additives.

And thus the task to hand was rather different from that set out so far in this chapter. Somehow some of the additives on a worthwhile segment from the additive table had been found. Presumably this would have been done by obtaining double hits on GATs encrypted from commonly used pages. The GATs apparently in alignment would have been stripped as before. If there had been enough depth this would have confirmed the alignment was correct. Various other intercepted messages were available. Some of these could be placed by the Mamba process into alignment with the known piece of additive. When this had been done for several such intercepts it would be possible to use the methods of Sects. 9.10 or 10.2 to extend that piece of additive. Eventually (with some luck) much bigger 'blocks' could be determined.

The project of constructing a working Mamba machine[15] was carried out in 1944. It turned out to require rather too much maintenance. When the IJN reduced the use of JN-25 code books around the start of 1945 the incentive for further work on the machine evaporated.

15.5 Weakness 3: Fourier Decryption

Fourier decryption is a mathematical process for guessing the characteristic of a potential decrypting additive from the characteristics of the GATs in depth. Once this is known, the number of groups that could be potential additives is much reduced. Its practical application is discussed in the next section.

The elements of the mathematical structure C_{10} are the ten remainders modulo 10 and will be denoted by the digits $0, 1, \ldots 9$. Ten real-valued functions $f_0, f_1, \ldots, f_9$ on C_{10} will be used. They are defined by:

---

[15]The machine was made in the NCR factory in Dayton, Ohio. This had been visited by Alan Turing in December 1942. The general story is described by Jim DeBrosse and Colin Burke in *The Secret in Building 26*, Random House, NY, 2004, which, however, does not mention Mamba. Considerable material on the Mamba project and various other such projects may be found in the RG38 and RG457 sections of the College Park NARA. Thus in RG457 there are Box 0584, item 1445 *Operations and maintenance schedule for Mamba*; Box 0591, item 1494 *Brief descriptions of RAM equipment*; Box 0616, item 1653 *Mamba (Scanner schematics)*; Box 0705, item 1742 *Mamba, 1944–1945*; Box 0949, item 2799 *Cryptanalytic machines synopsis, 1944*; Box 0950, item 2807 *Use of high-speed crypto equipment*; Boxes 1070 and 1071, item 3390 *NCR schematics and unidentified machine drawings*. There is also much material on the Copperhead I project.

The Dayton code breakers website includes a photograph of the Mamba machine. It is taken from RG457, Box 804.

$$f_0(j) = 1/\sqrt{10} \qquad\qquad f_5(j) = (-1)^j/\sqrt{10}$$
$$f_1(j) = \cos(\pi j/5)/\sqrt{5} \qquad f_6(j) = \sin(\pi j/5)/\sqrt{5}$$
$$f_2(j) = \cos(2\pi j/5)/\sqrt{5} \qquad f_7(j) = \sin(2\pi j/5)/\sqrt{5}$$
$$f_3(j) = \cos(3\pi j/5)/\sqrt{5} \qquad f_8(j) = \sin(3\pi j/5)/\sqrt{5}$$
$$f_4(j) = \cos(4\pi j/5)/\sqrt{5} \qquad f_9(j) = \sin(4\pi j/5)/\sqrt{5}$$

and satisfy the familiar orthogonality relationships:

$$\sum_{s=0}^{9} f_u(s) f_v(s) = \begin{cases} 1, & \text{if } u = v, \\ 0, & \text{otherwise .} \end{cases}$$

So any real-valued function g on C_{10} can be written as the sum of multiples of these ten functions:

$$g(s) = \sum_{u=0}^{9} a_u f_u(s),$$

where the coefficients (multipliers) are determined by the usual formula:

$$a_u = \sum_{s=0}^{9} f_u(s) g(s).$$

This applies in particular to the scanning distribution p (see above) for which the coefficients a_1, a_3, a_5, a_7 and a_9 are all zero[16] and:

$$a_0 = .31622, a_2 = -.00020, a_4 = -.00560, a_6 = -0.000002, a_8 = .17458.$$

The following 5-term formula for $p(s)$ is obtained:

$$p(s) = .10000 - .00007 \cos(2\pi s/5) - .00250 \cos(4\pi s/5)$$
$$- .00001 \sin(\pi s/5) + .07808 \sin(3\pi s/5),$$

and it yields the following remarkable approximate formula for $p(s)$:

$$p(s) \approx .10000 + .07808 \sin(3\pi s/5).$$

---

[16]The symmetry or functional equation $p(s) = p(5 - s)$ implies that five of the Fourier coefficients are zero.

The reader may wish to repeat these calculations for the function q of the 'weakness 2' section where $q(0) = .13051, q(1) = q(9) = .09056$, etc. This has the symmetry or functional equation $q(s) = q(-s)$ which implies that the coefficients of the five sine terms (f_5, f_6, f_7, f_8, f_9) are zero.

Now suppose a reasonably large random sample of N JN-25 book groups is being examined. The number with characteristic 0 is $c(0)$, the number with characteristic 1 is $c(1)$, etc. The previous calculation yields the following approximate formula for $c(s)$:

$$c(s) \approx .10000N + .07808N \sin(3\pi s/5).$$

The real aim of this discussion is to handle the case when the other side has subtracted an unknown additive from each of N JN-25 book groups. If this additive has characteristic x—also unknown—the formula for $c(s)$ has to be adjusted to:

$$c(s) \approx .1N + .0781N \sin(3\pi(s + x)/5),$$

and the right hand side may be expanded to yield:

$$.1N + .0781N \sin(3\pi x/5) \cos(3\pi s/5) + .0781N \cos(3\pi x/5) \sin(3\pi s/5).$$

This may be compared with:

$$c(s) = .1N + (a_3/\sqrt{5}) \cos(3\pi s/5) + (a_8/\sqrt{5}) \sin(3\pi s/5).$$

Thus $\sin(3\pi x/5)$ and $\cos(3\pi x/5)$ are determined by the approximate formulae:

$$\sin(3\pi x/5) \approx \frac{a_3}{\sqrt{a_3^2 + a_8^2}} \quad \text{and} \quad \cos(3\pi x/5) \approx \frac{a_8}{\sqrt{a_3^2 + a_8^2}},$$

where two relevant coefficients are given by:

$$a_3 = \sum_{s=0}^{9} (\cos(3\pi s/5)c(s))/\sqrt{5} \quad \text{and} \quad a_8 = \sum_{s=0}^{9} (\sin(3\pi s/5)c(s))/\sqrt{5}.$$

Here the integer x may be recovered from these approximations by use of the inverse trigonometric functions.

Suppose we have a depth of 25 with $c(0) = 3$, $c(1) = 0$, $c(2) = 4$, $c(3) = 3$, $c(4) = 1$, $c(5) = 2$, $c(6) = 4$, $c(7) = 2$, $c(8) = 1$ and $c(9) = 5$. One then calculates, on an anachronistic pocket calculator or otherwise[17]:

$$a_3 = (3\cos 0° + 0\cos 108° + 4\cos 216° + 3\cos 324° + 1\cos 72° + 2\cos 180°$$

$$+ 4\cos 288° + 2\cos 36° + 1\cos 144° + 5\cos 252°)/\sqrt{5} = .4472$$

---

[17]In an era before pocket calculators were introduced some rounded numerical values would have been used.

and:

$$a_8 = (3\sin 0° + 0\sin 108° + 4\sin 216° + 3\sin 324° + 1\sin 72° + 2\sin 180°$$

$$+ 4\sin 288° + 2\sin 36° + 1\sin 144° + 5\sin 252°)/\sqrt{5} = -4.4540$$

Returning to radians, one deduces that $\sin((3\pi x/5) \approx .100$ and also that $\cos(3\pi x/5) \approx -.995$ for some integer x. Thus $x = 5$.

Computer experiments with depths of 20 indicate that most of the time this method produces the correct value of x in two 'guesses'. For depths of 100 or more it is as strikingly accurate as the above somewhat contrived example and leads to the correct value of x immediately most of the time.

15.6 The (Non-historical) NC9

A device could have been constructed without excessive effort using the IBM technology of the 1940s to decrypt a depth when the characteristic of the decrypting additive is known. Special IBM devices made for the USN in WW2 were called NC1, NC2, NC3, NC4 (see Note 47 of Chap. 9), NC5, NC6, NC7 and NC8 leaving the name *NC9* for this non-historical device.

The first requirement is four decks of very special IBM cards. The first, with 1780 cards, would have each of the scanning groups of characteristic 2 punched in the first 5 columns of one card. These groups run from 00039, 00048 to 99987, 99996. It would be prudent to have some identification, such as 222, punched in, say, the last three columns. The next, also with 1780 cards, would have each of the scanning groups of characteristic 3 punched in the first 5 columns of one card. These groups run from 00003, 00012 to 99951, 99900. Identification, this time 333, would be punched in the last three columns. The third deck, with 925 cards, would give the scanning groups of characteristic 6 and be identified as before while the fourth would similarly store the 925 scanning groups of characteristic 9.

Consider the following example of 24 GATs in depth.

42400 51711 91452 76093 23408 69446 94117 82498 55936 52707 78028 17626
66833 68287 37278 94168 19015 70870 37905 43632 94643 44252 97203 47977

The characteristics of these are:

0	5	1	5	7	9	2	1	8	1	5	2
6	1	7	8	6	2	4	8	6	7	1	4

Suppose that it is known that the characteristic of the as yet undetermined decrypting additive is 6. Then the characteristics of the corresponding book groups are:

6	1	7	1	3	5	8	7	4	7	1	8
2	7	3	4	2	8	0	4	2	3	7	0

This list is examined[18] for (preferably) a 6 or a 9. If neither of these occurs, a 2 or a 3 is chosen instead. In the above example the first GAT comes from a book group with characteristic 6. This GAT is then subtracted from each of the original GATs in the depth to yield the following artificial depth:

00000 19311 59052 34693 81008 27046 52717 40098 13536 10307 36628 75226
24433 26887 95878 52768 77615 38470 95505 01232 52243 02852 55803 05577

This set of 24 decrypts to the same book groups as the original depth. So the first group, 00000, in the second list *decrypts to a book group of characteristic 6* which has to be scanning. Hence the additive needed is one of the 925 scanning groups of characteristic 6.

It would be quite feasible to enter the first nine or ten of the original groups in a subtractor machine and see what happens when the first group is changed to the various groups in a list on paper of 925 scanning groups of characteristic 6. The other groups on the machine would be changed 'in synch' each time and one would look out for groups among the 925 that decrypt all nine or ten on the machine.

An alternative method, which the NC9 is intended to mechanise, is to take the 925 cards punched with scanning groups of characteristic 6, and see which of these groups, when added to 19311, yield a scanning group. These cards would be put in one heap (heap A, containing about one third of the 925) and the rest in heap B. One then picks up the cards in heap A and sees which of these when added to 59092, yield a scanning group. These cards would by put in heap A, now about 100 cards, and the rest in heap B. This process would continue and eventually yield all the decryptions (in this case, the only decryption) of the original GATs.

This method would not work for depths of eight to ten. The characteristic of the required additive could not be worked out in advance and the decrypting additive would usually be far from unique. This method[19] is quite different from those described by Ely.

---

[18]The strategy is to choose a GAT that decrypts to a book group with characteristic 6, 9, 2 or 3 and there are then only 925, 925, 1780 or 1780 potential decryptions respectively.

[19]It is possible that the characteristic may help find the additive in a Phase 2 decipherment attempt that is not yielding a result quickly. It may well be possible to use a mixture of difference and Fourier methods to assist in the decryption of around 20 JN-25 GATs in depth when nothing is known about the more common book groups. However it would appear that in practice working out the optimal strategy would need modern electronic calculation capacity and so would be anachronistic.

The NC9 would need to automate this process. It could perhaps be programmed by the entry of one group, initially 19311, in a plugboard (jackboard, control panel) as in the diagram below:

1.	o	o 0	o 1	o 2	o 3	o 4	o 5	o 6	o 7	o 8	o 9
2.	o	o 0	o 1	o 2	o 3	o 4	o 5	o 6	o 7	o 8	o 9
3.	o	o 0	o 1	o 2	o 3	o 4	o 5	o 6	o 7	o 8	o 9
4.	o	o 0	o 1	o 2	o 3	o 4	o 5	o 6	o 7	o 8	o 9
5.	o	o 0	o 1	o 2	o 3	o 4	o 5	o 6	o 7	o 8	o 9

Here the little circles represent bronze sockets which can be joined electrically by plugging in simple leads. The group 19311 would be entered into the NC9 by joining the first socket in row 1 to that numbered 1 in row 1, joining the first socket in row two to that numbered 9 in row 2, and so on.

Once the plugboard had been programmed with a 5-digit group **b**, the NC9 would input a deck of cards in which the first five columns were punched with a group. It would sort these into two hoppers, A and B, according as to whether the false sum of the group on the card and the group entered in the plugboard was scanning or not.

Most likely any IBM technicians attempting to implement this idea would have used a collator with heavily modified wiring and then used a card to input the group **b**. The collator would then have had two feeds and two of its output hoppers would be used. An experienced user of IBM tabulating machinery has stated that such a device would handle about 200 cards per minute.

15.7   Weakness 4: Testing a New Cipher

Perhaps remarkably there is a method based on the scanning distribution for testing whether a new 5-digit additive cipher appears to have all book groups scanning. It needs preliminary calculation of certain means and standard deviations. As such calculation appears to require large-scale electronic sampling, the method is totally anachronous. But it is there. The details are set out in Peter Donovan's paper *The Flaw in the JN-25 Series of Ciphers II*, published in *Cryptologia*, 2012.

Appendix 1   Other Characteristics

Consider an additive system using 4-digit book groups. As an error-correcting device each 4-digit group $abcd$ could be extended to a five-digit group $abcde$ where e is chosen so that the false sum $a + b + c + d + e$ is 0. For example 5294 would be extended to 52940. The extended book groups are then encrypted by means of five-digit additives. Extending book groups in this way turns out to be insecure because it betrays alignment.

Earlier in this chapter, the characteristic of a 5-digit group $abcde$ was defined to be the (false) sum $a + b + c + d + e$. The following general algebraic relationship holds: *the characteristic of the (false) sum of two groups is just the (false) sum of the characteristics of the two groups.* Thus if 4-digit book groups are extended to become 5-digit groups with characteristic 0, the characteristics of the GATs obtained by additive encryption are the same as the characteristics of the corresponding additives. For example if the message consists of the eight book groups shown and the additives are as given on the third line:

Message	2360	6797	7499	7896	9640	9173	6687	3127
Extended	23609	67971	74991	78960	96401	91730	66873	31277
Additive	64575	13110	64590	59050	16157	53639	26042	57102
Characteristic	7	6	4	9	0	6	4	5
False sum = GAT	87174	70081	38481	27910	02558	44369	82815	88379
Characteristic	7	6	4	9	0	6	4	5

Here the characteristics may be calculated by the other side and used to put intercepts into depth. Suppose two messages in this code are being examined. The characteristics are taken to be 3-6-0-6-7-9-7-7-4-9-9-7-8-9-6-9-6 and 8-9-6-9-6-4-0-9-1-7-3-6-6-8-7-3-1-2 respectively. These two strings of digits have the segment 8-9-6-9-6 in common and so it is extremely likely that the correct alignment has the last five groups of the first over the first five groups of the second. That is all the leakage of information but it is quite serious enough. This simple redundant encryption allows the enemy to bypass the first defensive barrier (the indicators) of an additive system.

If, instead, the extra digit is inserted *after* encryption then *there is no loss of security*—nothing is given away. (Indeed, nothing would be given away if the same encrypted message were transmitted twice.) The calculations become:

Message	2360	6797	7499	7896	9640	9173	6687	3127
Additive	6457	1311	6459	5905	1615	5363	2604	˙5710
False sum	8717	7008	3848	2791	0255	4436	8281	8837
Extended = GAT	87177	70085	38487	27911	02558	44363	82811	88374
Characteristic	0	0	0	0	0	0	0	0

A different characteristic was mentioned in Sect. 14.5. An indicator group $abcd$ was constructed so that the false sum/difference $a + b + c - d$ was always 0. This was exploited. To return to book groups in additive cipher systems, almost any limitation on their choice is likely to be detected by the other side and quite possibly exploited to weaken security. Random choice of book groups is much safer.

Appendix 2   A Distribution for JN-11A

As noted before, JN-11 and JN-25 evolved in quite different ways in 1943 and 1944. Thus there is no historical interest in the observation that there would be an analogous process for an additive system in which the code book used 4-digit groups such that each book group had digits that summed to one more than a multiple of three. One can calculate the remainders upon division by 10 of each of the 3333 such book groups and confirm that 282 have remainder 0, 60 have remainder 1, 540 have remainder 2, 480 have remainder 3, 45 have remainder 4, 348 have remainder 5, 633 have remainder 6, 120 have remainder 7, 165 have remainder 8 and 660 have remainder 9. A Mamba process can be concocted for this context. The 'minimal' digits relative to modulo 10 false sums of JN-11A book groups are 1, 4, 7 and 8.

Appendix 3   Mamba-11

There are other possible characteristics. For example $a - b + c - d + e$ or $a + b + c + d - e$ could be taken to be 'the' characteristic and the additive system could be designed with this taking some fixed value on all code groups. Once the cryptanalysts detected such a practice was current, the above method of putting GATs into depth (alignment) would be available.

This chapter has dealt with a technique used to attack additive ciphers whose book groups were all multiples of 3. Suppose instead that all the book groups of the additive cipher being targeted were all multiples of 11. We recall that a 5-digit number $abcde$ is a multiple of 11 if the *alternating sum* $a - b + c - d + e$ is a multiple of 11. This alternating sum may well be negative. Thus for the number 29183 the alternating sum is $2 - 9 + 1 - 8 + 3 = -11$ and so 29,183 is a multiple of 11. One may check that $29,183 = 2653 \times 11$.

Then there is an analogue of the above processes available which may as well be called Mamba-11. It is presented here for 4-digit groups.

It is inappropriate to work with the characteristic $a + b + c + d$ of the digits of the group $abcd$ in the Mamba-11 context. Instead one works with the false alternating sum $a - b + c - d$. Then Mamba-11 works similarly to the original Mamba process. If instead (say) the book groups were chosen so as to leave remainders of 0, 5 or 9 upon division by 11, a modified Mamba-11 process would be available.

There might have been a Mamba-11 process available for an additive cipher system using 4-digit groups for which the book groups of an excessive proportion of the commonly used words were multiples of 11. Despite the comments on this matter in Parts F[20] and G of the *CBTR*, the authors know of no documentation of such a use of Mamba-11.

---

[20]Part F of the *CBTR*, apparently due to Professor Room, refers to special roles played by the prime number 11 in the construction of certain Japanese code books. According to the *CBTR* at least some early commercial codes had the feature that the more common words and phrases were encoded by numbers that were multiples of 11. This has not been verified and may be an error.

Appendix 4   Thwarting All This

In theory the IJN codemakers could have beaten all four of the methods given for exploiting the scanning distribution. All that was needed was choosing randomly 925 groups of each characteristic and then using the 9250 groups so obtained randomly as book groups for the 9250 most common words in new JN-25 code books. In practice any understanding of why this was necessary would have convinced any communications security unit that requiring all book groups to be scanning was a pernicious practice. So it is safe to assume that JN-25 book groups were chosen randomly, with about 2.7 % having characteristic 9, etc.

Part III
Ciphers and the Submarines

Chapter 16
Ciphers and the Submarines

The astonishing success of American submarines in the Pacific warrants greater recognition than it is often accorded. By the end of 1944 Japan had been brought to its knees, deprived of essential raw materials. The success of these submarines is even more remarkable for the fact that, until mid-1943, the USN Bureau of Ordnance forced them to operate with unreliable torpedoes. It repeatedly rejected complaints about torpedo depth control and failure of torpedo firing pins until incontrovertible evidence led to these major problems being fixed. This chapter provides data on the relative successes of submarine fleets in WW2 and then focuses on the American and German submarine fleets.

16.1 The Background Information

The startling data[1] given on the next page provides a basis for evaluating the relative successes of the major WW2 submarine fleets.

---

[1] The basic facts about submarines in WW2 have been set out by Mackenzie Gregory in *Underwater Warfare: The Struggle against the Submarine Menace, 1939–1945*, Naval Historical Society of Australia, 1997. Numerous other sources of information are available.

© Springer International Publishing Switzerland 2014
P. Donovan, J. Mack, *Code Breaking in the Pacific*,
DOI 10.1007/978-3-319-08278-3_16

Results Achieved by Sub. Fleets versus their losses in WW2

Navy	Total subs lost	Total tonnage sunk	Number ships sunk	Number sunk per lost sub	Tonnage sunk per lost sub
USA	52	5.2M	1314	23	101 923
UK	75	1.5M	697	9.3	20 266
Germany	781	14.5M	2828	3.6	18 565
Italy	82	1.0M	??	??	12 195
Japan	127	.9M	184	1.4	6 923
USSR	109	.4M	160	1.5	3 692

It seems reasonable to use 'tonnage sunk per submarine lost' as a measure of the success of a submarine campaign of the era.

16.2   Submarines in the Atlantic

A brief account of the corresponding Sigint/submarine conflict in the Atlantic is given for comparison purposes. The use of submarines against merchant shipping originated in WW1. German submarines inflicted considerable damage upon the British and French economies by sinking ships. Later, American ships became targets. By 1917, experience had shown that extra security could be gained by sending merchant ships in convoys, with some naval escort. Naval strategists had 20 years in which to absorb the strategic consequences of this. Life was not easy for the crew of a submerged submarine[2] detected by a hostile destroyer. After the Pearl Harbor raid, Germany declared war on the United States and was able to use its submarines to attack American coastal shipping. The initial reluctance of the USN to protect merchant ships by organising convoys defies rational explanation.

The Zimmermann telegram (Sect. 1.7) of 1917 expressed the view that changing from restricted use of submarines to unrestricted use would of itself win WW1 for Germany. This turned out not to be the case. However German Admiral Dönitz, a veteran of WW1 submarine warfare, pressed for major resources to be devoted to submarine construction from early 1939. He eventually won the argument. After 70 years, the submarine pens constructed at Lorient, France in 1940–1941 survive almost totally intact. The massive amount of steel and concrete used is the clearest evidence of the importance the Nazi regime assigned to the submarine in the war against the western Allies.

---

[2]This is why the I-124 was sunk in January 1942! The earlier submarines used electric power from batteries to operate under water. This was much slower. On the surface diesel engines could recharge the batteries and produce a much greater speed. In the middle of 1944 the schnorkel was introduced, allowing the submarine to travel just below the surface using this device for obtaining air and ejecting exhaust fumes.

The capture of the German submarine U-505 in 1944 yielded not only key cryptographic information but also a handbook issued by the German Naval Command in 1942. It stated that the continuous and successful use of submarines against Allied merchant shipping in the Atlantic was

> in the long run of strategically decisive significance for the outcome of the war, since the destruction of her sea communications means the loss of the war for a nation dependent upon sea-trade.

This has been published in English.[3]

One particularly strategic form of merchant shipping was the troop ship. The Allies in WW2 made extensive use of the high-speed ocean liners built for civilian use in the previous 20 years. Not one was sunk. The Japanese, not having such vessels, had to use low-speed merchant ships to transport troops.

Safe passage of the Atlantic by UK-bound convoys was greatly enhanced by finding routes which avoided contact with the U-boat wolf packs. Allied fleet control was able to use decrypts of communications between Admiral Dönitz and his U-boat captains for this purpose.

Up to the end of 1941, the code breakers at Bletchley Park were able to decrypt messages sent in the then current German naval Enigma cipher. So Sigint could assist convoys to avoid the worst hazards. In February 1942, a more secure encryption machine, the Triton Enigma, was introduced. This enabled the U-boats to operate with some considerable success in the mid-Atlantic, quadrupling the losses of Allied merchant ships. The so-called *Enigma blackout* lasted from February to December 1942, by which time Turing's new *bombes* and other improvements in cryptanalysis enabled useful operational intelligence derived from decrypted U-boat communications once again to be supplied to the RN, RCN and USN.

Several other improvements in technology and air power also combined with the Sigint information to render the Atlantic a dangerous place for U-boats. Better shipboard HF/DF equipment and improved airborne radar were gradually introduced. Operations research techniques[4] were applied to improve methods of attacking submarines from the air.

Better use of anti-submarine aircraft also helped. Permission was later obtained from Portugal to station aircraft in the Azores[5] and so to cover the mid-Atlantic gap out of range from Britain and North America.

---

[3]The U-505 was captured on 4 June 1944. It is preserved in the Chicago Museum of Science and Industry. The OKM document mentioned was published in 1989 as *The U-boat Commander's Handbook* by Thomas Publications, Gettysburg, PA.

The following comment is to be found on page 69 of the Potter-Nimitz book *The Great Sea War*: 'The German Navy poured vast resources into the construction of four major warships such as the *Bismarck*. Designing and constructing submarines would have served it better.'

[4]See C. H. Waddington, *O.R. in World War II*, Elek Science 1973.

[5]The reader studying naval aspects of WW2 needs to have an atlas available. It will display the strategic location of the Azores in the North Atlantic. A large globe is useful in revealing great circle routes.

16.3   German Sigint in the Atlantic, 1941–1943

The survival rate of the Allied merchant ships was reduced by the continued use of an additive system, British Naval Cipher 3 (BNC 3), which was employed by the RN, RCN and USN for convoy escort communications. It had been introduced in June 1941. The German naval intelligence was able to intercept, decrypt and decode these messages. Admiral Dönitz was kept informed and was able to send out modified instructions encrypted by Enigma. The enemy reading of it was detected[6] in March 1943 by Enigma decrypts revealing a change in the orders sent to U-boats at sea soon after instructions to convoy escorts were sent using BNC 3. An enhanced additive cipher system was introduced by John Tiltman which provided secure communications from mid-1943 onwards. See Appendix 4 of Chap. 13. This saga illustrates a remark once made by William Friedman: making one's own ciphers secure is usually much easier than breaking those of the enemy and is at least as important.

There was a *Target Intelligence Committee* (TICOM) set up in 1945 to examine captured German sites of cryptanalytical interest. It sent teams into Western Europe just behind the troops. The full story of the breaking of BNC 3 was available only after a TICOM team interviewing German cryptographers in June 1945 made its

---

[6]The March 1943 decrypt is well documented. See pages 114–115 and the note on page 239 of the useful book *The Secret in Building 26* by Jim DeBrosse and Colin Burke (Random House, New York, 2004). An additive system with a relatively simple vocabulary could have been made secure by use of some of the devices mentioned in Chap. 13. It should have been easy to issue new code books to ships in the relatively few North American east coast ports in use and thus to change the code book very frequently. A great deal was at stake. It should be borne in mind that one use of Enigma decrypts was to enable Allied convoys to be routed to evade the U-boats but apparently the Germans, who examined decrypts giving Allied instructions to merchant ships at sea, did not suspect that such instructions might have resulted from insecurity of their own codes.

The book *Die deutsche Marine-Funkaufklärung 1914–1945* by Heinz Bonatz (Darmstadt, 1970) quotes the full text of a letter from Admiral Dönitz of 3 November 1942 thanking the Sigint unit MND III for assistance in general and in particular in locating convoy SC107. Those who know German may read the following few lines of it:

'Der BDU hat bisher in hohem Maße die eigegangenen Funkaufklärungsergebnisse bei der Aufstellung seine Uboote verwertet und in mehreren Fällen auf diese Meldungen operierend Geleitzüge erfaßt. Eine besonders erfreulicher Fall ist die Erfassung des SC107 vom 31.10.42, wo der rechtzeitige Eingang der Funkaufklärungsmeldung über den Weg des Geleitzuges es ermöglichte, ... Durch diese Meldung hat die Abteilung Funkaufklärung hohen Anteil an den Erfolgen, die hier errungen worden sind.'

Curiously the British Admiralty sent out a message to various associates, including the ACNB, in December 1941 stating that 'The use of geographical names or latitude and longitude references in merchant navy code reciphered is considered insecure'. This may be found in the Melbourne branch of the NAA in MP1074/7 file for 11/12/1941 to 28/12/1941, barcode 4169371.

There is a quite striking message of 5 November 1942 in MP1074/7 that gives the daily positions of the liner *Ile de France* in a forthcoming voyage from Hobart to Auckland. Hopefully that one went by cable.

report. The following quotation is from a report written by R. T. Barrett in 1946, released[7] many years later. The title is *German Successes against British Codes and Ciphers*.

> The Germans, during the early part of the war, had been breaking our codes and ciphers with considerable success. Their attacks were thorough, persistent and highly ingenious. The success was in cryptography and they owed little to captured books and other kinds of leakage.
>
> During 1942, when the Battle of the Atlantic was at its height, the enemy was deciphering our own operational orders and the Admiralty's daily U-boat disposition signal within a few hours of dispatch.
>
> MINOR ECONOMIES IN THIS DETAIL OF CODE AND CIPHER SECURITY NOT ONLY COST US DEARLY IN MEN AND SHIPS BUT VERY NEARLY LOST US THE WAR.

Although the 'us' in this last sentence refers to the United Kingdom, the other Allies would have had dire prospects in 1942 if the U-boats had in fact won the war in the Atlantic.

16.4 Submarines in the Pacific

The strategic aspects of submarine warfare were scarcely a secret in 1941. It should have been evident that Japan was another nation dependent upon sea-trade. Indeed,

---

[7]The Barrett document is to be found in TNA ADM 223/284. It is reprinted as item 175 in *The Battle of the Atlantic and Signals Intelligence* edited by David Syrett (Naval Record Society, Aldershot, 2002).

Greg Mellen edited some undoubtedly genuine documents in the paper *Rhapsody in Purple* in *Cryptologia* 6(4), October 1982, 346–367. Tiltman is recorded as saying to an unidentified American in Washington 'You were right about Naval Cipher No. 3. It is no good and the Germans have been reading it all along just as you predicted. Our faces are very red and your stock is very high in London.' Disagreement about the use of Sigaba (Sect. 8.7) may have been part of the underlying cause of the use of Naval Cipher No. 3.

The well-known Churchill quotation 'The only thing that really frightened me during the war was the U-boat peril.' is from chapter XXX of *Their Finest Hour*, volume 2 of *The Second World War*. Elsewhere in that volume he states: 'The shadow of the U-boat blockade already cast its chill upon us. All our plans depended upon the defeat of this menace.' In volume 3 *The Hinge of Fate* Churchill states that 'For 6 or 7 months the U-boats ravaged American Waters almost uncontrolled and, in fact, almost brought us to the disaster of an indefinite prolongation of the war'. This refers to the initial failure of the USN to implement convoys for coastal and Caribbean shipping.

In volume 6 *Triumph and Tragedy* Churchill makes it clear that the loss of merchant shipping, of which the greater proportion was due to submarines, was one of the two major causes of Japan's defeat in the Pacific War. Admiral King confirmed this opinion. The other was the general failure of the IJN. The nuclear weapons used in August 1945 just brought the inevitable forward. Unlike Churchill, this book is able to clarify the substantial dependence of both of these causes on Comint.

TICOM also operated in the Pacific.

this should have been a considerable deterrent[8] to Japan from starting a war against countries with submarines in the first place. The oil embargo imposed upon Japan in mid-1941 would have drawn attention to the fact that it was dependent upon oil imports which in turn required shipping. The thrust into Malaya, the Philippines and the NEI would need shipping as well. And this shipping would be exposed to submarine attack once the opponents in any Pacific War became organised.

Because routes for merchant shipping between Malaya and Japan had to avoid Taiwan, the area needing to be covered by Allied submarines hunting Japan-bound convoys in the western Pacific was reduced.

Before 1941 the IJN appears not to have considered[9] the impact of submarine technology upon overall strategy. Its submarines had torpedoes that were much superior to those of the USN. These torpedoes had considerable success in sinking Allied warships. But although the strategy of early 1942 included cutting off the sea route between Australia and the United States, the IJN did not seriously attempt to use submarines to achieve this aim. In addition, it appeared unwilling to provide escort warships or aircraft in adequate numbers for convoy protection, preferring instead to maintain maximal strength for navy-to-navy encounters.

As early as February 1942 the Australian government received[10] intelligence reports showing that the shortage of merchant shipping was limiting the offensive operations of the Japanese Army. By April 1942 the strategic weakness of Japan was noted by the Allies. The mission of MacArthur's SWPA command included

exerting economic pressure on the enemy by destroying vessels transporting raw materials from the recently conquered territories to Japan.

There is a vivid contrast between this statement and the contents of a key relic of how some elements of the Japanese Government perceived the situation in May 1942. NAA Melbourne item MP1074/7 includes a copy dated 24 May 1942 of a Magic decrypt. The Japanese Ambassador to Chile had been enlarging upon the

---

[8]An invaluable account of the thinking behind Japanese naval strategy of that era was written in booklet form in 1946 by Masataka Chikaya, an IJN officer of 18 years experience. An English translation may be found in *The Pearl Harbor Papers* by D. G. Goldstein and K. V. Dillon, Brassey's, Washington 1993. Commander Chikaya, in total ignorance of Allied cryptology, asked 'Was not the foundation of our strategy completely upset by the activities of the USN submarines?'. He noted that the IJN had never properly studied marine transport capacity and commented that 'Submarines' roles and activities in the war that was to come had not been anticipated by the IJN clearly enough'.

[9]Commander Chikaya (see the previous note) made it clear that using submarines for attacking merchant shipping was not considered seriously by the IJN until the middle of 1943.

Sydney journalist David Jenkins' book *Battle Surface!: Japan's Submarine War Against Australia 1942–1944* (Random House, 1992) has considerable merit. It leaves the impression that the damage done by IJN submarines was much less than that done by German submarines in the Atlantic. Frumel played some role in defending Australian coastal shipping against submarines.

[10]The Canberra NAA file A7982 Z152 gives the February 1942 information. Further material about this issue may be found in A5954 517/27 and A5977 10/1943.

apparent victory in the Battle of the Coral Sea and had been saying that before long Japan would be able to carry products of Japan and the South Seas to Chile. This dream of profiting from the conquests was not to materialise.

16.5 American Submarines

The RAN dispensed with submarines as an economy measure around 1928. Although small numbers of Dutch and British submarines played a role in the Pacific War, the main contribution was American.

The USN did not need further reminding about the potential of these vessels. When the Japanese advanced early in 1942, the USN submarine tender *Holland* retreated along the coast of Western Australia to Fremantle (the port near Perth) and then to Albany on its south coast. The USN submarines based in the Philippines also had to retreat. But there was a big problem: as mentioned before, the USN torpedoes were defective. Tests carried out off Albany in Western Australia in mid-1942 established that they were travelling at too great a depth. An intercepted IJN message decrypted in Hawaii revealed[11] that these torpedoes also tended not to explode, and this was confirmed by USN tests in Hawaii. Both faults had previously been repeatedly reported by submarine commanders but their complaints had been rejected. Solving these problems took some considerable time.

As Darwin Harbour is shallow and exposed to bombing from Timor it did not become suitable for submarines until much later. From August 1942 Fremantle[12] became an active submarine base. Brisbane had received its first USN submarine in April 1942. There were some naval facilities in Townsville by that time, and these must have been useful for the submarines based in New Farm, Brisbane. A rather unsatisfactory refuelling base at Exmouth Gulf was used in 1943–1944. At one stage Midway was used as another refuelling base. The principal USN base for Pacific submarines was still Pearl Harbor.

An early reference to the sinking of a Japanese merchant ship may be found in the *Sydney Morning Herald* of 9 October 1942. The vessel was named as the

---

[11]The source is the interview (12 December 1963) by David Kahn with 'Ham' Wright formerly of the Hawaii USN cryptological unit. It is in the US National Cryptologic Museum as VF 35-11. The admirable account *Silent Victory* by Clay Blair, Jr (Lippincott, Philadelphia, 1975) makes it clear that the submariners deeply resented the defects in these torpedoes. The situation was finally fixed around September 1943.

[12]The WW2 submarines in Fremantle form the subject matter of *Fremantle's Secret Fleets* by Lynne Cairns (WA Maritime Museum 1995) and *Operations of Fremantle Submarine Base 1942–1945* by David Creed (Naval Historical Society of Australia, 1998). These give interesting background information on the facilities at Fremantle. They are complemented by *US Subs Down Under: Brisbane 1942–1945* by David Jones and Peter Nunan (Naval Institute Press, Annapolis MD, 2004). Numerous documents on the buildings adjacent to the Brisbane River and used by USN submarines may be found in the Brisbane office of the NAA.

Lisbon Maru and was sunk on 1 October. It remains a question whether the public identification of the ship constituted another breach of the security of Sigint.

A file[13] in the Australian War Memorial archives gives the losses to the Japanese merchant fleet inflicted by submarines based in Western Australia as being 25 in 1942, 48 in 1943, 168 in 1944 and 57 in 1945. Clay Blair, in his *Silent Victory*, states that submarines sank rather more than 1,000 Japanese merchant ships. The submarines deprived Japan of most of the benefit from its occupation of South-East Asia. Apparently no Japanese intelligence officer ever investigated whether so many ships could have been located by chance by spotter planes.

In 1944–1945 it was possible for the USN to phase down the use of relatively remote submarine bases in favour of Hollandia in Dutch New Guinea and later Subic Bay in the Philippines.

The magnitude of the losses of merchant shipping from Burma to Japan via Singapore was a reason for constructing the notorious Burma-Thailand railway.[14] This turned out to be rather futile, since the many bridges on the railway were prone to another form of attack.

16.6   The Sigint Factor

Several cryptologic achievements enabled intelligence on Japanese merchant shipping routes to be passed on to Allied submarines. American, Australian and British Sigint units were involved in breaking into a range of encryption systems. Some of these may be described as diplomatic or commercial code breaks. Thus in 1942, the Japanese used insecure diplomatic radio channels to convey information about merchant shipping. The most valuable sources of Sigint came from Japanese Army and Navy codes, and these major sources are described below.

16.7   JN-40

The invaluable Jamieson report (NAA item B5554) contains a very interesting account of the JN-40[40] series of ciphers, including the following extract:

---

[13] AWM124 4/260. A small proportion of these totals was due to British submarines. A few Dutch submarines were based there too. Clay Blair gives in *Silent Victory* quite detailed analyses of the sinkings, including of course those by submarines based at Pearl Harbor. As some were achieved by submarines in transit from one base to another, total figures such as those in AWM124 should be treated with some caution.

[14] It is worth noting that the railway line north from Trondheim in Norway was constructed in WW2 with comparable loss of life for much the same purpose.

The JN-40 series were basically double transposition of a two-figure code, the cypher text being transmitted in four-figure groups. It first appeared in September 1942, replacing the four-figure recyphered Merchant Shipping Code JN-39. No research was done on this system in Melbourne, as no traffic was available. Two months later, and after the system had been broken by the British unit at Kilindini, it was found that the traffic in this system, which was fairly heavy, had been retained in error by the USN Unit. The reading of this traffic then proceeded and good intelligence regarding the movements of shipping and convoys was obtained.

The second phase of JN-40 was introduced on 1st March (1943), concurrent with the introduction of JN-11. The bulk of the traffic relating to shipping movements, etc, now used the JN-11 system, while JN-40 was used for submarine, navigational and other general warnings, and for communicating with small vessels and units.

The 'four-figure recyphered code' mentioned above would have been an additive system in the terminology of this book. The 'USN unit' mentioned is Fabian's Frumel and the text quoted reveals that it was not making material available to the Australian-British unit under Nave in the 2 or 3 months immediately before the major Sigint re-organisation of November 1942. The twice relocated FECB, by then in Kilindini, Kenya, was by no means idle.

The B5554 report notes somewhat later that the breaking of JN-40 was made possible by its use of stereotyped[15,16] messages. The *GYP-1 Bible* (mid-1943) mentions the exploitation of messages transmitted in both JN-25 and JN-40.

In general the USN submarines underachieved in 1942. The problems included the defective torpedoes and a lack of appreciation by the higher command that the appropriate targets were the slower and less dangerous merchant ships.

---

[15]The Jamieson report may be in error in using the phrase 'double transposition' here. The matter needs further research.

A Magic intercept in 1942 had revealed Japanese disquiet with codes used for their merchant shipping. JN-40 was introduced in September 1942, replacing JN-39. The British team at Kilindini broke into it in November 1942, apparently using a plain text version of one message as a crib. This was processed at Frumel and elsewhere. The RAN report B5554 gives details of the series: JN-40A, JN-40B, JN-40C, JN-40D and JN-40E. The Melbourne work was handled throughout by Lieutenant Commander K. Miller assisted by J. Gray of the WRANS. It was 100 % readable, but the report says that it would have been secure if it had not been used for stereotyped messages.

John Winton in *Ultra in the Pacific* attributes the breaking of JN-40 to Brian Townend. The obituary of Townend in the London *Times* on 2 March 2005, confirms this, mentions the role of John McInnes and gives other information.

Page 292 of *Cryptanalysis of JN-25* (the *GYP-1 Bible*) notes that 'not infrequently' duplication occurred between JN-40 and JN-25 messages. This provided useful 'cribs'.

[16]The insecurity produced by stereotyped messages was mentioned in Sects. 8.18 and 8.19. An article by Ralph Erskine and Peter Freeman on Brigadier John Tiltman (*Cryptologia*, 27(4), October 2003, 289–297) refers to some theoretical analysis carried out at Bletchley Park by Gordon McVittie. This showed that use of stereotypes could reduce the secure life of a cipher system by two-thirds. At the practical level, NAA Canberra file A1196 40/501/271 contains instructions for supervisors of encryption activities. One important instruction is that stereotyping must be strongly discouraged.

This continued into the first quarter of 1943. Thus the early JN-40 intercepts did not produce any sensational effect. However the build-up of information about Japanese merchant shipping had begun.

16.8   JN-11

The introduction of JN-11 on 1 March 1943 must have provided a most welcome surprise to Allied cryptologists. By that time the consequences of having the book groups in the JN-25 series of ciphers all multiples of 3 had become clear. Presumably each new cipher system was examined for evidence of being additive and using such a code book. It turned out that the first of the JN-11 systems, JN-11A, used as book groups only those 4-digit groups for which the sum of the 4 digits was 1 more than a multiple of 3. Thus there were 3,333 possible book groups, beginning with:

     0001 0004 0007 0010 0013 0016 0019 0022 0025 0028 0031 0034 0037

and ending with:

     9961 9964 9967 9970 9973 9976 9979 9982 9985 9988 9991 9994 9997

A description of how the method of finding additives from JN-25 intercepts placed in depth extends to JN-11A intercepts has previously been given. In particular the machine invented by Turing and others in January 1940 may be used provided all the wheels in the first (say) column are set at 2 and then left alone.

There is even a Mamba process available for JN-11.[17] See Appendix 3 of Chap. 15.

As JN-11A had capacity for at most 3333 book groups it was a much simpler code than JN-25B. Recovering enough of the code book must have taken at most 3 months. So by the middle of 1943 a large number of very useful decrypts about Japanese merchant shipping were available. More information would have been gleaned from diplomatic and commercial codes. The IJN also sent messages about merchant shipping in codes of the JN-25 series.

---

[17]More on this matter is given in Appendix 3 of Chap. 15. The title is *Mamba-11*. However the mathematical background needed is fairly substantial. The story of JN-11 rather diverged from that of JN-25 in 1944 and will not be given here. NARA RG457 Boxes 143, 274, 275, 593, 817, 819, 811, 817, 819, 1108, 1301, 1375, 1469 and others contain material on JN-11. It was a very major cipher system.

16.9   The Water Transport Code 2468

As noted in Part G of the *CBTR*, a new cipher was introduced on 1 December 1942. Traffic analysis showed that it was likely to handle water transport messages for the Japanese Army. In fact it was fatally flawed by having two separate encryptions of its indicators. This was a somewhat surprising analogue of the duplicated encryption of Enigma indicators which had been exploited by Marian Rejewski a decade earlier. Technical details on these matters have been discussed in Chap. 14.

Perhaps the inventors of the system believed that frequent changes of its conversion square method of indicator encryption would provide security. The procedure described earlier would determine a new conversion square after the processing of about 150 intercepts using that square. Rejewski needed about 80 intercepts for his attack on the Enigma indicators, but since the Water Transport Code was heavily used, there was no problem in quickly collecting 150–200 intercepts using a new square. Reading of the Water Transport Code, also known as 2468 after its discriminant, turned out to be a wonderful source of intended noon positions, speeds and directions of merchant ships leased by the Japanese Army. It thus provided some of the information that the USN needed to direct its submarines into appropriate positions. It no doubt supplied much other useful intelligence too. There was a steady flow of messages with no value as intelligence but quite useful whenever a change of conversion square happened. Numerous boxes[18] of selected 2468 intercepts are preserved in the RG457 section of the American National Archives in College Park, Maryland. These show that 2468 had a relatively simple vocabulary.

The following text on the breaking of the Water Transport Code is taken from the Central Bureau Technical Record.[19]

> The effort applied towards reading the Water Transport System, a higher echelon army shipping system, was well repaid. Resulting intelligence gained concerning shipping movements was extremely valuable and the cryptanalytic experience and cross-references of data aided in the solution of other systems. A large part of the original work on the water transport problem was contributed by CBB, and it was a principal interest for about eighteen months, beginning early in 1943.

Abe Sinkov made it clear at a speech to CBB veterans in Sydney on 25 August 1987 that he saw this as being the second greatest achievement of CBB.[20] (He viewed the identification of the weakest points in the scattered territory occupied by the Japanese Army as the greatest achievement.)

---

[18]The 2468 intercepts are to be found in NARA RG4457, boxes 531–575. They are collectively item 1362.

[19]The *CBTR* is NAA file B5436 and may be read on line. The quoted text comes from Part G: *Main Line Systems*. Much more from Part G, as well as the quoted paragraph, is reproduced with annotations in Sect. 14.2.

[20]Archbishop Donald Robinson is to be thanked for supplying a copy of Sinkov's talk. David Horner's early paper on this subject (*Australian Outlook*, December 1978) makes essentially the same point on page 315.

16.10   JN-11, JN-25, JN-40, 2468 and Others Together

In 1946, IJN Commander Masataka Chikaya[21] wrote

> In the latter half of 1943 America's preparations for submarine warfare had become complete. They started to be active, very active. The mass attack by submarines was carried out in close concert with scouting operations that were absolutely superior. This was a system of attack that had been foreseen in the German submarine warfare in the Atlantic. This caused the losses of Japanese trade vessels to mount speedily and look endless.

The key phrase here is 'scouting operations'. In the main the shipping was located by Sigint. It was considered essential that no hint be given of this source. Thus, in many cases, spotter planes were sent to search for convoys on the routes where it was already known they would be found. The planes were not there to see but to be seen. This ploy appears to have worked. However an indication of the danger to which Japanese convoys were subject may be found in a German warning to German submarines dated 13 October 1944[22]:

> Consider whether there is not greater danger when cruising with a Japanese escort in some cases than in proceeding alone.

Part B of the *CBTR* notes on page 11 that the work of Central Bureau on Japanese Naval air/ground codes provided information on convoys of merchant ships. Instructions about air cover helped locate the convoys. This complemented information from JN-11, JN-25 and JN-40 and 2468 decrypts.

A unit was established to process intelligence about Japanese merchant shipping. For example, it played a role in the destruction of remnants of the *Take convoy*. This had left China late in April 1944 with some 20,000 troops withdrawn from Manchuria and destined for Halmahera, New Guinea and elsewhere. Its prospective noon positions were obtained by both Army and Navy Comint. After four large transports had been sunk, the remainder of the convoy was diverted to Menado in the Celebes. The CBB unit decrypted messages indicating that from there, landing barges would replace transports to complete the journey. This enabled the planning of bombing raids on the hapless barges en route. Much of the weaponry and supplies, and many troops, were lost as a result of the combined submarine/bomber attacks.

Intelligence about the movement of Japanese troop transports, derived from both Army and Navy Sigint, helped identify the bases weakly held by Japanese Army.

---

[21] A translation of Chikaya's account is given by Goldstein and Dillon (Note 8). Admiral Wenneker, German naval attaché in Tokyo, confirmed this in November 1945.

[22] This gem is taken from NAA Melbourne file B5553, which is a collection of Frumel intercepts. The date is 18 October 1944.

The National Security Agency booklet on *The Friedman Legacy* contains the text of the letter from General Marshall to Governor Dewey, the 1944 Republican candidate for the American Presidency. This was written in September 1944 and makes it very clear that Comint was a major factor in the attacks on Japanese shipping by USN submarines.

It helped assess current operations. More on this matter, which was of extreme importance in 1944, can be found in Sect. 23.2.

By contrast, American sea and air power was able to isolate completely the strongly held major Japanese base at Rabaul, denying it any further supplies and preventing any large scale removal of troops from it. The decision in August 1943 not to invade it but simply to leave it alone constituted a massive Allied victory, in that some 90,000 Japanese soldiers were virtually marooned there, becoming prisoners only after the end of the war.

The USN's submarines exerted significant strategic influence by cutting off various units of the Japanese Army, such as that in Rabaul, which then had to wait helplessly for something to happen. Even worse from the Japanese viewpoint, the cut-off units were forced to communicate in old codes, sometimes providing cribs to the Allied cryptologists. On 11 June 1944 General Marshall[23] wrote to General MacArthur:

> So long as 18th Army remains physically isolated and in radio communication with other Jap units it will continue to afford us valuable source for cryptanalytic assistance. To the extent that it will not interfere with your present operations it is highly desirable that this situation be preserved and fully exploited.

Another blockade was imposed on Japan as the war continued, in that contact with Germany became possible only by submarine. Even specially constructed cargo submarines had only limited capacity and were subject to harassment throughout the long voyage. For example, the I-52 was sunk in the Atlantic on 23 June 1944 carrying 300 tonnes of cargo, including two tonnes of gold. It had been intended that it would return to Japan with Enigma machines and essential commodities. Sigint had allowed the I-52 to be monitored throughout its voyage.

Appendix 1   The Hellships

The sinking of Japanese merchant ships had, from the viewpoint of the Allies, a most unfortunate consequence. Various of these, generally termed *hellships*,[24] were carrying Allied Prisoners of War (PoWs) back to Japan from sites, such as the

---

[23]The reference is taken from page 192 of Stephen Taaffe's *MacArthur's Jungle War: The 1944 NG Campaign*, which gives the source as RG4, Box 17, folder 1 in the MacArthur Museum archives. Marshall refers here to the IJA forces at Wewak and Hansa Bay, bypassed by the landings at Hollandia and Aitape. See Sect. 23.5.

[24]Greg Michno's *Death on the Hellships*, Naval Institute Press, 2001, appears to be the best reference on the deaths of PoWs in Japanese ships. To some extent it was written in co-operation with Linda Goetz Holmes, the author of *Unjust Enrichment*, Stackpole Books, 2001. Alistair Urquhart's *The Forgotten Highlander*, Skyhorse 2010, gives a first-hand account.

About 3 % of the large collection of Water Transport Code decrypts in boxes 531–575 of NARA RG457 has been examined without finding anything relevant to this issue. (See Note 18.) The other 97 % may provide further information about the movement of PoWs. It is also possible that there may be references to the movement of Korean women conscripted into prostitution by the Japanese Army. Spokesmen for the Japanese Government have denied that any written record of the 'comfort women' could be found.

Burma railroad or in Indo-China, where they had already been savagely treated, in order to put them back to work in mines and other hard-labour activities. Despite the Geneva convention, the hellships, as well as being grossly overloaded, were not marked by appropriate flags. Sigint information about Japanese convoys did include various details regarding their composition and cargoes, which in some cases revealed that certain ships were carrying PoWs. The dilemma facing the relevant Allied Command was how to react to this information. Selective sinking of ships in a convoy, even if this could be done under typical conditions of engagement in a submarine attack, could be used to leave ships carrying PoWs afloat but would risk the security of Ultra. As a result, such information was not always passed[25] on to those directing the submarine operations.

Some 20,000 Allied prisoners of all nationalities were lost due to submarine attacks on unmarked Japanese hellships, with the worst single disaster arising from the sinking of the *Junyo Maru* by a British submarine off the coast of Sumatra in September 1944. Some 5,500 of the 6,500 on board perished, including over 4,000 Javanese slave labourers.

An associated dilemma was caused by the Japanese practice in 1944–1945 of marking certain ships as hospital ships when they in fact were not. Sigint would occasionally identify fake hospital ships. If these were attacked there would again be a risk to the security of Ultra.

Around 1946 Irene Brion wrote down her memories of Sigint work. She published them in *Lady GI*, Presidio Press, 1995. On page 91 she notes: 'However I remember a message that told of a ship carrying an unusual cargo—women. Were they geishas? Whoever they were, they didn't rate the status of passengers.'

Not all Japanese deny that horrific war crimes were committed by the IJA in WW2. For example, IJN veteran Jintaro Ishida researched incidents in the Philippines during the Japanese occupation. His book has been translated as *Remains of War*.

[25]The article *Did Sigint Seal the Fates of 19,000 PoWs?* by Lee Gladwin, published in *Cryptologia* 30(3), July 2006, 212–235, gives evidence that the USN did its best to abstain from sinking merchant shipping known to be carrying prisoners of war.

Part IV
Organisation

Chapter 17
Central Bureau 1942–1945

The only satisfactory account of the operation of Central Bureau is that provided by its own report, the *Central Bureau Technical Records* (CBTR), written late in 1945 and from which this book's account of work on the IJA Water Transport Code has been painstakingly derived. This chapter consists mainly (Sect. 17.1—17.11) of an annotated partial text of Part A (entitled *Organisation*) of the CBTR and gives an overview of its wartime activities. The texts of Parts A to K are available on line via the Recordsearch option on the NAA website. The reference code is B5436. Another copy is in NARA RG457, Box 1086 item 3432. The original text is in conventional Roman type with additional comments in slanted type.

17.1 Formation, Early Development and Commitments

At the outbreak of war with Japan, the situation with respect to Signal Intelligence in the Far East was as follows:

(a) At Singapore the Far Eastern Combined Bureau, a British combined service organisation, operated Army and RAF field sections in addition to its Naval facilities. The Headquarters was long established and it was in active production.
(b) In India an organisation known as the Wireless Experimental Centre (WEC) was being developed.
(c) In the Philippines General MacArthur had a very small group whose major function was interception, for Washington, mostly of diplomatic material.
(d) In Australia there was a Naval 'Y' unit *(the SIB)* working in collaboration with GCCS and FECB Singapore on Naval and diplomatic problems, but on the Army and Air side nothing had been attempted.

© Springer International Publishing Switzerland 2014
P. Donovan, J. Mack, *Code Breaking in the Pacific*,
DOI 10.1007/978-3-319-08278-3_17

The CBTR rather surprisingly here omits reference to the USN team under John Lietwiler based at Corregidor. It also should have mentioned the Dutch team 'Kamer 14' at Bandoeng in Java. Indeed the first group of evacuees from Corregidor, led by Rudolph Fabian, went to Bandoeng and later on to Australia. The Melbourne COIC also deserves mention.

In March 1942, when General MacArthur established GHQ in Melbourne, one of the problems considered was the formation of a Signal Intelligence service. General Spencer Akin, Chief Signal Officer under General MacArthur, who had long been identified with Signal Intelligence and was well aware of its potentialities, was a prime mover in this connection. As an assistant, he had Colonel J. R. Sherr, who had a good technical background in cryptanalysis. He arranged for a first contingent of American personnel to be sent to Australia from Washington. *General Akin, irreverently called 'S. O. B. Akin' by many, became the (mostly absentee) commanding officer from January 1943. He was one of the 'Bataan Gang' of senior American officers who had been brought out of the Philippines with General MacArthur in 1942.*

Shortly before this, a large part of the AIF returned from the Middle East. With them came a section which had been operating in the Middle East and had performed valuable work in Signal Intelligence. *This was interception and traffic analysis work.* From this section was drawn the nucleus of the Australian Army Section of Central Bureau, under Captain A. W. Sandford. *Two other officers deserve a mention here. The first is Captain Jack Ryan, a veteran of WW1 and a radio expert. He served in the Middle East and progressed to leading the Army interception work for Central Bureau. The other was Stan 'Pappy' Clark, who developed vast natural talent at traffic analysis. Australian War Memorial file AWM52 7/39/5 mentions a report prepared by Sandford and Ryan on 8 March 1942 for Brigadier Simpson on a ship travelling back to Adelaide. It contained the first plans for what was to become Central Bureau.*

There were added to this AIF group some British personnel who had escaped to Australia from Java, an officer and a dozen other ranks from the Far Eastern Combined Bureau, Singapore; the men were experienced intercept operators.

The RAAF, on its part, had been investigating the general problem of Signal Intelligence. Their representative, Flight Lieutenant H. Roy Booth, who had been delegated to study the organisation of such a service by the RAAF, had recently returned from Singapore, where he had been sent to study RAF methods and Sigint organisation. *Apparently Flight Lieutenant W. C. Blakely accompanied him.* In addition, a small section had commenced interception at Darwin in 1941.

Roy Booth was a Sydney solicitor whose firm, Booth and Boorman, is still extant. By April 1941 he was serving as an intelligence officer in the RAAF assigned to the COIC.

In consequence of interservice discussions, Central Bureau was formed on 6 April 1942 as a combined United States Army, Australian Army and RAAF organisation. The name was chosen so as to convey no information whatever to outsiders as to the nature of the work being done.

The functions of the Bureau were defined as of a research and control centre to deal with all Army and Air intercepted traffic and relevant captured documents, this work to be carried on in close co-operation with London, Washington, India and other 'Y' centres. The Navy 'Y' organisation in Melbourne operated on a completely independent basis, and arrangements were made at this stage to expand that section to include United States Naval 'Y' personnel.

The Melbourne NAA file MP1074/7 8/3/1942 to 16/3/1942 contains a message to the British Admiralty Delegation in Washington dated 10 March. 'Special intelligence party ex Manila have arrived in Australia and are awaiting instructions from America. Subject concurrence of ACNB consider most helpful if they could remain in Australia to collaborate. Request this be represented by Capt Hastings to US Navy Dept.' In fact Frumel was up and running within 10 days.

The division of labour between the Army and Naval services was easily defined. Central Bureau was to concentrate on all military communications, the Navy group on Naval communications. The only problem consisted in determining how to divide the research on air-ground communications. It was generally agreed that the service which undertook the problem had to handle both Army and Navy air-ground systems. The Naval service indicated that they were quite happy to relinquish any interest in air-ground work and would like Central Bureau to take over the entire function. *In the heady days of April 1942 JN-25B mattered more than all the other ciphers put together. So Fabian would not have been particularly interested in lesser matters. The late General Akin left two drafts of an unpublished paper that have survived in the MacArthur Museum. His comment: 'the Navy component was apparently doing some work of this kind and declined to be included in the proposed activity' is fully compatible with Fabian's oral history account and must be closer than the CBTR account to the truth here.*

Each of the three services within Central Bureau was represented by its commanding officer and one or more technical assistants on a committee which determined Central Bureau policy and initiated action for the implementation of decisions made. *Some very interesting early minutes of the Committee that formed CB are available for online reading via the NAA website. The reference is A6923 SI/2.*

Arrangements were made to augment the headquarters staff, mainly from the United States Army and the AMF. A joint fund was established and orders placed for equipment to be secured from America. Plans were also made for the establishment of an IBM section, part of a general policy for the growth and expansion of the entire Bureau. *Ray Wyatt was an American Army radio specialist in Melbourne in 1942. He published an interesting account of his military experience in 1999 under the title 'A Yank Down Under'.*

The first operational steps taken by the committee consisted in preparing for the training and deployment of personnel to intercept Japanese communications. A RAAF unit was dispatched to Townsville to intercept enemy air-ground transmissions and to co-operate with operational formations there in the dissemination of tactical intelligence. They were assisted by 1 officer and 3 enlisted men of the

American personnel who had experience in the Philippines whence they had been evacuated by air. *The officer was Howard Brown, whose reminiscences survive in NARA in RG457 SRH-045.* Similarly AMF units were sent to Darwin and to Port Moresby. The Darwin unit was in operation by June, the Port Moresby party by September. With the establishment of these field units there began a steady flow of intercept material to the Bureau, and a start was made on its analysis. The early work consisted of the recording of all frequencies being used by the enemy, reconstructing Japanese radio nets, locating the transmitting stations, and identifying their call signs and frequencies. *The book 'The Eavesdroppers' by Jack Bleakley gives an excellent account of the work of the RAAF personnel who became '1WU', that is Number 1 Wireless Unit. It later sets out the achievements of 2WU, 3WU, 4WU, 5WU and 6WU. 7WU was at the training stage when hostilities ceased.*

At this stage the primary missions of Central Bureau related to field problems and low echelon material. Any facilities available after these functions were fully discharged, whether intercept or analytic, could be applied to the high command problem.

The amount of material available for study was necessarily very small at first. Moreover, it was being intercepted by inexperienced personnel and so involved many inaccuracies. However, as the training programme continued and some increases in personnel were obtained, the position became progressively brighter. The first successes in radio intelligence came from the gradual solution of the preambles of Japanese messages. *Stereotyped pieces of codes are easier to break!* From these, information was derived of sending and receiving stations as well of the units or individuals to whom messages were addressed. One interesting aid in connection with reading addresses came from the fact that some Army traffic was being relayed over Naval links. The conversion of the routing information to Naval procedure resulted in a different handling of the address. The system used by the Navy for enciphering addresses was much simpler than that used for the Army— it consisted merely of a mono-alphabetic substitution. Its solution was simple and provided valuable cribs into the Army system. *Cribs were all-important in WW2 cryptology! These three sentences confirm that IJN communications security was inferior to that of the IJA.* From this address intelligence, some information on the order of battle was obtained and the location of various Army units were traced. Thus began an appreciation of the organisation of 17 Army in Bougainville and the general order of battle of Japanese forces in the islands north of Australia.

The Shore Wireless Service of the RAN thus helped get both Frumel and CB established. Commander Newman had been instrumental in training the first RAAF interception team sent to Darwin in late 1941.

An early intelligence success came from the coverage of local Papuan circuits intercepted from Port Moresby. It was from these circuits, operators' chat and occasional items of low grade traffic that an estimate was formed of the size of the Japanese forces crossing the Owen Stanley Range towards Port Moresby. *This is rather modest. The Kokoda Track campaign (see Sect. 22.6) was of great importance. The Canberra NAA file A816 48/301/92 on RAAF mobile wireless units states that 1WU had been in Queensland since April 1942. It goes on to* state that 'The value of the unit for intelligence purposes is regarded as of the highest operational importance as, apart from its immediate operational utility, the information gained by the unit has been of the utmost importance in determining the location of enemy air bases, in detecting the movement of enemy aircraft and surface units in the area.' Hugh Melinsky states on page 30 of 'A Codebreaker's Tale' that 1WU intercepted in July 1942 information about the airstrip being built near the northern coast of Guadalcanal. Archival corroboration of this has not yet surfaced.

Melbourne NAA file MP721/1 W205/18 states that the Townsville remote receiving station was to be 'highly confidential' and that its site was pegged on 27 May 1942. File W205/2 describes how work on a conventional RAAF site in Townsville was deferred in favour of the 1WU station. According to Canberra NAA file A6923 SI/2 the RAAF issued the officer on the site with cash to 'obviate delays and to avoid disclosure of secret information'. Canberra NAA file A705 151/1/746 of 1946 describes the Townsville 1WU interception station as a 'rather fine re-inforced concrete air-conditioned building' designed for wireless unit work. There is a colour plan showing location of the concrete building and the 1WU camp site. The latter was made up of dilapidated houses which were left looking as such.

General MacArthur having moved his headquarters to Brisbane in July 1942, Central Bureau was transferred to Brisbane in September of that year. This move was necessary in order to be in close touch with GHQ and Allied Air Intelligence. It also had the effect of reducing the time lag in the passage of raw material from the field.

By the end of 1942 Central Bureau headquarters strength had grown to approximately 150 personnel. A small IBM section was functioning efficiently, traffic volume was increasing, and methods were being developed for the proper handling of traffic. The 126th (American) Radio Intelligence Company had arrived in the theatre to assist in the intercept problem, and in addition the strength of the field sections of both the AMF and the RAAF had been increased.

The diplomatic section was transferred from the Navy to the Australian Army as part of the November 1942 re-organisation. It was given support on occasions by Central Bureau.

Co-operation among the various centres was working most satisfactorily. Central Bureau was issuing a monthly report of technical progress; Washington, London and India were issuing periodic reports. These permitted each centre to keep informed of the developments of the others.

In September 1942 Central Bureau took over the house at 21 Henry Street, Ascot. This had been built in the previous century but by 1942 had lost most of its grounds to subdivision. Its usage in the 1920s and 1930s is difficult to establish, as are the processes that led to its selection for CBB. It is quite big with a wrap-around veranda upstairs and is now a private residence again. It was prepared for Central Bureau immediately prior to the move. The veranda was later wired in to prevent documents being blown away by the wind. It had a large and thus useful garage.

Near the house at the intersection of Kitchener and Lancaster Roads is a public park called Ascot Park. This was requisitioned early in 1943 and fenced in. Light huts, mostly 6 metres by 18 metres, were erected as required. The site ended up with 29 buildings. There were about three other sites used for other functions, such as garaging. The park has been restored to public use. An old fire station adjacent to the park, used for the IBM equipment from 1943 to 1945, survives.

As the huts were much more suitable for secret work than the house, eventually the Park took over as the principal site of CB. The layout did not resemble that at the British cryptological site at Bletchley Park, where the original huts were substantially replaced by the bomb-resistant permanent buildings that are mostly still there. The wartime Ascot Park survives in the form of plans (both 1943 and 1945) in the Brisbane branch of the NAA. Presumably these plans may be used to locate buried fragments of the footings for the huts.

Later, accommodation was provided for most of the American contingent, by that time quite massive, at the racecourse almost adjacent to Ascot Park.

The situation at the end of 1942 did show at least some promise. Some specialists of great skill, and in particular Abe Sinkov, had arrived. The Japanese Army mainline ciphers looked extremely formidable, but the first break-through was to happen in the first half of 1943, when the capacity existed to take advantage of it.

17.2   Assignment to GHQ

This part gives details of a re-organisation of January 1943. Central Bureau was placed under the control of GHQ, in fact under General Akin, who became its director. Its duties included cryptology, traffic analysis, direction-finding and supervising interception activities. The American Army, the Australian Army and the RAAF each provided an assistant director. Although General Akin did not intervene much in the running of CBB, he chaired a committee 'I' which made recommendations to GHQ on various matters. Arrangements were made for further reinforcements to be sent from the United States. Only the Mission Statement for CB is reprinted here. It indicates the position as in January 1943 as seen by the senior participants.

(1) Cryptographic

    (a) The cryptanalysing of the cryptographic systems employed for the secure transmission of enemy military traffic over point-to-point, air-ground and air-surface craft circuits.

    (b) The translation, preparation and dissemination of traffic, the cryptographic system for which has been solved.

    (c) Making available to superior and/or subordinated agencies, when directed by proper authority, the keys to solved systems.

    (d) The compilation of the codes and cyphers required for common use by the components of the S.W.P.A. *The Canberra NAA holds certain interesting documents dealing with WW2 communications security. One of these describes mechanical generating of one-time pads and tables of additives, originating from Bletchley Park. Another (vintage 1945) describes how the use of additive tables was monitored to prevent dangerous overuse. A description of how Central Bureau monitored Allied communications for insecure practices is given by Geoff Ballard in his 'On Ultra Active Service'.*

(2) Radiogoniometric and Identification

  (a) The reception, collation, evaluation and analysis of such intercepted military radio point-to-point, air-ground and air-surface craft traffic as may be necessary to provide for military intelligence purposes.
  (b) The identification and location of enemy aircraft, surface craft (carriers and tenders only) and military shore stations used for military point-to-point, and ground-air and/or land-aircraft carrier purposes.
  (c) The location and identification of enemy headquarters and other important enemy activities and establishments.
  (d) The movements and location of enemy aircraft, particularly those that may become engaged in raids against friendly areas or forces.
  (e) Data with regard to meteorological conditions in friendly and enemy territory. *Australian War Memorial file AWM124 4/132 makes it clear how much this mattered in March 1942.*
  (f) Distribution of the above data to the agencies designated by General Headquarters S.W.P.A. to receive the same.

(3) The Planning For and Co-ordination of Intercept Activities

  (a) Planning for the required intercept and direction finding personnel, equipment and stations.
  (b) Allocation of personnel, equipment and station facility requirements to the Allied Forces for the production of the properly trained and equipped personnel and required fixed intercept and direction finding stations and facilities. *In fact much of the training of American re-inforcements took place in the United States. See, for example, Irene Brion's book 'Lady GI'. Likewise the First Canadian Special Wireless Group arrived fully trained. Gil Murray formerly of the 1CSWG wrote an account of his experiences under the title 'The Invisible War'.*
  (c) Assignment and co-ordination of the missions of the various detachments engaged in direction finding and intercept work.

  The text notes that initial forecasts of personnel required were fairly accurate except for cryptanalysts. Various requests from 'Mic' Sandford to the Director of GCCS for more cryptanalysts survive.

17.3   First Steps in Solution

(a) Naval Air-Ground.

The systems which seemed to offer the best hope of solution at this time *(1942)* were the Naval Air-ground systems. *Details omitted.*

A program of concentration on Naval Air-Ground codes was initiated and a special section was formed. The operational code book was solved giving a much wider range of information. These Naval Air-Ground code books were the first solved by the Bureau.

From then onwards these codes were read continuously. *Details are set out in Part B of the CBTR. There is enough evidence extant to show that Nave was principally responsible for this achievement. See, for example, Canberra NAA file A6923 16/6/289 page 4. One interesting item in CBTR Part B is the statement that at one stage (after mid-1944) the messages for a day would be sent to the machine room at midnight to be punched onto cards with printouts being ready in the morning. Compare Sects. 1.14 and 1.15.*

(b) Army Air-Ground.

This topic forms the theme of Part C of the CBTR. The two sentences quoted show the importance of both captured documents and co-operation between cryptologic agencies.

Good progress was being made on the reconstruction of the code book when the book itself and some additive sheets were captured in India. In the light of the intelligence gained from this captured material enough was learned of the Army Air-Ground system to permit of continuous solution thereafter.

Page 7 of Part C acknowledges assistance from India, Guam (later), some USAAF units, 1WU, 3WU, 4WU, 6WU and three sections of the ASWG.

(c) Military Systems.

Very little progress was made, at first, in military systems, whether low or high echelon. Late in January 1943 an entry was made into a high command air point-to-point system, which used the discriminant 3355. The entry was made from the fact that a group of 17 messages was discovered having identical text except for slightly different endings. The last few groups of all 17 messages were sufficiently different to provide adequate depth for initial stripping. From this, some bits of additive were derived and two or three dozen code groups were identified on an arbitrary basis. In addition to this depth, three pairs of split duplicates were unearthed in which a fair amount of stripping was possible. Unfortunately the basic handling by the Japanese of their indicator system was so sound that no further progress was made and the problem was dropped shortly afterwards when other more significant developments took place.

This last extract reveals the problem that the jargon of the time causes for modern readers. It also shows the cardinal importance of indicator encryption systems in additive systems. (See Sects. 8.18 and 13.2.) Chapter 14 described the exploitation of an unsound indicator encryption system. Discussion of the work on military systems is resumed in Sect. 17.5 below.

(d) Weather Systems.

Some progress was also being made in the solution of enemy weather codes. These were technical meteorological reports of weather observations transmitted from numerous weather stations to a central headquarters which collated these data and used the information to prepare daily weather forecasts.

The basic principles of the system were solved and some regular stripping was possible. However, as the amount of material available for study was quite small, the amount of weather information derived was very limited. Nevertheless it often proved quite helpful to the *(Allied)* air forces in deciding whether or not the weather over selected targets was favourable for attack.

17.4 Intelligence Production in the Field

The success in the air-ground problem and the weather systems forced attention to the fact that this intelligence had to be built up in the field. Derivation of this intelligence at a rear headquarters, from traffic that was slow in getting down, and its passage back to the field involved delays which were quite unacceptable. Such information had to be derived and acted on immediately if it were to be of any real use. As a result, personnel trained in the handling of these systems were sent to the field sections, where they were able to treat intercepted messages and decypher them on the spot. The intelligence was then passed on at once to the local air force headquarters for immediate action. Central Bureau served as the research centre which kept feeding to the field sections code values, recovered additives and any other cryptanalytic information which would prove of assistance in field exploitation.

17.5 Further Steps in Main Line Solution

In April 43 the first successes on Army main line systems were achieved. *See Sect. 14.2.* This gave tremendous impetus to the entire cryptanalytic problem. It resulted in considerably increased technical communication among the various centres and paved the way for future developments. *Colonel Joe Sherr went to India to arrange communication and co-operation with the Wireless Experimental Centre (WEC) in Delhi. He died in a plane crash on the way back in late September 1943.* It constituted one of the reasons for calling a conference on inter-theatre co-operation and possible division of labour. From this time on until the end of the war there was always some success being achieved on high command material. At times, certain systems would remain unbroken for limited periods. On the other hand, there were occasions when practically everything transmitted was being read.

Early in 1944 the 9th Australian Division made the first significant capture of enemy cryptographic material in Sio, then the headquarters of the 20th

Japanese Division. The haul was somewhat accidental since the materials were discovered in a deep water-filled pit. Most significant of all was that the enemy was entirely unaware of this capture. Indeed, messages were read later on, signed by an officer, that he had overseen the destruction of the documents! For two full months we were in complete possession of current materials and read the Japanese messages immediately upon receipt. The intelligence supplied to G-3 was of inestimable value[1] and played a most important role in the Solomons and New Guinea campaigns.

Following on this fortunate occurrence the main line problem was handled with almost continuous success. In addition, the emphasis placed on the main line problem increased continually at the expense of low echelon studies. This change in emphasis, apparently contrary to the original ideas concerning the basic mission of Central Bureau, was forced on the organisation by the very nature of the low echelon problem, which was technically at least as difficult to solve as the high command material. Further, traffic volume was much lower since transmission was over circuits using low-powered sets. Field sections were seldom close enough to the front to be able to hear many of these circuits. Finally, in those cases where success was achieved on low echelon material, it was found that the intelligence was of relatively little value. The few items which had any worth at all had to be read currently because of their tactical nature, and this was very seldom possible. The result was that the low echelon problem was pushed more and more into the background.

Although important changes were made by the Japanese in procedure, frequently resulting in considerably increased security, information of these changes was regularly read in radio communications. So many of the Japanese headquarters were being cut off from any source of distribution of new materials that instructions with regard to changes in cypher procedure were passed to them by radio. In almost all cases the systems used to pass this information were being read. The changes thus due to come into effect were regularly anticipated, and methods for overcoming them were frequently available before the effective date of the change. The production of intelligence thus proceeded continuously in ever increasing volume. Derived intelligence was not always as current in the early months of 1944, at times some was as much as 3–6 months old. Nevertheless, the strategic information furnished permitted most effective overall planning. In addition, tactical information in varying volume was being continually supplied.

Inter-theatre co-operation on the Japanese problem became constantly more efficient. A joint conference in March 44 *(held in Arlington Hall)* defined a division

---

[1]Perhaps one should repeat what is stated here to point out what can be done with good Comint! 'The intelligence supplied to G-3 was of inestimable value and played a most important role in the Solomons and New Guinea campaigns.' See also Note 7 of Chap. 23.

A very detailed report on the use of intelligence by General MacArthur's SWPA command was prepared after the war in Tokyo. A copy has survived in the Australian War Memorial as AWM59. This report gives much general credit to Central Bureau but does not attempt to describe separate sources of intelligence for individual events.

of labour and served as a basis for future mutual assistance. Under the direction of Arlington Hall, arrangements were made for definite assignments *(of duties)* to each centre, which received from all other centres the traffic relating to its assignment. The results obtained by any one centre were radioed to all the others so that each was in possession of all available information. Regular progress reports issued by all centres provided safehand confirmation of radioed material, carried complete discussions of techniques and ideas for solution, described machine aids and ensured complete dissemination of technical information. *This book does not attempt to describe the techniques of 1944–1945.*

In connection with the very difficult main line problem, captured documents were most important. The handling of these documents was originally a function of the Allied Translator and Interpreter Section (ATIS) and all captured documents had to be passed to that organisation. It was discovered quite early in the war that the lack of officers in ATIS with a technical appreciation of cryptanalysis made its translations unsatisfactory and sometimes even incorrect. Further, the delays involved, by virtue of the tremendous amount of material that ATIS had to handle, could well negate the value of the documents altogether. An arrangement was therefore reached by which all documents of cryptographic importance were passed directly to Central Bureau from ATIS without any processing whatever. The Bureau's translation section then undertook the responsibility of handling these documents.

On several special occasions messages were read regarding the sinking of ships carrying cryptographic documents, or the loss in other ways of cryptographic material. On several such occasions officers representing Central Bureau were sent on special search missions to the scene and on at least two occasions considerable success was achieved. The most exciting of these was a salvage operation in connection with the sunken 'Yoshino Maru'. The ship had burned before sinking. The salvage party found nothing aboard except a tin box which had been jammed between two rungs of a ladder. *This was found in shallow water off Aitape, NG.* That box was found to be one of those which had contained cryptographic documents. They were badly charred and practically illegible. However by chemical treatment and photographic methods, a large part of the text was reconstructed.

17.6   Photographic Section

There were no fax or photocopying facilities in the 1940s. Hence cameras were used to copy documents and microfilm was used to enable quantities of information on paper to be sent by air to other centres. The following extract indicates the size of the communication problem.

The section handled a vast number of documents of various kinds, and microfilmed and reproduced captured cryptographic documents in the minimum possible time. One of its outstanding performances was the successful treatment and photographing of the charred additive books recovered from the 'Yoshino Maru' as recorded above. At one period approximately 100,000 negatives per month of traffic alone were sent overseas on microfilm.

The CBTR give minimal information about the CBB machine section. As already noted, Ronald Whelan, the former deputy head of the Bletchley Park IBM facility, wrote an account of that unit. This is available in the National Archives, London, as file HW 25/22. See also Sects. 1.14 and 4.3.

17.7   Personnel and Training

The US Army History of its Signals Corps notes that in February 1942 'good radio men' rated 'number 1 on the list of 181 shortages'. Getting together suitable interception operators in the numbers required was far from easy. The skill needed by an interception radio operator was quite high. Melbourne NAA file B5435 360 includes a letter from Sandford to Sir Edward Travis, the director of GCCS, of 8 November 1943 which states that 'it takes 12 months to convert even a trained operator to be efficient for our purposes'. Other archival references give somewhat shorter training periods, but the point remains. The attraction of having a large trained unit arriving from Canada must have been enormous.

The earliest system of training was the absorption method in which each section was responsible for the thorough instruction of personnel allotted to it.

In mid-1943 an Intelligence School was formed to cater for basic training in radio intelligence methods with emphasis on the air-ground problem. Instruction was undertaken by a permanent staff assigned to the task, and supplementary lectures were given by heads of sections and specialists in the various departments of the Bureau. The course of instruction extended over a period of approximately seven weeks, and the school provided initial training for all RAAF intakes and a proportion of U.S. personnel. A total of over 300 passed through the Intelligence School, being thereupon assigned for operational experience to field units as well as to Central Bureau. There were included in these 300 students several groups from U.S. Air Force units engaged in Radio Intelligence (Radio Squadrons Mobile assigned to the Pacific Theatre).

William Friedman had always practised documenting courses and procedures. No doubt his influence, albeit invisible, was taking effect here.

One difficulty experienced by the RAAF component was the fact that in the early stages of its reinforcement pre-selection of personnel had not been established as an essential measure towards ensuring requisite educational qualifications and aptitude for specialised intelligence work. Fortunately the major intake was effected after this principle had been accepted by higher RAAF authority.

One of several sources on this is the Canberra NAA file A11093 320/5K5, page 143. Undoubted integrity was required, as were matriculation standard education, research temperament, precision, initiative and 'classical or mathematical educa-tion'. The reader may wish to look again at Note 9 of Chap. 4.

All services found from time to time that their Tables of Organisation (War Establishments) were inadequate to cope with the demands for additional personnel and in particular to provide specialists in various departments (e.g., International Business Machine operators). The result was that frequent variations of establishment tables were necessary with consequential delays in the provision of personnel urgently required and in the conversion of staff from one task to another.

The continued expansion of the Bureau necessitated changes in internal organisation and consequently the Bureau was, late in 1944, divided into the following branches:

> A—Administration
> B—Solution
> C—Communications
> D—Photographic
> E—Traffic Analysis
> G—Machine Procedure
> H—Translation
> I—General Intelligence and Liaison.

At the cessation of hostilities the numerical strength of the Bureau and its field sections had reached a total of 4,339 personnel.

At the end of the war Central Bureau's scope took in the area from Okinawa to Brisbane in the North-South direction and from Guam to Borneo in the East-West. *Some material omitted.*

No record of the staffing of Central Bureau would be complete without a reference to the outstanding work performed by the Women's Services. WAAAF personnel were included in the RAAF component from the inception of Central Bureau. AWAS personnel joined after the move to Brisbane in 1942 and in May 44 a large contingent of WACs was added to the strength. A proportion of the female personnel was employed on the technical phases of the work, but a majority made themselves virtually indispensable as typists, stenographers, personal secretaries, IBM operators, and in the performance of many clerical duties in which they proved to be more patient and painstaking than men. Their contribution to the undertakings of Central Bureau was at all times a major factor in its operational efficiency—a circumstance which was fully realised when the transfer of the organisation from Brisbane to San Miguel, Luzon (without the Australian Women's Services) was undertaken.

Documentation survives of General Akin's attempt to persuade the Australian Government to allow Australian female staff of CB to go to Luzon. He described them in his unpublished documents as 'a most effective, loyal and highly effective group'. The records of the interception unit working for the Diplomatic Group near Melbourne include what may be a first in the Australian Defence Forces: two officers took leave on a Friday to get married on the Saturday.

17.8   Dissemination of Intelligence

Information disseminated to appropriate intelligence authorities by Central Bureau as a headquarters fell into three main categories as follows:

(a) Intelligence derived from the decryption of enemy high-grade traffic passed in the form of 'UBJ', which was in effect a file of the complete translations of decoded messages.
(b) A daily summary of Military W/T activity (BJ) based on traffic analysis.
(c) A daily report known as the 'CBW' Report covering Army and Navy air activity only and collated at headquarters from summaries signalled by all field units engaged on air-ground assignments.

In addition, daily reports were prepared by all field sections and passed by radio and safehand to intelligence formations in the field.

A comprehensive summary of intelligence distribution is not reprinted here. However page 109 of Canberra NAA file SI/2 makes it clear that by January 1945 the Signals Intelligence Board London made the big decisions about where information should be sent. Abe Sinkov in an oral history interview with the National Cryptologic Museum makes it clear that he kept Arlington Hall well informed on technical progress.

17.9   Liaison Officers

To achieve the closest co-ordination of effort in the field of Signal Intelligence, it became necessary for senior staff officers of Central Bureau to visit intelligence centres in London, Washington and Delhi. These liaison visits and the consequent interchange of technical material proved of considerable value in the overall study. There is no doubt that these tours of duty were also of great benefit to the centres visited.

A further result was the assistance rendered by overseas centres in the supply of specialist personnel urgently required in this theatre. In 1944 Central Bureau was reinforced by 20 British Army, 2 RCAF and 4 RAF translators; in 1945 by a further 12 RAF translators and phoneticians; and in the same year by the Canadian Special Wireless Group which was assigned to Australia to assist in the coverage of enemy main line circuits.

Similarly the Bureau welcomed visiting officers from overseas, it being appreciated that a great deal could be achieved in a short time by personal exchange of views.

In the early stages of the Bureau's development the demand for experienced Signal Intelligence officers was such that none could be spared for exchange duties with other similar organisations. In January 44 Washington took the first step in this direction by assigning liaison officers to Central Bureau for successive periods of duty, each of 6 months duration. These appointments were maintained until the cessation of hostilities. In July 1945 the stage had been reached where it was possible for Central Bureau to exchange liaison officers with Delhi (*the WEC*). The value of this interchange was just becoming evident when the war ended.

Another phase of the profitable employment of liaison officers was their assignment to headquarters formations in the field. As the production of current intelligence grew in volume and new Army and Air Force headquarters, inexperienced in the operational application of Signal Intelligence, were established in the theatre, such appointments became a matter of necessity.

The essential function of these liaison officers was to afford field headquarters the benefit of their specialised experience, to interpret in terms of operational significance the intelligence reports received from field sections and Central Bureau, and particularly to stress the limitations as well as the possibilities of radio intelligence.

The first appointment was made in 1943 to 6th Army which then had its headquarters at Finschhafen. Other similar appointments were made from time to time where necessary.

These officers performed another useful and highly important service, that of ensuring that appropriate measures were taken for the security of the intelligence provided. This duty was at a later date taken over by the United States SSO and British SLU detachments.

Frederick Winterbotham's well-known book 'The Ultra Secret' explains the Special Liaison Units set up initially for the European Theatre but later for the Pacific War too. The Americans had a similar system.

17.10 Communications

Early experiences demonstrated the absolute necessity for independent and efficient communications between all field sections and Central Bureau for the speedy passage to the Bureau of derived intelligence and texts of messages intercepted, and by the Bureau of operational instructions and technical information. Also in view of the complex nature of the cryptographic problems involved, a rapid exchange of technical information between Central Bureau and all overseas centres was essential to success. No one centre could cope with all problems and as time went on good communications became more and more important.

In both cases the need for security and the employment of special cypher systems dictated that all Signal Intelligence communication channels should be as far as possible separate from normal operational circuits.

So far as the Australian services were concerned the policy arrangement was made early in the history of Central Bureau for provision by the Australian Army of cypher personnel for internal and external communications, and for operation by RAAF of external circuits to British Empire terminals.

In order to meet this commitment a special cypher section was raised in December 43 staffed by Australian Corps of Signals. This section handled a total little short of two million groups per month.

Towards the end of 1943 a development of major importance took place in the installation by RAAF of the Brisbane–New Delhi radio link, together with the opening of Central Bureau W/T Station. This link also carried a substantial proportion of the Brisbane–London traffic. With the highly efficient U.S. Army radio teletype for Brisbane–Washington communications, the Bureau was well served for its overseas exchange.

Circuit diagrams showing the internal and external radio links in operation from time to time are shown in Central Bureau Technical Signal Intelligence reports issued monthly.

17.11   Forward Movements of Headquarters

With the development of the offensive in S.W.P.A., General Headquarters moved to Hollandia, Dutch New Guinea, in August 44. Consequently it became necessary to transfer an advance party from Central Bureau in order to serve GHQ and other forward headquarters formations with intelligence on the spot.

This party, known as Advanced Echelon, Central Bureau, Hollandia served a very useful purpose in carrying out traffic analysis locally. Traffic data were furnished by radio from all field sections and complete inference reports were regularly available within 48 h.

Similarly, following the invasion of the Philippines and the transfer of all major headquarters to Tolosa, Leyte, a Forward Echelon of Central Bureau was established there in November 44. This echelon worked in conjunction with the Wireless Units engaged in the air-ground mission and undertook the collation and dissemination of all air radio intelligence. The operational value of its work was later considerably enhanced with the establishment of direct circuits for the interchange of technical data and derived intelligence with all other theatres undertaking the same problem.

Upon the move of General and other headquarters to Manila in March 45, it became essential to plan for the transfer of Central Bureau to Luzon and consolidation there with its Advanced and Forward Echelons. These latter moved to San Miguel, Luzon, in May 45, and the greater portion of Central Bureau Brisbane was shipped to Luzon in July 1945. Movement of the remaining personnel, less a small Rear Echelon, was imminent when hostilities ceased.

Part K of the CBTR is entitled 'Critique' and gives comments jointly made by Booth, Sandford and Sinkov. They state that care was needed to avoid disruption in the relocation of Central Bureau. It was to become the signals intelligence unit in Operation Olympic, the proposed invasion of Japan. The nuclear weapons dropped on Hiroshima and Nagasaki forced the Japanese to surrender in August 1945. So Operation Olympic was cancelled and Central Bureau was disbanded.

Appendix 1   The American Citations

The following are all taken from Canberra NAA A816 66/301/232. They are the citations for high awards made by the American government for these three very senior figures in Central Bureau. The text dates back to 1946. The perceived need to keep the achievements of CBB secret led to the awards not going ahead.
US citation for A W Sandford SX11231.

> As an Assistant Director, Central Bureau, ..., Colonel Sandford displayed outstanding technical skill in organising and co-ordinating activities of the Australian personnel in effectively prosecuting the mission of radio intelligence. Under his distinguished leadership, much invaluable information was furnished in military intelligence. Through exceptional technical ability, resourcefulness and unremitting devotion to duty Colonel Sandford made a most significant contribution to the success of Allied operations in the SWPA.

US citation for Roy Booth 261929.

> As Assistant Director of Central Bureau ... Booth demonstrated conspicuous technical skill in organising and co-ordinating activities of the RAAF contingent of the Bureau. Under his skillful leadership, much meteorological data and vital information pertaining to hostile aerial activity was provided to allied air force elements. Through his comprehensive professional ability, resourcefulness and devotion to duty, Wing Commander Booth contributed materially to the continued effectiveness of air intelligence and aerial operations in the SWPA.

US citation for Stanley Clark for Medal of Freedom.

> For meritorious service which has aided the United States in the prosecution of the war in the South West Pacific Area from March 1942 to August 1945. As Head of the Radio Intelligence Section, Central Bureau, Office of the Chief Signal Officer, GHQ, SWPA, later United States Army Forces, Pacific, Major Clark exhibited exceptional technical ability and resourcefulness in the study of Japanese military and air-ground communications. He rendered consistently effective service in initiating and prosecuting projects of a highly specialised nature. Through his marked professional capacity, sound judgement and unsparing dedication to duty, Major Clark made a conspicuous contribution to the continued effectiveness of intelligence operations in the Southwest Pacific Area.

A file *Letters on Central Bureau Brisbane Liaison* survives in NARA as RG457, box 1027, item 3290. One 1945 letter opposes a proposed restructure of signals intelligence. It includes:

> Perhaps the best way to begin to describe CBB's point of view is to say that it is a small, but very proud, organization. It feels it has objective justification for this pride. General Sutherland has stated that its bulletin is the single most valuable item that there is in the theater—several personnel have been awarded the Legion of Merit—General Akin takes a tremendous personal interest in CBB and regularly obtains preferential treatment for it over other organizations in the theater—it can quote figures that per capita production at CBB generally exceeds per capita production at SSA [that is, Arlington Hall].

Appendix 2   Other Citations

The following material is taken from NAA file A6923 16/6/289, being the Army file on 'Administration of Central Bureau'. Sandford, writing about possible awards, states that:

> First and foremost is Colonel Sinkov. Apart from the fact that it would be a gesture on the international plane, it would be greatly appreciated by GHQ. Colonel Sinkov's phenomenal technical capacity and his untiring co-operation with the Australian component demand special recognition.

Sandford recommended awards for Captains Geoffrey Ballard and Charles Inglis but was not a whole-hearted supporter of Captain Nave:

> I have recommended nothing in the case of Capt Nave since I thought I should like to have your advice on this point. Technically, Capt Nave is by far the most brilliant officer in the Unit, but he is so lacking in initiative and appreciation of changing operational requirements of war forces (sic) that his efforts must be constantly guided by Major Clarke or myself. One cannot forget that it was he who 'broke' the first Japanese cipher handled by any of the military units during this war and on that account alone he should perhaps be specially considered.

Lt.-Col. Sandford would not have known about the JN-25 saga (Chap. 9) and Nave's role in getting the GCCS started on IJN codes and ciphers in the 1920s and 1930s (Sect. 3.3).

The Sydney University Archives hold the employee files on Professors Room and Trendall. Each contains a highly complimentary letter from the Department of the Army thanking the University for their services. See also Note 4 of Chap. 14.

Joe Richard (Sect. 14.3) was one of 11 Americans in Central Bureau awarded the Legion of Merit. He finished his career working for the National Security Agency. The others were Colonels Joe Sherr and Abraham Sinkov, Lt-Colonels Harry Clark, John Ernst and Hugh Erskine, Majors Charles Girhard and Zachariah Halpin, Captains Otto Marht, Robert Steinfurt and Clarence Yamagata.

Chapter 18
Organisation and Reorganisation

The previous chapter gave a contemporary overview of the development of one major Sigint unit, Central Bureau. There were several others, all of which have already been mentioned. The overall Allied Sigint activity was patently extremely expensive, particularly at a time when resources were stretched. Moreover there were various other units working on other sources of intelligence. Decisions must have been made at high levels that the cost was more than justified by the results.

18.1   Op-20-G in February 1942

This chapter does not attempt to tell the full story of how some 40,000 people[1] were employed in a global Sigint system by 1945. Instead a few brief comments on particular activities will have to suffice.

---

[1]The book *The Emperor's Codes* by Michael Smith gives much useful information on the various Allied cryptanalytic units in the Pacific War.

The names and functions of subdivisions of Op-20-G evolved quite rapidly throughout the Pacific War. The Dayton Codebreakers website has a glossary page which leads to a document setting out the structure in April 1944.

Those trying to confirm the estimate of how many people were involved in Allied Sigint operations in WW2 may begin with the figure of 4339 in Central Bureau in August 1945 quoted in Sect. 17.7 from the *CBTR*. The TICOM report into German Sigint in WW2 gives a total of 60,000 rather than 40,000. No contemporary breakdown of either total has been found.

© Springer International Publishing Switzerland 2014
P. Donovan, J. Mack, *Code Breaking in the Pacific*,
DOI 10.1007/978-3-319-08278-3_18

The USN cryptologic unit Op-20-G was re-organised[2] in February 1942. The various sub-units were:

GA	Administration	GP	Planning and Equipment
GB	Correspondence	GR	Training and Personnel
GC	Communications	GS	Machine Processing
GD	Deception	GT	Traffic Analysis
GF	DF Net for East Coast	GW	Traffic Receiving
GI	Combat Intelligence	GX	DF and Interception Control
GL	Collateral Information	GY	Cryptanalysis
GM	Math Analysis & Research	GZ	Translation & Code Recovery.

This list reveals that by 1942 the entire process involved elements of science and mathematics. The key 'combat intelligence' was being carried out in Hawaii and Corregidor. The Corregidor unit was transferred to Melbourne in three tranches.

The decryption and decoding of JN-25B in March, April and May 1942 made it clear that the JN-25 cipher systems were a principal target. This led to a decision to concentrate the work on new JN-25 systems[3] in Washington.

18.2 Use Made of Sigint

In general, it is quite difficult to evaluate the influence of any form of intelligence on the senior commanders. The special cases examined in this book are quite exceptional in that the role of Sigint is particularly clear.

18.3 Communications Security and Deception

A necessary feature of a Sigint war is keeping the communications of one's own side secure. The IJA and IJN simply could not match the Allies' successes in decrypting and decoding enemy messages. Section 8.7 gives an account of the American Sigaba/ECM machine and the British Typex.

---

[2]The reference is page 12 of *A Priceless Advantage: USN CI and Midway* by Frederick Parker. On page 18 Parker notes that the cryptanalytic unit GY took a few more months to have a JN-25 team up and running in Washington.

The introduction of the Op-20-GM 'Mathematics and Research' unit in February 1942 bore fruit later, both in Atlantic/European cryptology and in Pacific cryptology. For example, the recruitment of Marshall Hall, Jr. led to the Hall weights concept discussed in Sect. 15.3.

[3]The invaluable *GYP-1 Bible* records the arrangements in Washington for handling JN-25 in mid 1943. This was before the Mamba era (Chap. 15).

The equally invaluable *Jamieson Report*, being B5554 in the Melbourne NAA and available in digital form, deals with the various JN series with the exception of JN-25.

Although there was some deceptive use of radio by the IJN immediately before the raid on Pearl Harbor, there is little evidence of other major Japanese deception campaigns.

However Allied attempts were made in 1942–1943 to deceive the Japanese by sending occasional false information about the extent to which India and Australia were defended. Such a program was backed up by false radio traffic sent by spurious army units. Thus Australia had such a unit, in reality consisting of two or three radio operators, based at Hay, NSW.

A report in the Australian War Memorial (AWM92 3DRL 6201/86) of 8 May 1943 discussed methods of providing disinformation to the Japanese. One method was to provide fake indiscretions to the press. These would be effective only if they did become available to the Japanese. But it was known that the Chinese diplomats in Australia, who had access to the local press, were sending messages back home using insecure codes. In fact, this possible access route was further utilised. Low-level facts,[4] occasionally augmented by major pieces of disinformation, known as *purple whales*, were given to these well-meaning people.

Sadly there is little or no evidence that this enterprise influenced the Japanese to keep out of India, Ceylon and/or Australia. But two intriguing small paragraphs survive in the NAA. Canberra item A1196 48/501/171 entitled *RAAF policy on cipher* states that it was

> generally accepted by the Military Legation and by London that Chinese codes are not secure. Leakages do occur.

The date was 27 February 1942. Despite the conclusion that 'leakages do occur' apparently no effort was made to upgrade the security of Chinese diplomatic traffic. Presumably the decision to do nothing was made to allow purple whales to be transmitted when necessary. Canberra NAA item A10909 1 of February 1943 is a Magic intercept from the Japanese Ministry of Foreign Affairs to the Ambassador in Berlin stating:

> With the recent withdrawal of the Japanese forces from the areas of Guadalcanal and Buna, Australia is feeling a certain degree of relief and confidence in General MacArthur has increased.

This would appear to be a leakage of a report sent by Chinese diplomats in Australia being intercepted by some Japanese agency, passed on to the Ambassador in Berlin and intercepted again by those handling the Magic program.

---

[4]Churchill's aphorism about the truth being so valuable in war as to need protection by a bodyguard of lies works in reverse in a deception campaign. The major lie ('purple whale') is so valuable as to need protection by a bodyguard of minor truths.

18.4 Communications

Throughout WW2, the radio transmission of steadily increasing volumes of re-enciphered material from interception centres to the processing centres remained a challenge that required continual upgrading of communication channels.[5] The pre-war undersea cables could not carry such traffic but some automation of the radio transmission and reception of re-enciphered material was possible. As mentioned in Sect. 17.6, less urgent material was just written out on paper, photographed and carried by secure air mail in microfilm form.

It is possible that the quantity of heavily encrypted traffic sent between Sigint units could have attracted the interest of enemy security agencies.

18.5 Central Control

Conferences on Sigint were held as needed. Thus the finding in Sio, New Guinea of full documentation on Japanese Army cryptographic methods currently in use in New Guinea in January 1944 required a proper discussion a few months later. The situation was always in a state of flux. Abe Sinkov commented[6] many years later that the overall allocation of tasks was reasonably effective in stopping wasteful duplication of effort.

18.6 The GCCS in Ceylon and India

The FECB was withdrawn from Singapore to Ceylon (now Sri Lanka) at the start of 1942. Perhaps it would have been better moved on to Melbourne. Instead, when it was believed that Ceylon might be captured by the Japanese, its Naval component was transferred to Kilindini in East Africa. It later returned to Ceylon and became HMS *Anderson*.

---

[5]Communications channel capacity among Bletchley Park, Washington, WEC (Delhi) and CBB (Brisbane) continued to be an issue as the Sigint effort grew in size and complexity. For example, British Archives item HW 67/15 is a record of meetings mostly held at Bletchley Park in May 1943 involving Friedman, Tiltman, Sandford (CBB), Marr Johnson (WEC) and others. A considerable part of the discussion concerned the expansion of channels between the four centres and the allocation of more capacity on them to areas such as diplomatic intelligence.

[6]In his NCM oral history, Sinkov was somewhat scathing about the lack of co-operation between Central Bureau and Frumel in 1942–1943. It is possible that a visit by Admiral Halsey to General MacArthur early in 1943 resulted in much better co-operation between American Army and Navy cryptologic teams.

The Army component was reconstructed and expanded greatly at the Wireless Experimental Centre (WEC) near Delhi. A small American unit *US8* was stationed nearby.

18.7 US Army Sigint Work 1942–1945

The SSA (formerly SIS) ended the war in Arlington Hall Station, a former residential girls school with grounds sufficient for the major office buildings constructed for its massive staff. Its training facilities provided most of the American staff of Central Bureau. It worked on both European Theatre and Japanese material, with the Canadian Examination Unit being allocated some aspects, such as the communications of the Vichy Government.

18.8 Diplomatic and Commercial Intelligence

Up to 1939 diplomatic traffic had been a major target for GCCS. From the outbreak of the war in Europe its military sections expanded and kept on expanding. Eventually the diplomatic and commercial sections were moved from Bletchley Park back to London. Alastair Denniston was placed in command of this segment of GCCS with Edward Travis commanding the major component.

Up to 1942 the US Army and the USN had shared the task of working on Japanese diplomatic traffic. As Op-20-G had much more intelligence to process than the SIS/SSA, the diplomatic communications intelligence work was transferred to the Army. Following the *Holden Agreement*[7] of October 1942, under which the GCCS conceded overall control of work on Japanese Naval material to Op-20-G but took control for the Atlantic theatre, the Melbourne unit was re-organised.

---

[7]The British Archives item HW25/1 is Hugh Alexander's *Cryptographic History of Work on the German Naval Enigma*. The following is taken from sections 6 and 7 on page 57 of that report:

> In all they (the USN) produced well over 100 machines ('bombes') which were of the utmost value to us, not only on naval keys but also in air and army; indeed considerably more than half of the total American bombe time went on non-naval jobs. Their whole-hearted co-operation and readiness to use their bombes for jobs in which they, as an organization, had no direct interest was always very greatly appreciated by us; a great deal of our success on all keys (naval, air and army) in the last 2 years of the war was due to their help.
>
> In the cryptographic field they adopted from the beginning the clearly correct policy of supplementing our work rather than of attempting to cover the whole field themselves. With this end in view they set up a thoroughly efficient and businesslike organisation but did not put in it their best cryptographers. They were taking the lead in Japanese cryptography in which there was an immense field to cover and it would have been wasteful to have put their outstanding technical experts on to the Enigma in which the main problems had been solved and in which we had several years start.

So, in Alexander's view, the Holden agreement between Op-20-G and GCCS worked!

The Australian and British diplomatic group in Melbourne was transferred to the Army and closely co-operated with the GCCS diplomatic section. Professor Trendall continued as its leader for some time. The Australian Naval intelligence team and supporting staff in Melbourne were transferred to USN control from November 1942. Professor Room was transferred to Central Bureau and became its sole civilian. Commander Nave was not acceptable to Fabian and so went to Central Bureau too.

18.9 Postwar Reorganisation

In the USA the Army and Navy Sigint units were eventually merged into the *Armed Forces Security Agency* which later became the *National Security Agency* or NSA. The NSA—more correctly the NSA/CSS—has become the largest employer of mathematicians in the world.

In the United Kingdom the GCCS had used the cover name Government Communications Headquarters or GCHQ during WW2. In 1946 it adopted this name permanently. It moved out of Bletchley Park to one of its former outstations in North London and later moved to Gloucestershire. The tight relationship of the GCHQ with the NSA has survived into the new century. The British *Intelligence Services Act* of 1994 allows GCHQ facilities to be used 'in support of the prevention or detection of serious crime'.

The Examination Unit is now Communications Security Unit Canada and still works closely with the NSA, the GCHQ and the New Zealand GCSB. The Australian Defence Signals Directorate, formerly the Defence Signals Branch, was formed a few years after Central Bureau and Frumel had been disbanded. Again it works closely with the NSA, the GCHQ, the CSUC and the GCSB. There is a plethora of other intelligence and security agencies, some giving limited assistance to police work.

The publicly available information suggests that WW2 was the golden age[8] of Communications Intelligence.

---

[8]Or perhaps not. See the job advertisement inserted by the NSA in the *Notices of the American Mathematical Society*, December 2008:

'If you want to make a career out of solving complex mathematical challenges, join NSA as a Mathematician. At NSA you can bring the power of Mathematics to bear on today's most distinctive problems. We identify structure within the chaotic, and discover patterns among the arbitrary. You will work with the finest minds and the most powerful technology.' Presumably today's *most distinctive problems* include protecting the internet against attack from malicious governments, terrorists, criminals and nuisances.

The obituary of Andrew Gleason (*Notices of the American Mathematical Society* 56(10), November 2009 and mentioned in Note 29 of Chap. 1), describes an early large-scale recruiting of mathematicians for the NSA carried out by him.

Chapter 19
Security, Censorship and Leaks

Much of the Comint success described in this book arose from the skilled exploitation of mistakes made by the enemy. Such mistakes were likely to be detected and eliminated if there was suspicion that codes and/or ciphers had been 'compromised'. Discovery of a single major breach might even result in unrelated systems being made secure. This chapter describes examples of Allied security lapses involving the release of information that could have led to suspicions of this sort. These lapses caused consternation to the Allies but evidently had no other effect.

19.1 Sigint on Crete, May 1941

The British forces on Crete were under attack from the Germans in May 1941 and were forced to withdraw. By that time there was a system in place for delivering Sigint to commanders in the field. The GCCS field unit in Crete received Sigint via a secure link to Bletchley Park and delivered to the British commanders written accounts of this information with instructions that they should be destroyed immediately after being read. The use of the word *Ultra* to indicate the origins of the message had not yet been adopted. Instead the phrase *most reliable source* was used.

John Gallehawk[1] discovered a file entitled *Kreta: Dokumente des britischen Generalstabs*—Crete: Documents of the British General Staff—in the Imperial War Museum in London. The reference numbers are E120 AL 500/1/3. It was captured in 1945. Among innocuous material there is a hand-written message captured by the

---

[1]Gallehawk's booklet was published by the Bletchley Park Trust as Report No. 11 in October 1998. Geoffrey Ballard described the experience of the Australian traffic analysis group in Crete at the time in his *On Ultra Active Service*.

© Springer International Publishing Switzerland 2014
P. Donovan, J. Mack, *Code Breaking in the Pacific*,
DOI 10.1007/978-3-319-08278-3_19

Germans on Crete late in May 1941. A translation into German had been prepared
but it appears to have received no special attention. The text began:

> Telegram from London ZZZ
> No. 797   Desp 24.5.41
>   Rec'd 24.5.41   648GMT
> May 24th   797
>   Personal for General Freyberg
> O.L. 21/428
> According to a most reliable source, by midday May 23rd German troops ...

If the German intelligence service had given this the attention it deserved it would
soon have worked out that the *most reliable source* could only have been a decrypt
of a broadcast message encrypted by one of their own Enigma machines. It would
not have been particularly difficult to identify the original message in the files. The
Germans could and should have worked out that there were very serious problems
with Enigma and that the British had developed a very sophisticated cryptanalytic
unit. But nothing came of this.

Ironically the general British restrictions on the use of Sigint prevented Freyberg
from making much use of messages from Bletchley Park.

19.2   The Churchill Broadcast of 24 August 1941

A few months after the Crete incident and after the German invasion of Eastern
Europe had begun, Churchill received timely information from decrypts about
developments. Feeling obliged to make some protest, he addressed the issue on 24
August 1941 in a broadcast on the BBC. His text included the sentences:

> Since the Mongol invasions of Europe there has never been methodical merciless butchery
> on such a scale or approaching such a scale. . . . We are in the presence of a crime without a
> name.

This incident was rather different from the document captured in Crete. There were
other possible sources of information about the beginnings of what would now be
called genocide. Raphael Lemkin, a Polish Jew who escaped to Sweden and who
later went on to the United States, assembled all the information that he could locate
into a book entitled *Axis Rule in Occupied Europe* (late 1943). Lemkin got many
of the facts correct without access to Allied military intelligence. The Vatican must
have had some information from the priests in the field and so might have informed

the British government. Yet undoubtedly Churchill's broadcast did increase the risk[2] that the strength of British Comint would become obvious to the Germans.

19.3   The Capture of NZ Documents in Early May 1942

In April 1941 the loss of certain minor New Zealand vessels to German raiders was the subject of a report. A copy has survived in the Canberra NAA as item A7982 Z140. The 1940 *Automedon* incident, described in Sect. 2.3, is not mentioned in that report.

The NZ DNI, F. M. Beasley, visited Pearl Harbor in January 1942. (See Sect. 6.7.) He had previously agreed with Long to ask for more naval Sigint, particularly high-grade material, for both Melbourne and Wellington. This request was granted. The NZ archives reveal that in March 1942 the 'Captain on the Staff, Colombo' was on a distribution list for information from the NZ Naval Intelligence Department (NID). Despite the unfortunate experiences with the minor vessels and the *Automedon*, some Sigint obtained from Op-20-G, apparently including some from early reading of JN-25B8, was included in a sealed envelope sent to FECB in Ceylon.

This information was sent on the merchant ship *SS Nankin* which had arrived in Fremantle from Calcutta in March 1942 and which made its way via Melbourne to NZ and back. It left Fremantle again early in May 1942 and was captured, along with the relevant documents, by a German raider a week or so later.[3]

---

[2]See the book *Colossus: The Secret of Bletchley Park's Codebreaking Computers* by B. Jack Copeland and others, Oxford University Press, 2006. A chapter written by Stephen Budiansky (and available on his website) states on page 54: 'In July 1941, as the German Panzers thundered into Russia, Tiltman's group broke the hand cipher used by the German Police in the East and began to read the first hints of unimaginable horrors to come'.

In December 1942 Sir Anthony Eden told the House of Commons about 'receiving reliable information of the barbarous and inhuman treatment to which Jews were being subjected.' From the viewpoint of this book the use of *reliable* stands out.

[3]Quite a lot of material has been published on this incident:

(1) Hans-Joachim Krug and others: *German-Japanese Naval Relations in WW2*, Naval Institute Press, Annapolis MD, 2001 explains that in practice there was little rapid exchange of information between the two Axis navies;
(2) Edwin Layton, the fleet intelligence officer at Pearl Harbor, discusses the matter on page 418 of *And I was There* (1985). The book confirms that Comint was being sent to the NZ NID and that the capture, if handled promptly by the raider, could have resulted in great damage to the USN;
(3) Ian Gow and Y. Hirama: *Anglo-Japanese Relationships III*, Macmillan 2003, contains a paper by John Chapman entitled *from Allies to Antagonists* that covers the *Automedon* and *Nankin* incidents well;
(4) John Chapman has published a translation of parts of the war diaries of the German Naval Attaché Admiral Paul Wenneker under the title *The Price of Admiralty* but unfortunately the text for July 1942 has been lost.

Once again the story becomes a matter of what might have been. No radio warning was sent to either Berlin or Tokyo that IJN operational communications might be insecure. The relevant documents reached the German Naval Attaché (Admiral Wenneker) in Tokyo in July, by which time JN-25B had been scrapped and the Battle of Midway was well and truly over. He may well have expressed his worries to the appropriate authorities but no decisive action came of this.

19.4 Security Within Australia

Censorship of postal and telegraph services, and of radio and print media, was introduced into Australia immediately upon the outbreak of the European war in 1939. From the early days of the Pacific War, it was known that Australian radio broadcasts were being monitored by Japanese operators, but some careless reporting of information derived from Coastwatchers led immediately to swift Japanese reaction. A January 1942 Coastwatcher's report from Gasmata, which lies on the direct route from Rabaul to Port Moresby, advising that large numbers of planes were overhead flying towards Port Moresby (in fact to deliver the first Japanese air raid on it) was repeated in a news broadcast that led to a Japanese attack and subsequent landing at Gasmata and capture of the Coastwatcher there.

A little later, early in March, news broadcasts stated that Japanese warships had been sighted at Carola Haven on Buka Island, lying just north of Bougainville. Again, this information could only have come from a local source and Japanese troops promptly returned to the island and killed Coastwatcher Percy Good, unaware that another nearby Coastwatcher, W. Jack Read, had sent the original report.

We now turn to a sequence of breaches in security in WW2 cryptology in or involving Australia. In fact they were relatively inconsequential, but there was some considerable risk involved. The main damage done appears to have been that Central Bureau (CBB) lost much of the support Frumel could have given it. This was serious enough: with Naval code breaking running so far ahead of Army code breaking, the chance to look for possible cribs should have been taken. Commander Nave was forced out of naval code breaking in November 1942 but found an adequate field for his talents in CBB. The Melbourne NID, which had been responsible for the Coastwatchers, the Shore Wireless Service and the 1940–1941 start in Sigint, was reduced to a lesser role. Its close connection with British Naval Intelligence was thus weakened: the Admiralty had to use other routes to obtain the latest information from Pacific Theatre Sigint. Even then it kept a liaison officer in Melbourne (Alan Merry, RN) in addition to the British Admiralty Delegation in

---

Perhaps here too the myth of JN-25 documentation being recovered from the sunken I-124 clouded the issue.

Washington. When the former FECB was transferred back from Kilindini to become HMS Anderson near Colombo, it became a more effective ingredient in the Naval Sigint network and so was able to keep the British Admiralty informed.

The NID had been receiving occasional pieces of information flowing from Sigint from FECB and from the London DNI throughout 1940 and no doubt earlier. From January 1940 these would have carried the prefix 'Hydro', which was replaced by 'Ultra' in May 1940. Terminology was subsequently standardised to extend the use of 'Ultra' to the entire Allied intelligence community. One of these messages was distributed rather too widely and so in November 1940 the Secretary of the Defence Department intervened. Any such message circulated within Australian government units was henceforth to be prefixed 'From secret and reliable sources'. So the NID was insisting on proper security.[4] It is not clear whether in April 1942 all information received from Sigint sources was flagged.

19.5   A Minor Indiscretion

After the Pacific War began in December 1941 there was considerable growth in the office space needed for the defence forces. This had to be found quickly along the St Kilda Road arterial tram route in Melbourne. Much of this was done without publicity but one government Minister was partly responsible for a newspaper story about a school being thanked for patriotically making some accommodation available. This was drawn to the attention of Prime Minister Curtin who thereupon instructed all his Ministers to show discretion in making public statements.[5]

This was a minor matter. Anyone willing to put a week into observing passengers wearing military uniform on that tram route[6] would have been able to work out which premises were being used, and in particular which premises were being used round the clock. However getting the information to the enemy would have been much harder. In any case the Japanese were scarcely in a position to bomb Melbourne.

---

[4]The origin of the use of 'Hydro' and then 'Ultra' is mentioned on page 155 of Patrick Beesly's *Very Special Admiral*. The Australian file on the matter is Canberra NAA: A5954 2334/18 *Circulation of messages received from secret sources*.

[5]Canberra NAA file A5954 326/22.

[6]The late Mrs. H. Treweek recalled in 2004 that her late husband A. P. Treweek was once almost correctly informed on a St Kilda tram in Melbourne that 'down there they break all the Japanese codes!'. This may well have represented a mixture of rumour and inspired guesswork. Similarly, anyone in Bletchley who cared to do so could observe hundreds of staff arriving for the overnight shift at BP and draw the obvious conclusion. Since a considerable volume of intercepted enemy messages was re-enciphered in Australia and broadcast to cryptological units overseas, enemy intelligence agencies using TA should have been able to deduce that something was going on without help from St Kilda trams! But apparently nothing came from this.

Indeed, the transmission of information of this sort to the Japanese via Chinese diplomatic messages (Sect. 18.3) may have been tolerated to allow deception activities to continue.

19.6   The Curtin Leak of April 1942

Prime Minister John Curtin[7] is well known to have delivered background briefings about the military situation to certain senior representatives of the press. In March 1942 he directed the Defence Department to deliver to him daily at noon both a press statement on the current situation in the war and a confidential briefing on secret matters. He gave confidential briefings to about eight senior journalists most afternoons. This was in addition to his office releasing the open press statement for the day. A letter[8] of 9 June 1942 from Australian Broadcasting Commission (ABC) Canberra correspondent Warren Denning to the Acting Manager of the ABC is of considerable interest:

> With the arrival of Miss Wearne in Canberra it is possible for me to improve the means whereby you can receive copies of the confidential background information which comes to us from time to time from the Prime Minister's interview and other interviews with high authorities.
>
> Miss Wearne will be the central point in Canberra for this information. She will, from time to time, and as promptly as possible after receipt of the material send a copy to you.
>
> It is important that you should appreciate that this information is always of a highly confidential character not intended, in any circumstances, for publication in any form. It is intended to be a useful guide on the general situation and perhaps would help you to avoid making publishable comments which may be in conflict with the real situation.

---

[7]The title of NAA Sydney file SP286/16 8 suggests that it holds records of these private press briefings but in fact it holds nothing much prior to 1944 and the 1944 material has little interest. However one written briefing document for Prime Minister Curtin from 30 October 1942 on the Solomon Islands has survived in the NAA. See file A5954 333/7 page 108. This was forwarded to Curtin in Perth by teleprinter for 'a meeting with newspaper editors'.

Two relevant Australian books on this matter are:

Don Whitingdon, *Strive to be Fair*, ANU Press, Canberra, 1978.

Clem Lloyd and Richard Hall, *Background Briefings—John Curtin's War*, National Library of Australia, 1997. This has summaries of Curtin's briefings from 30 June 1942 onwards.

William Dunn in *Pacific Microphone* discusses how General MacArthur used the press.

The Google search engine when asked for 'Grattan Curtin' gives access to a a highly relevant paper based on a talk given by the senior contemporary journalist Michelle Grattan on John Curtin's relationship with and use of the press.

The NAA file about instructing Ministers to be discreet in making public statements is A1196 3/501/15.

Another memorable WW2 indiscretion by a senior figure was that of General de Gaulle in a broadcast immediately after the invasion of Normandy in June 1944. Unlike Churchill and Roosevelt he said that this was the real invasion rather than possibly just a preliminary step. Apparently the German High Command did not believe de Gaulle.

[8]The correspondence about the appointment of Miss Wearne is in the Sydney NAA as SP286/16 8.

In general these communications did not put Sigint at risk and merely resulted in a mild form of self-censorship being imposed on the press. However one such briefing late in April 1942 appears to have caused the publication in at least one American newspaper of information that could only have come from Sigint and thus implicitly have imperilled the process. Taken in conjunction with the *Chicago Tribune* leak of June 1942 (discussed later), there was considerable cause for concern.

The *Washington Evening Star* had published[9] on 27 April 1942 on its front page a story emanating from 'Allied HQ Australia' to the effect that Japanese warships were assembling in the Marshall Islands (north of New Guinea).

This information could scarcely have come from an observer on the spot. The age of the satellite was still decades off. So there is a fairly obvious inference that whatever Allied HQ Australia was, it had the capacity to obtain information from radio signals. Thus, if the above story had come to the attention[10] of the security staff of the IJN, there could have been a total re-examination of code and transmission procedures.

However the matter is even more complex than that. There were five methods of obtaining information from intercepted radio signals. The first was direction

---

[9]In fact this story was deleted in a later edition of the paper with the same date. Despite what has been published elsewhere the *Washington Post* did not carry the story at all. Apparently no systematic search of the American press of that day has been carried out. The same piece of information is in the Blamey papers 3DRL/6643 in the War Memorial Research Centre. The date suggests that Blamey was informed at least 24 h after the readers of the *Washington Evening Star*.

General Blamey somewhat improperly retained selected messages containing Sigint or about it in his papers. These have considerable interest. One may also find there a letter from Brigadier John Rogers, the Australian Director of Military Intelligence, dated 13 October 1943. 'In accordance with the promise given to overseas authorities could such memoranda as the attached be destroyed after perusal.'

[10]The paper *The Compromise of US Naval Intelligence after the Battle of Midway* by B. Nelson Macpherson, published in *Intelligence and National Security*, April 1987, quotes a 'Magic Summary' of 11 September 1942. It shows that the Japanese government asked its embassy in Lisbon for back copies of the *Chicago Tribune* and for American newspapers in general. Carrying out such a request would have at best been very time-consuming and expensive. An agent would have had to get a visa to travel to the United States, find transportation and then visit a major public library and take notes. Even if the FBI was not showing interest there would then be the finding of return transportation. At least the agent could have checked the *Washington Evening Star* at the same time, but only if attention had been drawn to it.

The Japanese embassy in Lisbon would have had even more difficulty in getting an agent to inspect back issues of the Brisbane *Courier-Mail*.

There is a related item in NAA Melbourne item MP1074/7, 1/9/1942 to 15/9/1942. It was a diplomatic message intercepted by the RAN on 11 September 1942, being a repeat of a message of 25 July. 'In view of extreme importance of material contained in *Times* and other British and American newspapers and magazines, as these can no longer be obtained in Kuibysheve [the temporary capital of the USSR], can Kabul obtain them including other English language newspapers published in India for dispatch to Japan.'

Section 8.19 lists various steps needed for security. One of these is: *Diligently monitor enemy documents, announcements and actions for evidence that your own systems are being read.*

finding (DF). As noted in Sect. 1.13, this could detect the location of the source
with reasonable accuracy. A chain of DF stations had been set up well before 1942.
This was not particularly secret. The second method was traffic analysis or TA.
Skilled TA operators could obtain some information about who was communicating
with whom without breaking any codes. Once again this was not particularly secret.
The third method consisted of identifying operators by their style (or 'fist' for
Morse operators) and was not particularly secret either. The fourth method was
identifying individual radio transmitting sets by their peculiarities. This is discussed
in Sect. 1.13. It was kept secret from the Axis powers throughout the war. The fifth
method was code breaking, particularly of JN-25B. It was also extremely secret.

It is clear that the precaution advocated by the NID could have avoided all this
trouble. Information prefixed 'secret and reliable sources' should not have been
mentioned to the press at all. The Melbourne paper *The Age* carried an editorial
on 5 May 1942 about it being improper to divulge to anyone else information of
interest to the enemy. Special mention was made of Americans not complying with
this principle.

Apparently the published article came from an American reporter C. Yates
McDaniel working from Australia.

General Marshall in Washington raised the matter with Owen Dixon,[11] the
newly-appointed Australian Ambassador in Washington. Marshall took the matter
very seriously and on 30 April 1942 sent a message[12] to General MacArthur:

> Marshall to MacArthur.
> Publicity purportedly from your HQ gives important details of recent Coral Sea action.
> Cincpac considers premature release of info re action his forces jeopardises successful
> continuation of fleet action. Info re action of Cincpac forces will be released only thru
> Navy. Please bring this matter to attn of Aust'n authorities and secure their co-operation.

The response was:

> MacArthur to Marshall. 15 May 1942.
> I said any corrective measures would have to be taken by Aust govt. He has agreed
> in principle and directed members of govt to be completely restrained, and has privately
> assembled country's editors and obtained promise of restraint, giving me partial control of
> Aust'n censorship. This has been done so that our govt is in no way involved so no possible
> allegation that it is attempting to impose its will. Please inform President in view of his
> comments your radio 31, 6 May.

Marshall responded a little later:

> Marshall to MacArthur 24 May 1942.
> Information in my 109 (22 May, about coming attack on Midway) not to be passed
> beyond yourself.

---

[11] Owen Dixon left a diary which Philip Ayres has used in his biography *Owen Dixon*, Melbourne
University Press, 2003. Dixon was married to the sister of Walter Brooksbank, the senior civilian
in the Naval Intelligence Department in Melbourne. However there is no evidence that he had any
information from Brooksbank about Sigint.

[12] These quotations are from files AWM54 423/11/202 Part 1 pages KA26 and KA32 in the
Australian War Memorial Research Centre.

This last came somewhat late. Commander Long, the DNI in Melbourne, had already informed the British DNI of the Midway decrypts on 20 May 1942. The text survives in the British National Archives as file ADM 223/868 as well as in the Melbourne NAA and is well worth quoting:

> (To) Admiralty (For D.N.I.).
> IMMEDIATE
> Your 1536B 19th to D.N.I. Melbourne only.
> From Special Intelligence it is considered certain that operations against Midway Island are certain against Aleutian Island very probable and against Hawaii Island probable.
> 2. Air attacks against Midway Island to begin on 'N minus 2' day from a Carrier force to North Westward from about 50 miles. . . .
> . . .
> 7. All the information from which the above was extracted has already been passed to Washington via U.S.A. naval channels.

No doubt this overstates the share of the credit earned by Frumel. Equally clearly it passed information to those who had no need to know it but who had a general need for full background knowledge.

19.7 The Battle of Midway as a Security Leak

As well as protecting Sigint-derived information via the 'need to know' rule, the basic principle underpinning the use of such information to execute an operation against the enemy was that 'cover' be provided, usually by arranging some type of visual observation via search planes or ships. The most sensational violation of this principle was in fact the Battle of Midway. Patently, the USN must have had a reason[13] for moving its aircraft carriers to that minor island at just the right time.

---

[13]The Australian War Memorial file AWM54 423/11/202 Part 1 on page KA36 gives a message from another American admiral. 'Our intelligence points to concentrations of Jap Flt at Truk later part of June but gives no indication of attack either Alaska, Midway or Hawaii. Obviously, however, you would not have redisposed your forces without good reason which it would be helpful for me to know.' The NAA Melbourne file B6121 119 gives an interesting 'good reason' concocted a few days after the Battle.

Part of Note 5 of Chap. 21 bears duplication. The regulations on the use of Comint given in Sect. 1.17 included: 'In general, if any action is to be taken based upon Ultra information, the local Commander is to ensure that such action cannot be traced back by the enemy to the reception of Ultra Intelligence alone. . . . No action may be taken against specific land or sea targets revealed by Ultra unless appropriate air or land reconnaissance has also been taken.'

As discussed in Note 34 of Chap. 9, the loss of the I-124 gave the IJN communications security people another misleading 'good reason'.

Admiral Yamamoto, generally believed to have been responsible for the Pearl Harbor raid, was shot down in April 1943 by an ambush that patently needed prior knowledge of his flight plans. Rudolph Fabian of Frumel was against this exercise. The Oral History interview with him held by the American National Cryptological Museum records the sentence 'It could have compromised the hell out of us.'

To make things worse, this happened only a few weeks after the Battle of the Coral Sea. The IJN would have been aware that at least some of these carriers would have needed to be moved quite a distance.

19.8   The Chicago Tribune Affair

A better known leak occurred in the *Chicago Tribune* on 7 June 1942. It was copied in the *New York Daily News*, the *Washington Times-Herald* and certain other newspapers. A headline on page 1 stated that the USN had possessed foreknowledge of the Japanese intention to strike at Midway. As already discussed, this was almost obvious but there was much turmoil before the matter subsided. No explicit mention of code breaking was made. A grand jury investigation was commenced in Chicago in early August 1942 but no charges were laid. However the difference between this and the April leak is clear: the April leak required the reader to make a somewhat subtle inference. The June leak did not. Representative Elmer Holland of Pennsylvania stated in Congress on 3 August 1942:

> It is public knowledge that the *Tribune* story ... tipped off the Japanese high command that somehow our Navy had secured and broken the secret code of the Japanese Navy.

The *Tribune* reporter involved, Stanley Johnston, was an Australian.

The OP-20-G team took this leak very seriously. Arthur McCollum,[14] later based in Brisbane, recorded the reaction.

> My goodness, the place was shaking.

Subsequent analysis of changes in Japanese Sigint practice after Midway has not supported any suggestion that the IJN, through its own review processes or via awareness of the US leaks, seriously considered the possibility that JN-25 had been compromised.

Even though the USN had caused much of the ruckus subsequent to the publication in the *Chicago Tribune*, the inevitable clamp-down followed. The Frumel team led by Rudolph Fabian would have been fully aware of this. They had every reason to believe that their work on JN-25 had already had enormous effect and was likely to have much more as the war progressed. They would also have been aware that John Lietwiler had been considering shooting all USN cryptologists (including himself) left on Corregidor rather than letting any be subject to Japanese torture.[15] From the beginning of the interaction between Frumel and MacArthur's

---

[14]Arthur McCollum played a key role in the USN 7th Fleet intelligence team in Brisbane. An oral history interview with him is in the US Naval Academy at Annapolis. See also pages 677–678 of D. Goren's *Communications Intelligence and the Freedom of the Press*, in the *Journal of Contemporary History* vol 16 (1981) 663–690.

[15]The reference here is the interview with Duane Whitlock carried out for the National Security Agency and available from the National Cryptologic Museum in Fort George Meade, Maryland.

GHQ in Australia, Fabian had insisted on delivering Sigint by making an oral report to General MacArthur alone. This may have improved security a little but it aroused antagonism from MacArthur's senior staff and prevented the proper use of professional staff in assessing the Sigint information.

The unwillingness of the US Army and Navy to share fully information derived from their respective intelligence services affected the use of Sigint in Australia. While, initially, MacArthur's new Central Bureau had received some co-operation from Frumel, this was to stop. The minutes[16] of the CBB executive meeting of 16 June 1942 show that Lieut Gilbert Brooksbank of NID (brother of the long-time chief civilian staff member Walter Brooksbank) announced that Admiral Leary, the senior USN officer in Melbourne, had instructed Commander Newman to pass information only to certain specified authorities. Central Bureau was not on the list. Thus a major new activity for the Army and the RAAF lost the advice of the RAN officer with the greatest relevant experience. The situation deteriorated over the following months. The RAN declined to attend any more Central Bureau meetings. Newman wrote to CBB's Sandford saying that he could give no more help officially but could perhaps respond to private letters. In 1943 CBB complained to General Akin about lack of co-operation from Frumel. Even more striking is the case[17] where CBB asked General Akin to request officially assistance from a USN interception officer in Adelaide River. The assistance had been forthcoming anyway but it was believed to be unfair to have the officer imperil his career.

Security appears to have been tightened after the leaks of April and June. A circular[18] to senior Australian Army intelligence officers from the Chief of the General Staff, signed by then Colonel Rogers on his behalf on 30 July 1942, states:

> Information obtained as a result of the interception of traffic passing on enemy communications ('Y' information) is of vital importance AND CAN BE LOST BY US FOR THE DURATION OF THE WAR BY A SINGLE SLIP ON THE PART OF ANY CARELESS OR THOUGHTLESS PERSON. Circulation must be limited to those officers who are directly concerned with using the information. . . . 'Y' information should never be passed on or repeated purely for information.

In particular this edict would appear to prohibit the passing of Sigint ('Y' information) to Australia's Prime Minister, John Curtin.

The circular contains a few other aspects of lesser but still substantial interest. The indicator that the information was 'Y' had now become 'from usual sources'. The danger of cribs was recognised: unless one-time pads were being used for encrypting any message containing this phrase, the transmitting operator was to move it to a randomly chosen place in the interior of the message while the receiving operator was to restore it to the beginning. There was a general directive to send 'Y' intelligence by radio as little as possible, perhaps because the Allied success with JN-25 reduced general confidence in all encryption systems.

---

[16]The early Central Bureau minutes survive in NAA file SI/10.

[17]This comes from a collection of letters from CBB to General Akin in NAA file B5435 222.

[18]The NAA reference is SI/10. See page 218 in the digitised version.

19.9 The Courier-Mail Affairs

Page 1 of the Brisbane *Courier-Mail* of 2 March 1943 displayed a headline 'Japs building up all forces in Nth islands'. The article which followed included 'Japanese land forces now in the SW Pacific Area are known to include troops believed to have been drawn from other areas'. Worse was to come on Monday 8 March 1943. A story on page 1, attributed to 'yesterday's communiqué from General MacArthur's headquarters', was entitled '136 Allied planes fought Jap convoy' and included the following: 'The ground forces which the enemy attempted to land, the strength of which is estimated at 15,000 are now identified probably as of the 51st and 20th divisions with certain other special troops.'

The NAA file B5435 222 contains a letter to General Akin, the absentee director of CBB, from Squadron Leader Roy Booth, head of the RAAF component. Booth drew attention to the issues of the *Courier-Mail* of 5 and 8 March 1943. The former included a reference to the arrival of fresh troops in NG having been anticipated and preparations made before the convoy was sighted. As Akin was based at the SWPA headquarters he was in a good position to curtail press releases of this sort.

Careful reading in Japan of a newspaper report to the effect that SWPA forces had destroyed much of its 51st and 20th divisions would have shown the Japanese intelligence organisation that the Allies had access to the details of military movements. This would have been a strong indication that Comint was being exploited. The report could have been picked up from insecure Chinese codes or elsewhere. However nothing much appears to have come of this incident.

On 10 December 1943 'Archie' Cameron, then an intelligence officer and a Member of the House of Representatives but previously Minister for the Navy wrote[19]:

> We have given them gratuitous information of our ability to read through the veil. Our publication in January last (1943) of their order of battle in the Owen Stanleys and Buna was one instance. Our declaration in March (1943) that part of 20 division was in the Lae convoy clearly disclosed to the Japs that we were getting beyond the veil. The 20 division was not in Rabaul and never has been there. We had not been in contact with it. It looked to me like a victory for our signals interception. It was a revelation to the Japanese Army. The safest, in fact the only safe course in these circumstances, is to say nothing.

One key word here is 'gratuitous'. There was no possible gain in publishing details that could have come only from Sigint. However Cameron appears to have erred in believing that the IJA or, for that matter, the IJN, properly analysed the matter.

---

[19] A small file of letters sent by Mr. Cameron survives in the Australian War Memorial Research Centre. The reference is AWM54 225/2/3.

However it is clear that 9 months later Mr. Cameron was not totally satisfied. He had the wit to see the point behind the regulation (same source as at end of previous note) 'Names of enemy ships revealed by Ultra sources may never be quoted'.

According to Admiral Halsey's 1947 *Admiral Halsey's Story*, stories eventually appeared in the Australian newspapers to the effect that the shooting down of Admiral Yamamoto had been initiated on the basis of Comint. 'But the Japs evidently did not realize the implication.'

The shooting down of a plane carrying Admiral Yamamoto in April 1943 also violated the regulations on the use of Sigint. The exercise depended totally on precise intelligence. In this case the information from JN-25 intercepts was backed up by the reading of lesser local cipher systems by RAAF 1 Wireless Unit (1WU) at Townsville. It seems that the Japanese blamed the lower-level messages for the breach of security.

Another piece of evidence of the failure of Japanese intelligence to work out what was really going on is the following intercepted message[20] from Manila to Tokyo of 23 February 1944.

> The present situation is such that the majority of tankers returning to Japan are being lost. Although every effort is being made by high command to develop counter measures, we request that the further effort should be exerted in training of crew members and in studying methods of escort.

19.10   Cobra and Lagarto

Another major leak was still to come. In March 1944 there were two small Australian guerrilla groups, Cobra and Lagarto, operating in East Timor. A message[21] to the Cobra group on 7 March 1944 read:

> Your six big relief. Col Major C and all here sick at heart past week due intercept Jap cipher naming you personally and apparently claiming your capture January 29.

So this at the very least provided men highly liable to capture with some information about code breaking. It may well have gone direct to the Japanese. It may have imperilled the benefits from the Sio windfall. But again nothing happened. One NAA file on Lagarto contains a message to the Japanese just after the surrender asking them to care for the prisoners of war, including the survivors of these two operations. There is also a letter to those managing the operation reprimanding them for leaking Sigint information.

19.11   The 1944 Presidential Election

In 1944 General Marshall was informed that the Republican candidate for the Presidency, Governor Thomas Dewey of New York, was considering raising the matter of the use of Comint in 1941 as a major issue in the campaign. Marshall wrote to Dewey stating that such action had every prospect of damaging the American campaigns on all fronts. The importance of Comint was stressed. It was stated that

---

[20]The quote is from the Blamey papers (Note 9) 3DRL/6643, folder 2/65.

[21]Barbara Winter's book *The Intrigue Master* on DNI Long gave the reference to the Cobra and Lagarto affairs.

the 'Purple' diplomatic cipher system was providing intelligence needed by General Eisenhower in Europe. Dewey agreed not to mention the matter in public and in fact did not do so. President Roosevelt was re-elected.

19.12 Post-War Leaks

The post-war imposition of security appears to have started in London on 31 July 1945 with the issuing of a long-term ban on releasing information about Sigint in WW2. The precedent was set with the publication of Churchill's *The Second World War*, which contains some mention of Magic, the interception of Japanese diplomatic material, but none of other Sigint in the European Theatre. Yet the chapter on Midway in volume IV *The Hinge of Fate* (1951) of the six-volume work mentions 'intelligence' in four places, leaving little room for doubt that it must have been Comint. For example:

> One other lesson stands out. The American intelligence system succeeded in penetrating the enemy's most closely guarded secrets well in advance of events. Thus Admiral Nimitz, albeit the weaker, was twice able to concentrate all the forces he had in sufficient strength at the right time and place. When the hour struck this proved decisive. The importance of secrecy and the consequences of leakage of information in war are here proclaimed.

As the text of the letter from General Marshall to Governor Dewey of September 1944 had become public knowledge in December 1945 courtesy of the Congressional Inquiry into the Pearl Harbor Raid, this gave away little more than was currently available to all. A crunch came in Australia with the writing of the Australian War Memorial's official history of WW2.[22] The Editor, Gavin Long, sought the same general right of access to the official files as Charles Bean had been granted for the WW1 history. This was denied as Britain wanted to keep some parts secret. The War Cabinet decided in 1945:

> The exercise of censorship by the Government is to be limited to the prevention of disclosure of technical secrets of the three Services which it is necessary to preserve in the post war period.

G. Harmon Gill's volume *Royal Australian Navy 1942–1945*, published in 1968, did not mention the breaking of ciphers. However the one-volume summary *The Six Year War* by Gavin Long was published in 1972 and contained some brief but inaccurate references to Naval code breaking but none to Army code breaking.

The Australian veterans of Central Bureau were told on discharge not to assemble as a group for 30 years.

Various leakages happened in September 1945. Thus the *Canberra Times* of 6 September 1945 reported a speech made by Defence Minister Beasley in the Sydney

---

[22]NAA Canberra item A816 35/301/146 is the file on the AWM Official History. British National Archives (TNA) has a file in the HW series on the deletion of references to Sigint from the Churchill memoirs.

suburb of Manly in which he said that the cracking of the Japanese Naval code by an American Naval officer just prior to the Coral Sea battle played a big part in saving Australia. This was reprinted on the front page of the *New York Times* of the same date. On 10 September 1945 the London *Daily Telegraph* stated that 'Japan's defeat was greatly helped by the fact that all Japanese Naval code messages were read and that we therefore knew all her plans beforehand'. This was an exaggeration. On 11 September 1945 the Melbourne *Argus* recycled this story under the headline 'USN had Japan's Secret Code—Knew Enemy Plans for Battle of Midway'. The text stated that 'the Japanese-coded radio messages were unmistakable'. (So they were—decryption and decoding were the problems.) By this time the prospects of the JN-25 series of additive systems being revived were minimal. These and a few other such press stories seem to have been forgotten.[23]

Admiral Nimitz was co-editor (with E. B. Potter) of *The Great Sea War*, published in 1960. This was based on the WW2 material in their *Sea Power* published at much the same time. Both versions make it clear that the Battle of Midway was the fruit of Comint.

In 1974 Frederick Winterbotham became the first participant in WW2 Sigint to write a book on the subject. Jozef Garlinski recorded in his book *Intercept* (Dent, 1979) that when carrying out the research into Bletchley Park in December 1977 he had interviewed one of its veterans, Mrs Ruth Thompson, He then learned for the first time that Bletchley Park had a Japanese section. By that time an account of the new RSA encryption system that rendered the codes of the Second World War obsolete had been published in *Scientific American* in August 1977. David Horner and Desmond Ball were able to unearth enough to publish the first accounts of code breaking in Australia about that time. The record of the congressional inquiry into the Pearl Harbor raid was one source. Rather later Geoff Ballard and Jack Bleakley received some moral support in writing their books on Central Bureau from the Australian Army and RAAF viewpoints respectively. The Australian Navy kept, and still keeps, its silence.

The American National Security Agency arranged oral history interviews around 1980. These have been declassified.[24]

---

[23] A few press cuttings survived in the Defence Department records and are now in the NAA Canberra as item A5954 560/4. William Friedman later collected various other leaks in a document now in RG457 in the College Park NARA.

[24] Amusingly the interview with Rudolph Fabian has the descriptions of the special tricks used with JN-25 blacked out. The full original text was finally declassified in 2006. A booklet *The Quiet Heroes of the Southwest Pacific Theater* edited by Sharon Maneki and published by the NSA in 1996 has considerable overlap with these oral histories. The oral history material is available in the National Cryptological Museum.

Mavis Batey's book *Dilly: The Man Who Broke Enigmas* (2009) mentions the failure of the British GCHQ to declassify WW2 vintage material, such as HW 43/33. See pages 1 and 223. This 'Part I' covers techniques used to break JN-25 systems, and presumably duplicates American material on public access in NARA RG38. The adjacent 'Part II' HW 43/34 was declassified in February 2013. It has 100 pages on the JN-25L53 system.

Part V
Conclusion

Chapter 20
Conclusion

Our description of the key technical aspects of the Allied cryptanalytic successes against the principal cipher systems used by the IJA and IJN in 1939–1943 is based essentially on archival documents, including oral history interviews of a number of the most important participants. Understanding all this required mathematical expertise, research skills, and comprehension of the communications and information technologies of the period. These technicalities explain why the provision of Communications Intelligence in particular was irregularly available to Allied operational commanders, depending upon the nature of each cipher system change made during the course of the Pacific War. Nevertheless, the quality of the provision of Signals Intelligence to those commanders, and their use of it, strongly influenced the timing, direction and ultimate outcome of this war, from early 1942 onwards.

20.1 General Conclusion

Intelligence derived from decoded JN-25 messages was, as we have shown, of overwhelming importance in 1942 and 1943. Major Army ciphers were unreadable until the Water Transport Code (2468) was broken in mid-1943 and all others remained so until discovery of coding materials at Sio in January 1944. This enabled a huge number of previous intercepts in several Army cipher systems to be read and for a short time allowed current traffic to be broken, greatly assisting General MacArthur in his advance across New Guinea towards the Philippines. Of even greater significance, steadily from mid-1943 onwards, was the pooling of convoy data from 2468 and various IJN codes, including JN-25, enabling the US Navy's submarine fleet to wreak havoc on Japanese merchant shipping and troop ships, resulting in Japan's military and economic capacities facing terminal decline from 1944 onwards.

© Springer International Publishing Switzerland 2014
P. Donovan, J. Mack, *Code Breaking in the Pacific*,
DOI 10.1007/978-3-319-08278-3_20

On the other hand, our conclusion that no intelligence of operational significance could have been obtained from JN-25 sources by late 1941 (and so could not have influenced judgements at that time about a raid on Pearl Harbor) is also soundly based on technical reasons. The important process of stripping additives from JN-25 message GATs by use of data on common book groups was by then only recently producing dividends. The consequent building up of entries in the code book certainly assisted those charged with translating those entries into plain language equivalents, but attempting to gain the sense of a message was still in its infancy.

These processes developed rapidly after the Pearl Harbor raid because the rapid provision of more Sigint resources and the availability of many more operational messages enabled the Allies to exploit a huge blunder made by the IJN when it changed the JN-25 system on 4 December 1941: it retained the B code book, which immediately meant that all work over the previous year, devoted to building it up, remained useful to our code breakers and gave them a totally unexpected boost.

While that blunder alone was extremely costly to Japan, the original decision by the IJN to restrict its code words (book groups) in JN-25 to those 5-digit numbers which were multiples of 3 turned out to be of even greater cost. Once discovered, it so reduced the complexity of the cryptanalytic tasks required to break it that this system remained insecure throughout most of the war. The whole JN-25 saga provides an example par excellence of the precept that one's own codes must be subject to the strictest attack by one's own expert cryptanalysts before being used.

The importance of the separate decisions made by both the UK and US governments in the 1920s and 1930s to develop expertise in cryptography and in particular in the monitoring and breaking of Japanese communications systems is clearly demonstrated in this book. Accumulated knowledge and experience lay behind the 1939 breaking of the Japanese Purple cipher machine by Friedman's SIS team and the breaking of the first JN-25 system by Tiltman at GCCS. The fruitful 1941 collaboration on JN-25B between the FECB and Station Cast was possible only because of previous decisions to increase Sigint activity against Japan. Without the knowledge of JN-25B available at the end of 1941, it is highly unlikely that useful Comint would have been obtained from it before mid-1942. By that time, the course of the Pacific War would have been very different.

20.2 Acts of Genius?

The success of Allied Sigint in WW2 was heavily dependent upon the education and intelligence of those involved with code breaking, and with their ability to work together under considerable stress. Teamwork was essential throughout this activity, but it was unusual to attribute success to specific teams or individuals within the larger code breaking units. This book has included accounts of selected successes where such attributions can be made. However, the known contributions of certain individuals were so remarkable that they may reasonably be described as acts of genius, deserving specific recognition. For example, the construction by the GCCS of the Robinson and the subsequent Colossus machinery was described by an American observer at Bletchley Park on 1 December 1944 in the following terms:

> Daily solutions to Fish (the SZ42) messages at GCCS reflect a background of British mathematical genius, superb engineering ability and solid common sense.

William Friedman, John Herivel, Dilly Knox, Marian Rejewski, Frank Rowlett, Alan Turing, John Tiltman, Bill Tutte, and the engineers Joseph Desch, Tommy Flowers and Harold Keen played leading roles in Atlantic and European Sigint. (Rowlett's breaking of the 'Purple' diplomatic cipher machine was much more relevant to the European theatre than to that of the Pacific. He also played a major role in ensuring the security of high-level American communications.)

Turing's work on statistical methods in cryptology is seminal in much WW2 cryptology yet seems to be minimally appreciated. Jack Good of GCCS and Solomon Kullback of SIS made useful contributions in statistical techniques and elsewhere.

In the Pacific area, the names of Joe Desch, William Friedman, John Tiltman and Alan Turing recur, along with Joe Rochefort and a few in his team. Marshall Hall, Jr, found the statistical peculiarity in JN-25 traffic that led to the *Hall weights* and so must have been a key cryptologist for Op-20-G.

The breaking of the indicators of the Water Transport Code was a very significant achievement that may have had indirect input from Rejewski and almost certainly had substantial input from Abraham Sinkov and T. G. Room.

The modern American NSA 'where intelligence goes to work' described itself as 'the nation's largest employer of mathematicians'. This is taken from the advertisement quoted in Note 8 of Chap. 18. 'In the beautiful, complex world of mathematics, we [the NSA] identify structure within the chaotic and patterns among the arbitrary.' The technology has changed out of sight but the type of talent needed seems to be much the same.

Around 480BC Sun Tzu wrote a short account of *The Art of War*. One of the last sentences in Lionel Giles' translation of 1911 reads:

> Hence it is only the enlightened ruler and the wise general who will use the highest intelligence of the army for purposes of spying and thereby they achieve great results.

Perhaps the best contemporary summary was made by a Canadian signals intelligence officer in May 1945:

> So we leave our task with a feeling of legitimate satisfaction.

Part VI
Background

Chapter 21
From Pearl Harbor to Midway

This chapter outlines the critical 6 months from the Pearl Harbor raid of 7 December 1941 to the end of June 1942. Twenty years earlier Winston Churchill had referred to 'the incomparable advantage of reading the plans and orders of the enemy before they were executed' (already quoted in Sect. 1.6). This applies with great force to the crucial Battle of Midway, 4–6 June 1942.

21.1 The Onset of Hostilities

The first acts of the Pacific War occurred on Sunday 7 December, the day before Pearl Harbor, in the South China Sea. Japanese aircraft attempted to shoot down a British reconnaissance plane and successfully shot down an American plane, each having been sent out to try to locate the exact whereabouts of the Japanese invasion force. Japanese landings took place at Kota Bharu in the early hours of the next day—Monday 8 December—in fact, some 2 h prior to the surprise Japanese carrier-born aircraft attack on the USN base at Pearl Harbor. (The International Date Line complicates the description of some events, including this one: the earlier action in Malaya did occur on 8 December, local time!)

The Japanese also invaded Thailand in a number of places with the Thais initially providing armed resistance. But the Thai Premier declared a cease-fire on 8 December, allowing the Japanese free passage towards Malaya and Burma. Later in the month Thailand signed a treaty of military alliance with Japan and in January 1942 it formally declared war on the UK and the USA.

The raid on Pearl Harbor occurred at dawn local time on 7 December 1941. It resulted in massive damage to a dozen capital ships, the loss of 2,400 American lives, injuries to another 1,200 American servicemen and the destruction of numerous warplanes on the ground. The Japanese had suffered quite disproportionately small losses. The immediate consolation for the USN was that its aircraft carriers

© Springer International Publishing Switzerland 2014
P. Donovan, J. Mack, *Code Breaking in the Pacific*,
DOI 10.1007/978-3-319-08278-3_21

had not been caught at the same time and, as the oil storage and the ship repair facilities at Pearl Harbor were overlooked in the raid, a fast recovery was possible. Despite the enormous damage to ships and the loss of personnel suffered by the USN through the attack on Pearl Harbor, the fact that the carriers of its Pacific fleet were at sea at the time would prove to be decisive in the unfolding of the war in the Pacific over the subsequent 6 months.

Also, on 8 December, immediately after the above attacks, air raids occurred on Singapore, on the US islands of Guam and Wake in the Pacific, and in depth on US air bases in the Philippines (which bases were also unprepared, despite a radio message advising of the Pearl Harbor raid), where great damage and loss of aircraft occurred. Australia, the UK and the USA separately declared war on Japan on 8 December.

The air raids on the Philippines continued on 9 and 10 December, with the Cavite naval base being seriously damaged on 10 December. Initial Japanese landings on the Philippines also occurred on that day. But the most shocking news for Australia on this day was the sinking, by Japanese aircraft, of the Royal Navy capital ships HMS *Prince of Wales* and HMS *Repulse*, off the coast of Malaya. This occurred soon after their arrival in Singapore where they had been sent to strengthen a weak British fleet. The RN must have been aware of the vulnerability of warships, without air support, to aerial attack given the successful British raid on the Italian fleet at Taranto 13 months earlier. The great damage at Pearl Harbor was magnified by having these two ships exposed to attack. Because the USN battleships sank in shallow water near excellent dockyards most of them could be repaired. The two British ships were lost forever.

Japanese forces invaded Guam on 10 December in such strength that the US military stationed there surrendered. The small USN interception and radio intelligence station there was totally destroyed by its staff, who thus managed to avoid being identified by the occupying Japanese forces. The initial Japanese attempt to invade Wake on 11 December seriously underestimated the strength of the US presence there, and in fact landings on Wake, and its capture, were delayed until 23 December.

On 11 December, Germany and Italy, as co-signatories of the tripartite pact with Japan, declared war on the USA, eliciting an immediate counter-declaration.

General Marshall immediately summoned to Washington a promising, recently promoted, Army officer, Brigadier-General Dwight D. Eisenhower. Eisenhower produced within a few days what turned out to be reasonable plans[1] for a fall-back position, based on holding Australia and maintaining safe surface communications with it across the ocean from California and Hawaii. This interim strategy was implemented.

---

[1]The Canberra NAA item A816 14/301/559 contains some very interesting correspondence on this from Admiral King and General Marshall to Prime Minister Curtin and from Curtin to President Roosevelt. The dates are the first half of 1942.

21.2 The Japanese Thrust

The speed with which the Japanese managed to achieve their immediate objectives in the South-West Pacific Area (SWPA) and in SE Asia, was truly frightening. The Japanese troops in Malaya fought their way steadily southwards towards Singapore. On 15 December 1941, Japanese forces crossed the Kra Isthmus in Southern Thailand and captured the airfield at Victoria Point on the southern tip of Burma. This exemplified a key strategy of the Japanese in their later invasions of the southern Philippines, British NW Borneo and the East Indies: initial targets were airfields from which Japanese fighters and bombers could support naval and military attack groups.

The first American servicemen landed in Australia, at Brisbane, on 21 December. Hong Kong surrendered to the Japanese on Christmas Day, and two days later, Australia's Prime Minister, John Curtin, announced that Australia now looked to the United States as the major player in any plan for the defence of the Pacific region.

On 4 January, Japanese planes raided Rabaul on New Britain, and these raids continued almost daily afterwards. On this same day, General Wavell was appointed head of all Allied SW Pacific forces and in mid-January was stationed in Java in charge of ABDACOM—the combined Australian, British, Dutch and American forces. The next day, the British evacuated by sea to Colombo almost all the staff of its Singapore Sigint Station, FECB.

On 10 January Japan declared war against the Netherlands and soon afterwards its forces occupied Tarakan in Dutch Borneo and Menado in the Celebes, again sites hosting usable airfields. Japanese forces occupied Balikpapan in Dutch Borneo on 23 January. Ambon in the Moluccas was under their control by early February.

The air raids on Rabaul were continuing. On 21 January, they extended to Kavieng on New Ireland and to Madang and Salamaua on the New Guinea coast. On 23 January, the Japanese landed on and captured both Rabaul and Kavieng. Soon afterwards, the first air raid on Port Moresby in Papua took place. These raids continued steadily thereafter.

On 1 February, planes from the USN carriers *Enterprise* and *Yorktown* attacked IJN targets in the Marshall Islands, retreating eastwards before any effective resistance could be mounted. These attacks forced the IJN to begin to reappraise its plans for further consolidating its position in the SW Pacific.

The worsening situation in the Philippines caused the USN to order the progressive removal of the staff of its Sigint unit, Station Cast, to Western Australia and then on to Melbourne. The first batch left by submarine to Java on 4 February.

The IJN submarine I-124 was sunk in January 1942 in relatively shallow waters 50 km off Darwin. This minor incident may have later clouded for the IJN the source of the USN foreknowledge of the attack on Midway: it might have been possible for the Allies to recover cryptographic materials from the wreck. (In fact they did not. The foreknowledge of the June attack on Midway came from a quite different route. See Note 34 of Chap. 9.)

The Japanese forces in Malaya overcame all resistance on the peninsula by early February and crossed the Straits of Johore into Singapore on 9 February. Singapore surrendered on 15 February, with many Australian, British and Indian troops becoming prisoners of war. This came as an enormous shock to Australia's citizens and brought with it the realisation that Australia itself was vulnerable. Sir Alexander Cadogan, Head of the British Foreign Office, noted in his diary[2] that the fall of Singapore was the darkest day of the war. Winston Churchill made similar remarks.

The bombing of Darwin on 19 February, in part from some of Admiral Nagumo's Pearl Harbor attack carriers en route to the Indian Ocean, killed several hundred civilian and military personnel, sank ships in the port, damaged the town and the airfield, and gave Australians an even greater shock.[3] (Darwin suffered a further 63 air raids, many of which were not widely reported within Australia. The final raid occurred in November 1943.)

On 20 February, the US carrier *Lexington* was off Rabaul and engaged in air battles with Japanese planes, providing the IJN with a further reminder of the ongoing danger from the USN carriers in the Pacific. The favourable progress made by the Japanese in their attack on Singapore enabled them to invade Sumatra on 14 February, and to land on Timor on 19 February, thus continuing the encirclement of the main NEI island of Java.

Later in February, carrier-based planes of the US Navy attacked Japanese bases in the Gilbert and Marshall Islands. This action again focused the IJN on the threat remaining from the USN carriers, a threat that was reinforced within a few days by a US carrier-based attack on Wake Island and a subsequent raid in early March on the small Japanese-held Marcus Island, lying in the western Pacific within bomber range of Tokyo. This certainly caused concern among the Japanese High Command, which had confidently assured the Emperor and his citizens that enemy air raids on mainland Japan were impossible.

On 25 February, Wavell's ABDA Command was dissolved, and Wavell flew from Java to India via Colombo. Two days later, his remaining British, American and Australian staff left by sea for Colombo, the Dutch personnel remaining behind.

The Battle of the Java Sea, 25–28 February, was a last-ditch, unsuccessful attempt by remaining Allied naval forces to thwart the invasion of Java, which began on 28 February. A number of Australian, American, British and Dutch warships were sunk Around the Java coast in a 3-day period.

Japanese aircraft attacked Broome and Wyndham in Western Australia on 4 March. On 8 March, Japanese troops entered Rangoon in Burma, Java surrendered, and Japanese troops landed at Lae and Salamaua on the northern coast of New

---

[2]The Cadogan diary entry is quoted by Robin Denniston on page 14 of *Thirty Secret Years*, the biography of his father.

[3]Since the war there has been some controversy as to the extent to which the Australian Government was contemplating abandoning much of the northern part of the mainland. More attention should have been paid to NAA Brisbane item BP243/1 MS62 *Most Secret Works for Defence—Various Locations in Queensland*, which contains contingency plans to destroy the ports and airstrips in Northern Queensland. See also NAA Canberra item A7942 Z76.

Guinea in order to capture airfields from which further raids on Port Moresby could be made. The invasion force had been spotted en route, and information conveyed to a USN task force, including the carriers *Lexington* and *Yorktown*, in the Coral Sea. A decision was made to launch an attack on the invasion fleet, moored in the Huon Gulf, before the airfields were repaired sufficiently to enable planes from Rabaul to use them. On 10 March, a combined air strike from the two US carriers, over the Owen Stanley Range, the central East-West mountain spine of East New Guinea, sank several ships and damaged others, thus inflicting losses directly on the IJN.

General MacArthur left the Philippines on the 12 March, declaring 'I shall return', and arrived in Australia on 17 March. Around this time it was agreed that the United States would take principal responsibility for defending the Pacific but not the Indian Ocean and the Middle East.

Japanese forces were occupying sites on Bougainville at the end of March as a prelude towards advancing into the Solomon Islands.

21.3 April 1942

The IJN's carrier force, sent into the Indian Ocean in late March 1942, raided[4] Colombo and Trincomalee in Ceylon (Sri Lanka) in early April, and sank some British warships off the coast. The main British naval squadron, under Admiral Somerville, failed to engage with the enemy. As the British had a distinctly inferior force, this was fortunate. Concerned over the vulnerability to further attack of his ships and bases in Ceylon, Somerville on 25 April removed his ships to Mombasa in Kenya, on the east coast of Africa, arriving there on 3 May. He took with him most of the previously evacuated FECB naval Sigint staff. They set up a new station at Kilindini, East Africa (now Kenya). Japanese warships remained active in the Bay of Bengal, disrupting shipping along the coast of India.

So, for the first 4 months of the war, Japan had enjoyed unbroken success. One minor exception was its loss of the submarine I-124 already mentioned.

At the insistence of Prime Minister John Curtin, the Australian Army's 7th Division was recalled from fighting in the Middle East at the end of March 1942. Unfortunately some units were diverted to Java against Curtin's wishes and subsequently captured there. The 6th Division was also recalled. In April, the first major contingent of US troops—the US 41st Army division—arrived in Australia. The presence of these units provided a timely and valuable morale boost to Australia's citizens.

The Japanese captured Bataan on 9 April, taking some 75,000 prisoners of war and effectively ending American resistance in the Philippines except for the defenders of Corregidor. This island, in the entrance to Manila Bay, was the final location of the important USN Sigint station, Cast.

---

[4]Theories that the choice of Sunday for the raid on Pearl Harbor (see Appendix 1 of Chap. 7) was significant are supported by the raid on Trincomalee also occurring on a Sunday.

The defence of Corregidor ended on 6 May, with the surrender of some 10,000 Filipino and US servicemen. It has been suggested that the lengthy resistance mounted against the Japanese in the Philippines delayed further Japanese advances towards New Guinea and the Solomons by 1–2 months. This delay was very significant.

General MacArthur, now appointed Supreme Commander, established General Headquarters for his South-West Pacific Area (SWPA) command in Melbourne on 18 April. At this time he initiated his own intelligence group, Central Bureau (see Chap. 17), needed because of his isolation from other US Army intelligence operations. It was to grow steadily as the roles of military and air force personnel in the SWPA became more and more important in the ongoing war, and moved with MacArthur to Brisbane later in 1942. However, because of the threat posed by the IJN during 1942, Allied Naval Sigint was far more important than military Sigint in 1942 and most of the Comint came from the breaking of the IJN's JN-25 operational code.

On 18 April, the US Army Air Force (now the USAF), in collaboration with the USN, launched a daring seaborne raid on the Japanese homeland by deploying 16 B-25 bombers from the carrier *Hornet*, part of a naval force that went to within 1,000 km of the Japanese coast. This is now generally called the *Doolittle Raid*. The bombers were only just able to take off from the carrier flight deck but could not return. The crews had been instructed to fly on to China and most of them were able to do so.

The physical damage caused by this raid was small but the psychological shock, both to civilians and to the Japanese High Command, was not. It was resolved that the Japanese protective barrier in the Pacific, provided by its bases in the Mandates and at Rabaul, would be enhanced in two ways. The first of these was to be the capture of Port Moresby, strategically located on the southern coast of Papua and from which Rabaul could be raided by Allied planes. The second was to be the seizing of Midway Island, from which Japanese heavy bombers could attack Pearl Harbor. Moreover, the Midway operation would be planned so as to lure the USN's Pacific carriers to their destruction. Following that, the islands of Fiji and Samoa would be conquered. All this, if achieved, would have given the IJN complete control of the shipping lanes across the Pacific between the USA and Australia, and dominance over the waters off the eastern Australian coast. Combined with the IJN's apparent mastery of the Indian Ocean, Australia would then be isolated from its Allies.

21.4   The Battle of the Coral Sea

There had been an apparent lull in Japanese activity in the SWPA since their successes in early March. From early April onwards, the code breakers in Colombo, Hawaii, Melbourne and Washington had been piecing together intercepted signals indicating preparations for a major new thrust involving a naval task force as well as troopships. By the end of April, it was known that Port Moresby and Tulagi in the Solomons were the targets.

In early May, the IJN practised large scale radio deception in order to disguise eastward movement of troops, materiel and warships. It altered shore station call signs and those of ships involved in the Port Moresby operation. There was also a massive increase in its priority level radio traffic. These combined to increase substantially the difficulty of provision of accurate Sigint to Admiral Nimitz in the first half of May.

Nevertheless, the composition of the intended Japanese naval invasion force, with troopships coming from the Marshall Islands and Rabaul, via the Coral Sea, was known, and also its approximate date of departure. The upshot of this Comint was that the USN was able to plan to intercept the Port Moresby invasion force en route, in what was to become named as the Battle of the Coral Sea.

On 5 May an Australian reconnaissance plane[5] sighted this force heading across the Coral Sea towards Port Moresby. A combined US-Australian attack group engaged with it on 7–8 May, sinking *Shoho*, a small new IJN carrier, seriously damaging *Shokaku*, a medium-sized Japanese carrier, and destroying most of the planes of the larger carrier *Zuikaku*. Fear of further aircraft attacks, with reduced defences, caused the Japanese force to return to Rabaul. Though the Allies suffered greater losses, with the carrier *Lexington* sunk and the carrier *Yorktown* damaged, the failure of the invasion force to complete its mission was a major setback to the Japanese military. Moreover, neither of the two large IJN carriers was available for active service until after the coming conflict at Midway, where they might well have made a difference to the outcome. Within months of the battle, Port Moresby had been reinforced and any possibility of an easy Japanese victory there had evaporated.

The Battle of the Coral Sea was the first naval battle in history in which the opposing fleets engaged solely via use of their carrier-based aircraft, with no ship of either fleet ever sighting an enemy vessel. Naval warfare had entered a new era.

An immediate consequence of this battle, and one of lasting operational significance, was the recognition by Admirals King and Nimitz that the radio and

---

[5]The date of the IJN incursion into the Coral Sea was not obtained from JN-25B intercepts. There was a tricky but systematic extra encryption used for dates in that additive cipher system. This was worked out in the last week or so before JN-25B was supplanted and so the date of the attack on Midway could be and was determined. Compare Note 27 of Chap. 1, in which Mackenzie King's diary entry on the coming Battle of Midway is quoted. The IJN ships approaching the Coral Sea were observed by the RAAF plane sent out to monitor the approaches. Presumably Comint lay behind this use of the plane. The IJN would have seen it and then later attributed any apparent foreknowledge of its plans to an apparently chance event. One reference for this is NAA Canberra item A9695 489: *RAAF Participation in the Coral Sea Battle*.

Attention must be drawn again to the regulations on the use of Comint given in Sect. 1.17, which included: 'In general, if any action is to be taken based upon Ultra information, the local Commander is to ensure that such action cannot be traced back by the enemy to the reception of Ultra Intelligence alone. ... No action may be taken against specific land or sea targets revealed by Ultra unless appropriate air or land reconnaissance has also been taken.'

communications intelligence now available to them in the Pacific theatre had proven itself to be of great value and had to be included in all future intelligence briefings on activities of the IJN. Thus they were in receptive mode for the growing amount of Sigint relevant to the IJN plans for Midway. There was still some considerable disagreement about the interpretation of this intelligence between the Op-20-G units Hypo in Hawaii and Frumel in Melbourne, and the high level intelligence groups in Washington.

21.5   The Sigint Factor

One of the most significant outcomes of the Japanese plan to invade Port Moresby and the resulting Coral Sea battle was the enormous increase in the number of radio messages sent at that time by the IJN using its main (JN-25B) operational code. This substantially increased the number intercepted by the Allied Sigint stations. They were able to use and enhance their traffic analysis, call sign and direction-finding skills, which would be of critical importance following the expected major change to the JN-25B system that would completely stop the flow of Comint[6] until the new system was broken. The large increase in the number of intercepted signals also provided much more material to the code breakers. They were now proficient in adding new book groups to the current JN-25 code book and these helped the linguists to assign valid meanings to more book groups. This increased the flow of accurate operational intelligence to Admiral Nimitz.

The situation at that time was described by Eric Nave (see Sect. 3.3) in his memoir, now lodged with the Australian War Memorial. He wrote[7]:

> However there was at this time *one new factor, a dominating one* [our italics], the naval code JN-25 was now being read by the US cryptographers. A routine change was postponed twice and this, with a new code book, would have given a virtual blackout. Distribution had become a real difficulty. In order to avoid having to send the same message in two different codes at the same time, with disastrous security consequences, two delays took place with welcome benefit to the USN. Without Sigint (information on the raid on Midway) two USN carriers would have been in the Coral Sea and one under repair.

Nave's reference to the postponement of the replacement of the JN-25 system by a new version, initially expected on 1 April and then on 1 May, but not in fact effected until late in May, is most significant. These delays resulted in sufficient information coming from successfully decrypted IJN operational messages for the

---

[6]NAA Melbourne archive series B5555, however incomplete, is the key RAN record of Comint, particularly in 1942. From the viewpoint of this book, and perhaps from the more general viewpoint of WW2 history, it is the most interesting series in the Australian NAA. It confirms evidence from other sources that JN-25B was being increasingly read in March, April and May 1942.

[7]Nave's account of the timing of the reading of JN-25B8 is in stark contrast with that given in the book *Betrayal at Pearl Harbor* by James Rusbridger and Eric Nave. Rusbridger appears to have written much of that book without consulting Nave.

USN to set up both the Coral Sea and Midway battles. General MacArthur wrote to General Marshall in Washington suggesting that every effort be made to exploit this 'good fortune', which could not be expected to last. Had the replacement occurred even as late as mid-April, the war would at the very least have lasted much longer and perhaps there would have been adverse effects on the Allies in other theatres of the war. But this was not to be.

The change in JN-25 from the B version to the newly named C version, on 27 May, was complete in that the code book, the additive table, and the indicator encryption details all changed. Decryption of old JN-25B intercepts continued after 27 May and provided further useful intelligence to Admiral Nimitz. More importantly, the fact that the IJN Midway and Alleutians fleets were continuing to observe radio silence after the change[8] was seen as providing further evidence that the full operation against Midway was continuing.

By then, Op-20-G had gained much experience in the special methods used to attack JN-25 cipher systems. Additional staff had been trained. Had the change from JN-25B to JN-25C included a new code book in which the book groups were random 5-digit numbers, it could have set Op-20-G back many months before the extraction of useful Comint could commence. There was great relief when the code breakers confirmed, early in June, that the new code book again consisted of 5-digit groups, again all multiples of 3, and that some useful decrypts could be expected within 2 or 3 months. Likewise, the new indicator encryption system was open to the special methods of attack previously developed.

21.6 Comint about Midway

From early May, Sigint showed that a major operation was being planned by the IJN. As data built up, there was disagreement within the USN as to the principal target of this operation, identified as 'AF' in the intercepted messages. This was resolved by a ruse from Station Hypo: the Midway base was ordered to send a plain language radio message advising that a problem had developed with its fresh water supply. As expected, this message was intercepted by Japanese operators on Wake Island, who then advised Tokyo of a water problem at AF. This confirmed Midway as the principal target, as the Hypo and Frumel code breakers had thought.

The IJN intended to launch a diversionary attack on some US islands in the Aleutians simultaneously with the major assault on Midway, with the dual objectives of capturing them and luring the remainder of the US Pacific fleet and especially its

---

[8]The replacement of JN-25B8 by JN-25C9 on 27 May 1942 led to the most memorable piece of radio intelligence of the whole Pacific War. The ships involved in the attack on Midway kept radio silence but there was plenty of other traffic in the new 5-figure cipher. The reference is NAA Melbourne B5555 4: 'There is considerable volume of traffic being broadcast for units in the Midway Operation forces but no traffic originated by them. The operation has apparently begun and forces at sea are keeping radio silence.' (1 June 1942.)

carriers to their destruction. The final intelligence briefing given to Admiral Nimitz by his intelligence officer, Commander Layton, relied heavily on the hard work of Rochefort's Hypo team and Fabian's Frumel team.

NSA historian Henry Schorreck[9] commented on this as follows:

> The Comint documentation presented here ... is more than enough to demonstrate the amount, timeliness and accuracy of the Comint provided Admiral Nimitz: thanks to Comint, he knew more about the Midway Operation than many of the Japanese officers involved in it. He knew the targets; the dates; the debarkation points of the Japanese forces and their rendezvous points at sea; he had a good idea of the composition of the Japanese forces; he knew of the plan to station a submarine cordon between Hawaii and Midway; and he knew about the planned seaplane reconnaissance of Oahu, which never took place because he prevented their refuelling at French Frigate Shoals.

Two particular items of Comint are deserving of special mention amongst the wealth of such information. The first is that, as previously mentioned, its use in setting up the Coral Sea Battle resulted in two IJN carriers being unavailable for the Midway operations. The second is that the decrypt, concerning the decision of the Japanese High Command to abandon any further sea-based invasion of Port Moresby in favour of a land-based assault from the north coast of Papua, enabled Nimitz to order Halsey's two carriers patrolling the Coral Sea, *Enterprise* and *Hornet*, to return *undetected* to Pearl Harbor. They were able to do so after ensuring that their presence in the Coral Sea had been observed by Japanese reconnaissance planes. This sighting caused the IJN to recall an attack force en route to invade Nauru and Ocean Island, east of the Solomons.

Hiding from the IJN the return of *Enterprise, Hornet* and the damaged *Yorktown* to Pearl Harbor was an important aspect of Nimitz' plans. He noted that Japanese radio interception stations were picking up air-ground messages between Port Moresby and nearby Allied planes and air-ground chatter between carriers and pilots coming into land, and insisted upon strict radio silence in such circumstances.

MacArthur suggested that knowledge of the Japanese Sigint activity could be exploited by using radio deception, whereby two USN warships in the Coral Sea would correspond with selected shore-based stations so as to create the impression that a USN carrier task force continued to operate there. Nimitz agreed and subsequently USN Sigint confirmed that Yamamoto had been advised of this USN presence while he was steaming towards Midway.

Sigint was also able to provide Nimitz with dates for the marshalling of ships for each constituent group in the Aleutian and Midway operations and, via detailed traffic analysis, with almost complete lists of every ship in each group.

---

[9]Frederick D. Parker's *A Priceless Advantage* gives a connected, detailed analysis of the Comint background to the Battles of Coral Sea, Midway and the Aleutians. Henry F. Schorreck's *Battle of Midway: 4–7 June 1942: The role of COMINT in the Battle of Midway* (NARA RG457 SRH230) is a shorter, less detailed but thorough account.

Canberra NAA item A5954 530/2 contains the Advisory War Council Minutes for 17 June 1942 and also uses the word *advantage*: 'As it was, we had the great advantage in each action of knowing the Japanese plan beforehand.'

It was also essential to Nimitz that his carrier force departures towards Midway not be known to the IJN. Thus, again based on Comint, he ordered the stationing of a warship at French Frigate Shoals, between Pearl Harbor and Midway. This was the obvious location at which an IJN submarine could refuel seaplanes spying on Pearl Harbor. This action frustrated the IJN's plan to use French Frigate Shoals in this way in late May. Had the IJN seriously expected unusual activity occurring at Pearl Harbor, it could have employed an alternative method of surveillance—its midget submarine attack[10] on Sydney Harbour on 31 May was preceded by observation using a submarine-carried demountable float plane. Such an exercise could have been carried out at Pearl Harbor instead and possibly exposed the prospect of a counter-ambush by the USN carriers!

The IJN submarine cordon intended to detect and attack USN warships heading westwards from Pearl Harbor was not fully positioned until early June, by which time the USN carriers—*Enterprise, Hornet* and a hastily repaired *Yorktown*—were on their way to Midway. Although Admiral Yamamoto was provided by the Sigint unit on board his flagship with intelligence reports containing information relevant to both the Midway and Aleutian operations, he apparently chose not to pass them on to his fleet Commanders because of the strict radio silence order currently applying to them.

21.7 The Aleutian and Midway Battles

Consistent with Sigint predictions, IJN carriers in the Aleutians operation bombed various sites in the island chain on 3 June, including Dutch Harbor, despite extremely bad weather. Japanese troops were landed on Kiska on 6 June and on Attu on 7 June. Admiral Theobald, in command of the USN Task Force sent north by Nimitz to frustrate these landings, had apparently decided that the real intention of the IJN forces was not to support landings on these extreme western islands of the chain, but to isolate USN forces and island bases from the Alaskan mainland. He chose to ignore all the Comint-based briefings received from Nimitz and deployed his ships well to the east of the focus of invasion, thus being unable in any way to disrupt the Japanese landings. The outcome of this was that the two islands remained under Japanese occupation until Attu was recaptured in May 1943 and all Japanese forces were evacuated safely by the IJN from Kiska in late July 1943.

---

[10]New Zealand radio intelligence had informed the Melbourne Naval Intelligence Department of IJN submarine activity in the Tasman Sea in May. It is possible that Frumel may have had information from JN-25B decrypts about the coming attack on Sydney Harbour by midget submarines. It is also possible that no warning was given to the naval forces there in order to preserve the secret that JN-25B was being read. One USN cruiser in the harbour was not damaged. The three midget submarines involved were all sunk after they had killed 19 RAN and two RN personnel.

On 3 June, the Japanese Midway occupation force was, as expected, spotted by a Catalina about 700 miles west of Midway and attacked by high flying bombers from Midway the same day, without suffering any damage. This sighting convinced Nimitz that Admiral Nagumo's carrier strike force was indeed likely to be found in the predicted location northwest of Midway. Pre-dawn on the morning of 4 June, Nagumo launched his planned attack on Midway, sending a fleet of some 100 aircraft to destroy the runways and other facilities, prior to invasion by the occupation force. The additional fighters and ground forces sent there by Nimitz managed, despite heavy losses of aircraft, to limit the damage inflicted sufficiently for the attack leader to recommend that a second attack would be necessary prior to any landings. This meant that planes on the carriers, that were being armed with bombs and torpedoes for use against a possible but unexpected USN counter attack, had to be reloaded with incendiary and fragmentation bombs destined for Midway.

Midway had launched a mix of torpedo, dive and high altitude bombers towards Nagumo's strike force. Although no damage was inflicted and many were destroyed, the continual attacking by these aircraft disrupted activity on board the carriers, slowing both the re-armament process and the preparations for the return of their planes sent to Midway.

In the midst of this, Nagumo received news of a sighting nearby of USN warships, followed some time later by advice that they appeared to include a carrier and later still that torpedo bombers were heading towards him. He ordered a change of course, which resulted in one dive bomber group from the USN carriers missing his force altogether and landing on Midway. The other two dive bomber groups flew well beyond their target before turning back. The three torpedo bomber groups all carried out low-level attacks but again inflicted no damage and suffered heavy losses.

At this stage, it looked as if the USN ambush had failed and that an intensive carrier to carrier confrontation now loomed, with the odds favouring the IJN. But then, the returning USN dive-bombing groups spotted the wake of a single IJN destroyer. It was correctly guessed that it was speeding to catch up with the Nagumo force. So the dive-bombing groups turned in that direction. (The destroyer had been delayed because it had been searching for the USN submarine *Nautilus*, which had been trying to attack the occupation force.)

Aided by their approach from the rear of the IJN strike force and also as most of the IJN fighters were still at low altitude from defending against torpedo bombers, the dive bombers had, within minutes, inflicted fatal damage on the IJN carriers *Akagi*, *Kaga* and *Soryu*. The carrier *Hiryu* was not attacked and, later that morning, the *Hiryu* sent attack waves towards the USN carriers, resulting in the *Yorktown* eventually being abandoned. In return, that afternoon, dive-bombers from the *Enterprise* found and damaged the *Hiryu* so badly that it sank next day. The carriers *Enterprise* and *Hornet* remained unscathed. *Yorktown*, still afloat, was being towed away when it was torpedoed by a Japanese submarine and finally sank on 7 June. The destroyer towing the *Yorktown* was also sunk.

Admiral Yamamoto, ensconced in his super-battleship *Yamato* near the bulk of the invasion force some 1,000 km west of Midway, eventually decided to order the warships escorting the occupation force to head full speed towards Midway and attack it, only later to realise the futility of this and order a withdrawal. In executing this, two heavy cruisers, *Mikuma* and *Mogami*, collided and *Mogami* was so badly damaged that it could sail westwards only at reduced speed, with *Mikuma* accompanying. Next day, they were attacked by US planes and *Mikuma* was sunk. Thus ended the Battle of Midway.

21.8   Consequences of the Battles of the Coral Sea and Midway

The enormous loss of seaborne air power suffered by the IJN at Midway led Admiral Yamamoto to order an immediate withdrawal of the IJN forces back towards their bases in the Mandated Islands and Japan, even though the IJN retained a clear numerical superiority over the USN in terms of conventional warships operational in the Pacific theatre.[11] The Japanese High Command, reviewing its strategy, placed greater importance on gaining control of the Solomon Islands. This was already in train and had advanced a step further with a landing on Guadalcanal on 8 June. By 13 June, it had been decided to construct an airfield there, moving in the men and equipment needed without delay.

Following the Japanese move into Rabaul and their continual further penetrations into the Solomons, the US had appreciated the developing threat to the sea-lanes linking it with Australia, which would compromise the use of Australia as a major base for its forces in the SW Pacific. A plan by the Americans to launch offensive actions through the Solomons and New Guinea, once further Japanese advances had been stopped,[12] led to the South Pacific Command being set up under Vice Admiral Ghormley. It covered an area previously shared by Admiral Nimitz and General MacArthur. The stage was being set for a particularly important and costly conflict over the next 6 months.

An unexpected consequence of singular importance to the future of the war was the finding in July of a crashed but redeemable Zero fighter on Akutan Island in the Aleutians. It was transported to San Diego and by September was restored to flying condition. Experiments with it enabled modifications to be made to the newly

---

[11] Although there have been numerous major land wars, the title of the Potter and Nimitz book *The Great Sea War* points out that there was only one major sea war.

The June 1942 Battle of Midway may be compared with the March 1941 Battle of Matapan. Both were set up using Comint. Each had great strategic consequences.

[12] As noted earlier, the Japanese Army had another fundamental problem: its activity in China consumed much of its resources. Occupation of the newly captured territories consumed more. In 1944 Sigint would provide General MacArthur with information about which Japanese-held places were weakly defended.

developed Grumman F6F Hellcat fighter as well as enabling US pilots to develop tactics to overcome the Zero in battle. Hellcats accounted for some 5000 of the 6500 enemy aircraft shot down by carrier-based planes in WW2.

The evident value of the Sigint available to Nimitz through the Coral Sea and Midway battles accelerated the acceptance of Sigint and especially of Comint by the USN and the other services. USN commanders were ordered to attend to all Comint briefings sent to them. Op-20-G received an enormous boost in personnel and equipment later in 1942. An intelligence centre set up in that year by the USN became JICPOA—Joint Intelligence Center Pacific Ocean Area—in 1943.

21.9   The Burma Campaign, March–June 1942

During this period, the British situation in Burma steadily deteriorated. The IJA increased its strength there, eliminating any effective Allied air power.[13] An American Military Mission had been based in Chungking from September 1941 and was responsible for co-ordinating the US lend-lease supplies to China through Burma. After Pearl Harbor, the US, conscious of the enormous numbers of Japanese troops tied up fighting in China, appointed General Stilwell as its liaison officer with Chiang Kai-shek and head of the US China-Burma-India (CBI) Command. He found himself having to negotiate with both the Chinese and the British regarding his role in Burma, and was unable to take effective command of the Chinese forces in Burma before they were routed. The Japanese captured Lashio, the terminus of the Burma Road, in late April, thus cutting off the last land supply route to the Chinese. Stilwell in early May led a small group of military and civilian personnel west through the Burmese jungle to safety in India. The British in the south had also been forced back and by late May the Japanese had gained control of all important targets in Burma, and the prospect of an attack into India had to be contemplated.

There is a curious item of Sigint associated with Stilwell at this time. The US Tenth Air Force had been created to assist operations in the CBI theatre, but was barely in existence by the end of March, when its commander, General Brereton, had assured Stilwell that he did not expect to be ready for combat until early May. Stilwell was therefore mystified to learn that it had carried out two raids—one on the Andaman Islands, and one on Rangoon, on 2 April. Brereton advised that he had been ordered directly from Washington to do this, 'to assist the British'. Now, Ensign Chamberlin, one of the small remaining group of Navy personnel at Station Cast in the Philippines, in late March, recalled the following incident:

> The only enemy traffic being translated was that for which we had a cipher key. One day a long series of messages were intercepted for which much of the key had not been recovered. What was readable seemed to indicate that a large invasion convoy was being readied by

---

[13]URL http://www.army.mil/cmh-pg/brochures/burma42/burma42.htm and Donovan Webster's *The Burma Road* are recommended.

the Japanese. It appeared to be of such importance that Honest John sent off a 'priority' to Washington requesting assistance. The response was negative, so Lietwiler became a 'cryptanalyst' for about thirty-six consecutive hours and as a result, Rufe Taylor could translate the entire series which turned out to be the Japanese Invasion Plan for India. This was immediately transmitted to British General Wavell. General Wavell borrowed planes and pilots from the Middle East to augment his own forces and hit the convoy off the Nicobar Islands with such force that the elements which remained limped back to Singapore and India was never invaded. Later, in Melbourne, we learned General Wavell had said in a message to Washington that 'under no circumstances must the Corregidor Unit cease to function'.

Perhaps the true story is that the Station Cast decrypt was responsible for the Washington directive to General Brereton.

The Americans, determined to show continued support for the Chinese, set about arranging an airlift of supplies over the Himalayas, from airfields in Assam, to Kunming, thus adding incentive to the Japanese plans for invading India. For his part, Chiang Kai-shek continued to agitate for greater Allied military activity in Burma and in the Bay of Bengal.

Worried about the vulnerability of their merchant shipping to US submarines on the long journey from the South China Sea, through the Strait of Malacca and then around the Malay Peninsula and eventually to Burmese ports, the IJA in June commenced the construction of the infamous Burma Railway, designed to link existing rail sectors in Thailand and Burma by a new line running north-west from near Bangkok to just south of Moulmein in Burma.

21.10   Events Elsewhere

The United States was suffering massive losses of Caribbean and coastal shipping due to German submarines operating in the Western Atlantic Ocean. This went on for 7 months following the German declaration of war. Inexplicably the USN took a great deal of time to absorb the lesson from 1917–1918 about the use of convoys. Chapter 16 describes the immense damage caused by German submarines elsewhere in the Atlantic in 1942.

Although Britain had been saved from immediate invasion, the situation in Eastern Europe remained critical throughout much of 1942. The coming of warmer weather in March 1942 enabled further German advances to happen. However, Moscow was held. A long siege of Leningrad (now named St Petersburg again) continued all year, being finally relieved only in January 1944.

A fortnight after the Battle of Midway, the British forces in Tobruk (now in Libya) were captured when that key port fell to the Germans and Italians. Thus the British army was forced back to El Alamein in Egypt. The decisive turning of what Churchill called the *Hinge of Fate* was still some months off. At least the sinking of the key Japanese aircraft carriers almost totally eliminated the prospects of any invasion of Australia and so made a build-up of British forces in Egypt easier.

Chapter 22
Guadalcanal and Papua

Two campaigns dominated the second half of 1942 in the Pacific, both being finalised only early in 1943. These were the battle for Guadalcanal in the Solomons and the Japanese attempt to attack and capture Port Moresby, via an overland offensive from Gona and Buna on the northern coast of Papua (the eastern part of New Guinea) across the central spine formed by the Owen Stanley Range (the *Kokoda Trail* campaign). These were certainly not the campaigns anticipated by the Allies as a consequence of the Coral Sea and Midway battles. Their thinking was already directed towards the recovery of strategically important captured territory. Towards this end, and also from the experience gained in the Coral Sea engagement, the strategic benefit to be obtained from having available an operational airfield close to the eastern tip of New Guinea was appreciated. Following surveillance, the Allied Command selected a coconut plantation at Milne Bay. This had the advantage of providing for the development of a deepwater port in the bay. So, by late June, airstrip construction had commenced, supported by a strong contingent of troops.

22.1 Allied Planning in Mid 1942

A comprehensive regional plan, designed to capitalise on Midway, and accommodating the separate priorities of his Army and Navy commanders, was released by General Marshall on 2 July. It specified a triple of offensive operations, aimed ultimately at Rabaul:

Task 1. Seizure and occupation of the Japanese occupied Santa Cruz Islands, Tulagi, and adjacent islands in the southern Solomons.

Task 2. Seizure and occupation of the remainder of the Solomon Islands, of Lae, Salamaua, and the northeast coast of New Guinea.

Task 3. Seizure and occupation of Rabaul and adjacent positions in the New Guinea/New Ireland area.

© Springer International Publishing Switzerland 2014
P. Donovan, J. Mack, *Code Breaking in the Pacific*,
DOI 10.1007/978-3-319-08278-3_22

The scheduled date for starting Task 1 was 1 August 1942. Realising that an Allied airbase on the northern coast of New Guinea would be essential, a reconnaissance group from Port Moresby recommended a site at Dobodura, near Buna on the north coast. Around 10–12 August, as part of *Operation Providence*, detailed planning for landing a contingent of engineers and soldiers (Buna Force) at Buna, hopefully without Japanese knowledge, was already in train when the first indications of relevant Japanese movements were detected.

Intercepts, picked up soon after the Coral Sea battle and communicated[1] to MacArthur, had revealed that the Japanese were exploring the possibility of combining a land-based assault on Port Moresby, from the northern coast of Papua southwards over the Owen Stanley range, with a smaller amphibious assault from the Coral Sea. It seems that Allied HQ did not seriously regard this as imminent and likely to interrupt its Buna force schedule.

However, news from the Solomons could not be ignored. The Japanese had established a seaplane base at Tulagi in the Solomons in May and had landed exploratory parties on Guadalcanal just after the Midway battle. They had soon commenced the building of an airstrip there. Then, on 6 July, a 12-ship convoy landed heavy construction equipment and some 2,500 construction workers near Lunga Point on the northern side of the island. Within a month, the strip was close to completion.

The build-up of shipping in Rabaul and sightings of troop-laden smaller vessels off New Britain had been monitored since mid-July. Fears that a Japanese landing on the north coast of New Guinea was in train were discounted in favour of a view that this activity was also directed at the Solomons. By the time the real destination of these ships was confirmed by sporadic sightings of their convoy east of Salamaua, bad weather, plus a shortage of Allied aircraft, allowed most of its troops and equipment to be successfully landed close to Buna on 22 July. Troop movement towards Kokoda began immediately, while additional landings helped to set up a strong Japanese base covering Gona, Sanananda and Buna.

22.2 The Threat to Australia

The implications of the Japanese capture of Port Moresby, and of their establishing an airbase on Guadalcanal, were evident, since either would substantially strengthen the Japanese threat to isolate Australia. Air raids on Townsville and Port Hedland in

---

[1]Melbourne NAA file B5555 5 *Kokoda Trail: First Information* has Commander Newman's comment that 'After the Coral Sea Battle, C-in-C Combined Fleet ordered postponement of attempted occupation of Moresby until Allied Sea and air strength could be reduced. Very valuable early information of an intended alternative move (the Kokoda Trail) was then obtained as shown in the attached intercepts.' Newman adds that this information was passed on to Admiral Leary, USN, who would have deduced that the IJN carriers must have been sent elsewhere, and to General MacArthur, who was responsible for organising resistance along the Kokoda trail to protect Port Moresby.

July, and the ongoing Darwin air raids, had heightened the serious concern already felt in Australia after the first air raids on Darwin.

Loss of Port Moresby as an Allied forward base in New Guinea would have itself been a serious setback, but a Japanese base there would have rendered the north-east coast of Australia, as well as shipping in the Coral Sea, vulnerable to constant attack. A Japanese airbase on Guadalcanal would have a number of adverse consequences for the Allies. It would have strengthened the defensive shield of the major base at Rabaul. It would have provided powerful opposition to any USN carrier-based attacks on Japanese bases in the region. It would have provided additional air cover in support of Japanese landings on Pacific islands such as Ocean Island, Nauru, New Caledonia and Fiji. It would have directly threatened the sea communications routes from the West coast of the USA to Australia, thus compromising any prospective reconquest of the South-West Pacific Area.

22.3 The Battle for Guadalcanal

The speed of the Japanese progress at Guadalcanal forced the USN to focus immediately on capturing and holding the airstrip under construction before it was activated, and then turning it into an Allied airbase.

An invasion force was assembled in New Zealand, and US marines landed on Tulagi and Guadalcanal on 7 August. Japanese reaction was swift, with waves of bombers sent from Rabaul to destroy the transports and warships and so crush the invasion immediately. The flight path from Rabaul followed the island chain from New Britain down to Guadalcanal, and the Coastwatchers[2] stationed along these islands provided invaluable warning of incoming air attacks. These warnings enabled the USN's carrier-based fighters and warships to optimise their use of resources in defence, and generally to overcome the attackers.

The Japanese also determined on an immediate naval attack on the invasion fleet. A strong force of cruisers and destroyers was sent down on 8 August for a night-time attack on it. A combination of superior night-fighting skills, surprise, and an uncoordinated response led to a resounding victory for the IJN at the *Battle of Savo Island*, the worst US naval defeat of the entire war. For various reasons, the Japanese commander, Admiral Mikawa, decided to return directly to Rabaul after this battle. Had he decided to attack the transport ships unloading supplies and equipment for the 13,500 US Marines already onshore, he would have found them virtually undefended. As it was, the shock of his attack led to all the transports being withdrawn before unloading was completed, so Mikawa was more successful than he realised!

Japanese air and sea attacks on the Marines intensified and continued unabated once it was apparent that no air cover defence was available, the carriers of the US

---

[2]More on the Coastwatchers is to be found in Sect. 6.6.

invasion force having disappeared. The first US aircraft to land on the incomplete but by then usable Guadalcanal airstrip (now named Henderson Field) arrived on 20 August. An absolutely essential aircraft presence was maintained there throughout the campaign, despite horrendous naval, air and land-based assaults by the Japanese for the next few months. Sometimes, the only US aircraft left standing ready for action were those that had been diverted to Guadalcanal from sinking or damaged US carriers in the Solomons area!

Several major naval engagements followed the Savo Island battle. On balance the net result in ships destroyed favoured the IJN, but this was overshadowed by its loss of planes and trained aircrew. The importance of Henderson Field both to naval engagements and to all-out attacks on Guadalcanal may be better appreciated when it is realised that during these, the USN lost its carriers *Wasp* and *Hornet*, while the carrier *Saratoga* was torpedoed and put out of action until the end of the year. However the USS *Enterprise* suffered damage several times but emerged triumphant to continue throughout the war. There were times when only Henderson Field could supply air support for any engagement with Japanese forces attacking or attempting to reinforce Guadalcanal. In fact, almost a year passed before there was another direct carrier-to-carrier confrontation in the Pacific.

Early counterattacks on Henderson Field by freshly landed Japanese infantry groups might have been successful, had the commanders in Rabaul correctly estimated the size of the US contingent and reacted accordingly. Instead, the Japanese considered that the US force was a small exploratory one. They grossly underestimated its strength until several thousand of their troops had been lost in their original attempts to recover the airstrip. US radio and communications intelligence also assisted the defenders in these early forays by the Japanese military. In particular an early attack by the Ichiki detachment on the Tenaru River edge of the land held around Henderson Field by the US Marines was beaten off. Sigint was most helpful in this action.

The US Marines of Guadalcanal used for the first time an extremely valuable operational 'encryption system', namely specially trained Navajo[3] 'codetalkers'. They were essential for two reasons: rapidly moving battle fronts meant that the standard, slow encryption/decryption systems then in use simply could not cope and plain language could not be used because some IJA interception operators were proficient in English.

By late October troop numbers on Guadalcanal, on both sides, had grown dramatically and further unsuccessful attempts had been made by the IJA to regain the airfield. Huge losses were suffered by the Japanese. By December, with some 30,000 IJA troops on the island, and with the US now regaining naval and air superiority, it was clear that these troops could no longer be supplied with even

---

[3]Chester Nez' *Code Talker*, Berkeley Caliber, New York, 2011, written with Judith Schiess Avila, is the only first-hand account. Page 132 contains the memorable sentence 'The message that would take 4 hours by Shackle took only two and a half minutes by Navajo code'.

minimal food or equipment. In a daring and well-planned evacuation some 10,000 Japanese troops were taken off the island and successfully returned to Rabaul in early February, to the surprise and chagrin of the American forces.

Guadalcanal was a costly and brutal battleground for both sides. The US ground forces lost some 1,700 killed and 6,000 injured or debilitated by tropical illnesses, while the USN and its allied navies lost 29 warships and some 5,000 personnel. 600 aircraft were lost, with 400 aircrew. Japanese troop losses alone were some 25,000 dead. The IJN lost 38 ships, including transports, and some 3,500 personnel. Also lost were some 800 aircraft and 1,200 aircrew. The forced evacuation of the Japanese gave a major morale boost to the Allies and was a serious loss of face to Japan.

Admiral Yamamoto was well aware that American industrial capacity would outstrip Japan's in the rate of provision of new ships and planes from early 1943 onwards. This knowledge continued to fuel the IJN's belief that it should, despite Midway, strive to engineer an early and victorious 'Decisive Battle' that would force the USA to negotiate for peace. So the IJN was somewhat reluctant to commit its capital ships and carriers to smaller-scale operations against the USN. This reluctance influenced the Guadalcanal campaign in favour of the US forces.

22.4   Use of Intelligence in the Guadalcanal Campaign

The 'real time, on the spot' intelligence provided by the Coastwatchers was invaluable throughout the Guadalcanal campaign, as it enabled optimal use of the embattled fighters and bombers on Henderson Field against both air and sea attacks. As previously mentioned, JN-25, the main IJN operational code, was of limited tactical use to the Allies for much of the campaign. This was because JN-25C, the version that had been introduced by the IJN at the end of May, was only just beginning to yield useful decrypts by August, the time of the USN landing on Guadalcanal. The capture of JN-25C documentation there at this time resulted in its prompt replacement by JN-25D and cryptanalysis of JN-25 was back to square one again. Nevertheless, DF and TA radio intelligence information regarding the disposition of IJN units proved of extreme value to the USN in the later naval battles leading up to the battle of Guadalcanal in November.

In late January 1943, towards the end of the campaign, the Japanese submarine I-1 was forced aground on Guadalcanal by two RNZN corvettes, *Kiwi* and *Moa*. The USN recovered much cryptographic material destined for Japanese units. This disrupted IJN communications and also led to the IJN changing only the additive table for JN-25D, a much less drastic setback for Op-20-G than the introduction of yet another code book.

22.5   The Papuan Campaign

Rapid construction work at Milne Bay produced a usable landing strip by late July. Allied fighters arrived immediately and were later reinforced. From this site it was possible to launch attacks on the Japanese forces that had just been landed at Buna and Gona on the north coast of Papua, without having to fly over high sections of the Owen Stanley range. Also, the closer proximity of the air strip to the Solomons and the Bismarck Archipelago, including Rabaul, made the Japanese bases in these locations more vulnerable to air raids from General Kenney's Air Force,[4] which was just beginning to grow in numbers of bombers and fighter planes.

The Japanese Rabaul Command decided immediately to send troops by sea to capture the Milne Bay airfield. Sightings by Coastwatchers enabled one contingent, sailing from Buna on 24 August, to be located the next day at Goodenough Island and destroyed by fighters from Milne Bay itself. Bad weather prevented attacks on the main force coming from Kavieng and on later follow-up forces until after all Japanese troops had landed on the northern shore of the Bay. A break in the weather enabled a successful attack on the stores brought to the Japanese base camp at the first landing. As at Guadalcanal, the Japanese Rabaul Command underestimated the size of the Allied contingent. After some days of intensive and brutal combat, and having lost many troops killed in action or exhausted by the jungle conditions, the surviving Japanese retreated and were evacuated on 6 September.

22.6   The Advance Towards Port Moresby

Soon after their first landing,[5] the Japanese forces at Buna had launched exploratory patrols towards the village of Kokoda, at the foot of the Owen Stanley range. The Allied force there at this time consisted of a Papuan group and a battalion of Australian militia without much formal training and no experience of battle. Some reinforcements were eventually sent, and, although still heavily outnumbered, the defenders managed to confine the Japanese advance to the Kokoda area for nearly two weeks, before being forced to retreat into the hinterland.

The main Japanese forces arrived at Buna via convoys from Rabaul over the next few weeks and by mid-August the order to advance towards Port Moresby

---

[4]The United States Army Air Force was renamed the United States Air Force after WW2. There was no separate Air Force signals intelligence unit. Air Force General George Kenney, Commanding General of the Fifth Air Force, wrote the book *General Kenney Reports* in 1949.

[5]Attention has to be drawn here to Melbourne NAA item B5555 5, being part of the Frumel documents selected by Commander Newman for preservation.

'After the Coral Sea Battle, C-in-C Combined fleet ordered postponement of attempted occupation of Moresby until Allied Sea and air strength could be reduced. Very valuable early information of an intended alternative move [presumably the Kokoda Trail] was then obtained as shown in the attached intercepts.' . . . 'Passed on to Admiral Leary USN and General MacArthur.'

was given, while construction of an airfield at Buna was also commenced. General Willoughby, MacArthur's chief intelligence adviser, remained firm in his belief that the Japanese build-up in strength at Buna was to protect its airfield and did not presage an overland assault. It was not until early September, after the Australian defenders had been forced to withdraw steadily, that MacArthur's GHQ accepted that substantial additional troops were necessary and began to supply them. But by then the advancing Japanese had reached the southernmost ridges of the Owen Stanleys, at Ioribaiwa on the Kokoda Trail.

The situation now changed dramatically. By bombing and strafing the Buna airfield and key points along the Trail, General Kenney's planes had successfully prevented the supply of food and equipment to the Japanese front line. The troops were forced to scavenge for food. Also, the worsening situation at Guadalcanal, and the failure of the Milne Bay operation, led the Japanese High Command at Rabaul to subordinate all activities to the recapture of Guadalcanal. A staged withdrawal back to the beachhead at Gona and Buna was ordered. This coincided with a coordinated Allied assault, supported by aerial attacks and supply drops, which eventually resulted in Kokoda being recaptured on 1 November. The threat of an immediate successful Japanese overland (or amphibious) assault on Port Moresby was removed.

22.7 Final Victory in Papua

Removal of the threat to Port Moresby by no means meant that the Papuan campaign was effectively over. The retreating Japanese forces had been told to continue to fortify Buna and Gona and to *fight to the death*, in order to delay Allied military forces for as long as possible. As the earliest possible return to the Philippines remained an almost obsessive commitment for General MacArthur, he instructed his commanders to clear the area of Japanese quickly and irrespective of cost. The cost was indeed high in Allied casualties. Frustration over the time taken to complete this task seemed to prevent MacArthur gaining a real understanding of the conditions under which his troops were fighting. Unlike Milne Bay and Guadalcanal, where Allied forces were repelling Japanese attacks, here his forces were attempting, for the first time, to overcome an extremely well fortified position. This position was manned by troops who did fight to the death, in appalling tropical conditions that resulted in serious illness and fatigue among Allied troops. Moreover, despite the realisation by subordinate commanders that heavy ordnance support would be critical in helping to destroy the interlocking bunker and pillbox defences, MacArthur did not consider this necessary. This decision undoubtedly prolonged the campaign and added considerably to the overall casualty list. When the battle was over, in late January 1943, Allied battle casualties were some 3,100 killed and 5,500 wounded. But in addition, there were some 20,000 cases of disease[6] (principally malaria) and almost all those who fought returned from the area in a very poor physical state. Of the 17,000 strong Japanese force, at most 4,500 survived, withdrawing into New Guinea.

22.8   New Guinea and the Wau Campaign

Having decided in November not to reinforce further their beleaguered enclaves at Gona and Buna, strategic considerations now required the Japanese commanders to provide a defensive line protecting Rabaul from the west. Convoys from the Carolines moved down the New Guinea coast, occupying Wewak, Madang and Finschhafen in force. Other convoys from Rabaul landed troops at Cape Gloucester and Arawa on the western edge of New Britain.

Lae and Salamaua had been under Japanese control since March 1942, so there were now a series of Japanese bases on the northern coast of New Guinea, roughly 200–300 km apart, stretching westwards from Salamaua to Aitape and then Hollandia (in Dutch New Guinea). These, with Japanese bases along the northern perimeter of the Dutch East Indies, gave Japan an outer defensive barrier all the way from Halmahera to Rabaul. The Allies, expecting further Japanese reinforcement of the Lae/Salamaua area, were by late 1942 able to monitor shipping movements in the Huon Gulf and mount attacks on them from their operational airfields at Port Moresby and Milne Bay. A convoy from Rabaul to Lae in early January was tracked and attacked, resulting in a loss of a quarter of its troops and half its supplies.

The surviving Japanese troops were then ordered to move inland and seize Wau, an Allied airfield in the highlands behind Lae. *Kanga Force*, a small Australian group based at Wau, had been effective in harrying the Japanese around Lae and Salamaua. The airfield itself was both a threat to and a strategically desirable target for the Japanese, and its value was well appreciated by the Allies. Alerted by intelligence reports, reinforcements from Milne Bay were flown into Wau just in time to repulse the attackers and preserve the airfield.

22.9   The Battle of the Bismarck Sea

The depletion of their forces at Lae as a result of the failed attack on Wau, and the resolve of the Japanese High Command in Rabaul to hold this coastal area at all costs, soon led to a major Allied victory. This victory is attributable to

---

[6]In the later years of the war in New Guinea research into tropical medicine paid off. The Allied troops benefitted from access to appropriate drugs. The Japanese troops did not. Indeed, the *Official History* of WW2 published by the Australian War Memorial has four volumes on medical aspects but none on intelligence or communications.

a combination of vital Sigint and the increased Allied air strength derived from the newly constructed airfield at Dobodura (the original purpose of the aborted Operation Providence).

The formation of a large convoy in Rabaul, intended for the Japanese base at Lae on the north coast of New Guinea, appears to have first been identified by Frumel. MacArthur was notified[7] on 19 February 1943. On page 87 of *The Eavesdroppers*, Jack Bleakley states that 1WU at Townsville also obtained useful corroboration of the naval intelligence. On page 320 of *The Signal Corps—The Test*, George Thompson and others claim that information was received from radar as well. Although the authors have not seen documentation of these other sources of intelligence, there is no reason to doubt them.

The convoy of eight transport ships carrying some 7,000 soldiers, and escorted by eight destroyers, set sail on 1 March and was spotted by an American plane later that day. Presumably, this plane had been sent on that route less to observe what was already known to be on the water than to cover up the source of information that led to the subsequent attack. Such deceptions became almost standard practice.

The convoy, which had ineffective air cover, was attacked by bombers on the following two days, in the *Battle of the Bismarck Sea*. Eventually all eight transports and four of the destroyers were sunk, with about 3,000 troops killed. The venture into the Coral Sea, in May 1942, must have seemed a distant memory to the Japanese now that air and sea control of at least part of the north coast of New Guinea had been taken by the Allies.

A few days later, a surprising consequence of this battle occurred on Goodenough Island, one of the three major islands in the d'Entrecasteaux group, roughly 150 km north of the eastern tip of New Guinea. Goodenough Island had been seized by the Japanese Army in its initial advance along the north coast of New Guinea. It was attacked on 23 August 1942 by Australian troops, with the Japanese withdrawing a few days later. The small Australian garrison maintained on the island managed to ambush eight Japanese who had landed in two flat-bottomed boats. These boats contained large quantities of documents in sealed tins. It is not clear why these and other small groups of battle survivors had moved so far off the route from Rabaul to Lae, or why documents of considerable sensitivity were not torn up and thrown overboard. These documents made their way back to Brisbane and turned out to contain lists of names of officers in the Japanese Army and their units as of October 1942. Details of units of the Japanese Army were also found in the boxes.

The importance of this information was considerable. Bletchley Park had recognised the value of assembling and collating all the information about a specific

---

[7]The reference here is a letter from E. E. Stone prepared on 9 March 1943 to support the opposition from Op-20-G to any proposed merger of signals intelligence with that of the US Army. It claims that MacArthur was receiving all potentially useful information in the hands of Frumel. In particular it mentions MacArthur received the key information on the Lae convoy on 18 February 1943. This is to be found in SRH 200–201, being the OP-20-G file on Army-Navy co-operation in Signals Intelligence 1935–1945. It is in RG457 at NARA.

military command to have been derived from Sigint. One such data collection—
Elsie's Index—related to the Japanese situation. Such card-based data banks were
extremely useful in interpreting other intercepts. Although Central Bureau was
not then reading any mainline IJA traffic, the Goodenough Island find meant that
General MacArthur's HQ had now received a valuable contribution to a data base
for use in processing later Comint. Credit for obtaining this bonanza lies with the
Australian soldiers on the ground, the relevant (mostly naval) Sigint, good fortune,
and, of course, with those responsible for the successful aerial attack on the convoy.

The Goodenough Island haul was copied in Brisbane by ATIS, the Allied
Translator and Interpreter Service, and then forwarded to other relevant Allied units.
ATIS was responsible for processing captured documents other than cryptographic
material and grew steadily in size as the advancing Allied forces sent it vastly
increasing quantities of such material.

The Japanese Army, unwilling to accept that communications security had
been breached, suspected that certain German Lutheran missionaries working in
the area around Lae had betrayed the destroyed convoy to the Allies. A few of
these missionaries had been arrested by the Australian authorities in 1939 and
taken to a camp for holding enemy aliens on the mainland. The others, including
children, were considered not to pose a security risk, and remained. They[8] were
now massacred by the Japanese in retribution.

22.10   The Global Situation in Early 1943

In China, Japanese forces were fighting to overcome Chiang Kai-shek's Kuomintang
alliance. They focused in the latter half of 1942 on capturing Chungking, the city
in Western China harbouring his headquarters. Heavy bombing of this city had
continued unabated throughout 1942. The already extensive Japanese ground forces
prepared for an all-out attack. Reinforcement, involving the transfer of troops from
the SWPA, was planned. But the unexpected outcomes of the Battle of Midway
and the bitter campaign in Guadalcanal thwarted these plans. Chungking was not
captured.

The situation in Burma remained disastrous for the Allies, with the IJA threat-
ening to attack India's eastern border (including modern Bangladesh) and foment
anti-British activity there. A successful assault would also have threatened the
airlifting of military supplies to the Chinese Army from airfields in that part of
India. There was no real improvement in the Burmese situation until later in 1943.

---

[8]The reference here is pages 101–103 of Gregory Michno's *Death on the Hellships*, Naval Institute
Press, Annapolis MD, 2001. Some of those responsible were taken to the appropriate War Crimes
Tribunal in 1947.

The Axis occupation[9] of North Africa was ending. The Soviet Army had achieved an immense victory at Stalingrad. The 4-rotor Enigma used by U-boats had been broken at Bletchley Park, the insecure RN cipher assisting the U-boats was being replaced and the Germans were losing the Battle of the Atlantic. The Japanese Army had been ousted from Guadalcanal and Papua, and, after the Battle of the Bismarck Sea, would never again successfully use troopships in convoy to reinforce its bases in New Guinea.

President Roosevelt[10] was able to say, quite correctly, that the Axis had to win the war in 1942 or lose it later.

---

[9]Gastao de Freitas Ferraz was radio operator on the Portuguese fishing ship *Gil Eannis* and also a German agent. Bletchley Park had detected this. He was apprehended at sea on 1 November 1942 to prevent warning being given of the American *Torch* fleet then on its way to North Africa. Another victory for Ultra! See the *International Herald Tribune* of 3 March 2009.

[10]State of the Union address, January 1943.

Chapter 23
Rabaul and the Philippines

By mid-1942, the senior Allied Commanders in the Pacific had begun to consider the military pathways[1] that would initiate the journey back through occupied territories and to Japan's shores. The positioning of the major Japanese bases within the land and sea territories over which Japan held sway in mid-1942 determined that Rabaul and the principal bases in the Mandated Islands would have to be dealt with, prior to any advances either further westwards towards the Philippines or directly towards the Japanese mainland. The absolute importance of air support superiority was clearly recognised. This necessitated that both carrier-based and land-based aircraft be available in sufficient numbers and be suitably disposed to sustain naval or military operations. Critical changes in planning in 1942–1943 finally resulted in General MacArthur being granted permission to advance towards the Philippines via a westward thrust along the north coast of New Guinea.

23.1 Changes in the Strategic Plans

General MacArthur had agreed to a three-stage seizure and occupation approach, as previously noted:

Task 1. The Santa Cruz Islands, Tulagi, and adjacent islands in the southern Solomons.

Task 2. The remainder of the Solomon Islands, of Lae, Salamaua, and the northeast coast of New Guinea.

Task 3. Rabaul and adjacent positions in the New Guinea/New Ireland area.

---

[1]Chapters 21–23 can scarcely be read without constant reference to maps. As they do not constitute original research, general reference is made to the US Army and Australian War Memorial histories of the New Guinea and Solomons campaigns.

© Springer International Publishing Switzerland 2014
P. Donovan, J. Mack, *Code Breaking in the Pacific*,
DOI 10.1007/978-3-319-08278-3_23

The drawn-out Papuan and Guadalcanal campaigns placed these tasks on hold, but by the end of 1942, with their successful termination pending, plans for progressing the Pacific War were part of the agenda for the coming January 1943 meeting in Casablanca of Churchill and Roosevelt and their senior advisors. A strategic overview was critical to determine a feasible campaign strategy in the Pacific during 1943. The eventual outcome for the Pacific region was defined as follows[2]:

> Specifically, the Allies intended to capture Rabaul, make secure the Aleutians, and advance west through the Gilberts and Marshalls in the Pacific towards Truk and the Marianas. The Central Pacific advances were supposed to follow the capture of Rabaul.

Intensive discussion over the next few weeks reduced these expectations, as far as operations in the southern Pacific region were concerned, to what was effectively Task 2. Explicitly added to the Task was the seizure of Woodlark and Kiriwina islands, north of the eastern tip of New Guinea, to provide additional air bases. This was specified in a directive issued on 28 March 1943. Hence previous draft plans, under the code name Elkton, were modified to produce *Operation Cartwheel*, a general plan to move towards Rabaul via a two-pronged assault. One prong was directed northwest from Guadalcanal, up the chain of Solomon Islands and Bougainville, towards New Britain and New Ireland, containing respectively the major Japanese bases of Rabaul and Kavieng. The second prong would use the recaptured Lae and Salamaua on the northern coast of New Guinea as a springboard for landings on the Admiralty Islands to the north and directly onto New Britain. The invasion and capture of Rabaul was to be deferred pending successful completion of these tasks.

General MacArthur's principal focus remained a determination to return[3] to and liberate the Philippines en route to Japan. Admirals King and Nimitz preferred to move towards Japan via the Japanese Mandates, some key western Pacific islands and Formosa, utilising Luzon, the island in the northern Philippines, as well. This required that the USN hold total naval superiority over the IJN, because the vast distances (often of 1,500 km) between successive stepping-stones along this pathway required carrier-based air dominance protecting each invasion force in turn. MacArthur favoured advancing along the northern coast of New Guinea to attain a satisfactory coastal point from which he could invade Mindanao. He would use both existing and new airfields to provide the air support needed for this operation, and would also need USN support for successive sea-based landings along the coast. (The coastal terrain prohibited consideration of a westwards land-based advance.)

The previously mentioned occupation by the IJA of Wewak, Madang and Finschhafen compounded the setback MacArthur had experienced in Buna and

---

[2]The quotation is taken from page 8 of *Cartwheel: The Reduction of Rabaul*, by John Miller, jr. appearing as part of the *US Army in WW2* history series, subseries *The War in the Pacific*, published by the Center of Military History, United States Army, Washington D.C.

[3]The Google search engine gives many thousands of references to General MacArthur's statement 'I say, to the people of the Philippines whence I came, I shall return'.

Gona. The western prong of Operation Cartwheel suddenly faced much stronger defences. Moreover, additional Japanese forces, including those evacuated from Guadalcanal, were used to provide extra defensive strength in the central and northern Solomons.

The lull in operations in this sector, following the conclusion of the Guadalcanal and Papua campaigns, reflected the large losses of men and equipment suffered in those campaigns. But the USN had new warships and carriers in production, with delivery beginning later that year and steadily increasing in number. So fighter and medium bomber strengths would also increase. The IJN, severely depleted in planes and aircrew as a result of its Guadalcanal campaign, knew that its relative strength would deteriorate over time and that the Allies would gain air and naval superiority.

23.2 Cartwheel Commences

Initial operations in each prong of Cartwheel were scheduled for late June. General MacArthur was in overall command and directed the western attacks, while Admiral Halsey directed the Solomons campaign from October 1942. Meanwhile, during April, Admiral Yamamoto, conscious of the build-up of Allied forces at Guadalcanal and in New Guinea, decided to mount a major series of air strikes against targets in both areas, using enhanced numbers of fighters and bombers at Rabaul. Yamamoto came down to Rabaul from Truk to direct this activity. Sigint and Coastwatcher intelligence helped the Allies minimise the damage inflicted by these attacks and helped to inflict further losses of aircraft and aircrew on the Japanese. Yamamoto, mistakenly led by his subordinates to believe that much greater results had been obtained, decided to make a morale-building flying visit to the Buin area. Arrangements for this, sent via JN-25 messages, were intercepted and decrypted by Allied codebreakers, and confirmed courtesy of a message sent in a lower level code. This latter was intercepted by 1WU in Townsville. After obtaining approval at the highest level to act on this information, aircraft from Henderson Field shot down Yamamoto's plane over the Bougainville jungle on 18 April. This was a profound shock both to the IJN and to the people of Japan, and a major intelligence success for the Allies. The IJN must have suspected that their codes were compromised, but apparently blamed the use of the low-level code, rather than its JN-25 system, for the disaster. However, an Allied intelligence breakthrough of much greater importance was also in progress at the time.

The Imperial Japanese Army introduced several new major (*mainline*) cipher systems during 1942. One of these, the Water Transport Code, usually referred to by Allied codebreakers by its discriminant 2468, was the new system used to organise and control the shipping employed for the movement of its troops and stores across its whole area of operations. It, and other important mainline systems, were 4-digit additive cipher systems. Sustained attempts by cryptanalysts to break into indicator systems of other mainline ciphers were unsuccessful except that the system for one such code, known as 3366, was broken but the code was changed soon after and so became unreadable again. A major effort (see Sect. 14.4) at Central Bureau in Brisbane and at Arlington Hall in Washington found that the indicator system used for 2468 involved a form of redundant encryption that was susceptible[4] to attack. As a result, by mid-1943, intercepted messages in this system were being routinely decoded. The information from these, combined with Comint derived from several IJN systems and air-to-ground systems, had an ongoing critical impact on the future course of the war in the Pacific. As well as assisting the Allies in planning specific operations against the enemy, this information greatly increased[5] the effectiveness of the USN's submarine fleet in sinking ships in convoys from the resource-rich captured territories of SE Asia and the NEI attempting to reach the Japanese mainland. The supply of oil and other essential war materials to Japan deteriorated steadily from late 1943 onwards and became a critical factor from late 1944.

The main initial target of the Solomons campaign was the capture of the Japanese air base at Munda Point in New Georgia. The operation began in late June. After a relatively easy beginning, strong defences near the airstrip and determined fighting by the Japanese troops resulted in a costly and protracted engagement, but the Allies gained control of the airstrip on 5 August. Construction work on the strip began immediately and it was available for use by fighters from 14 August and by bombers from mid-October.

The New Georgia campaign cost the Allies over 1,000 dead and some 4,000 wounded, with thousands more suffering from disease or combat fatigue. Immediately, as a result of these numbers, a decision was taken to bypass the next heavily defended targeted island, Kolombangara, leapfrogging over it to take the almost undefended nearby island of Vella Lavella, which also had a good operational airstrip. This decision was feasible only because of the Allies' confidence in their ability, via air and sea power, to contain the bypassed forces and control anticipated air and naval assaults on the advancing forces. Once these latter conditions prevailed elsewhere, the policy of bypassing and isolating enemy strongholds, via landings on weakly defended sites considered suitable for airfield construction and as ports, was to be effectively applied throughout the rest of the Allied campaigns in the Solomons and New Guinea.

Vella Lavella landings occurred in mid-August and by early October all Japanese resistance had ended in the New Georgia Islands, with some 2,500 IJA troops killed and 9,000 evacuated from Kolombangara to Bougainville.

---

[4] The insecurity in the indicators of cipher 2468 is discussed in detail in Sect. 14.4 and elsewhere. A full understanding requires some mathematical sophistication.

[5] Much more on this matter is given in Chap. 16.

23.3   The Reduction of Rabaul

After high-level discussions had been held in July, General Marshall proposed to General MacArthur that Rabaul, with a garrison of over 100,000, now be neutralised, rather than invaded. This was to be achieved via the capture of Kavieng on New Ireland, Manus Island in the Admiralty Islands, and Wewak in New Guinea, using sea and air power to isolate Rabaul from Truk and destroy it as a base for attacks on Allied forces. MacArthur argued that Wewak was too strongly defended, so that it would be preferable to seize a base further west on the New Guinea coast after which Rabaul had to be captured. Finally, at the August 1943 meeting of Roosevelt and Churchill in Quebec, the Combined Chiefs agreed that MacArthur should isolate, rather than capture, Rabaul and the north coast of New Guinea as far west as Wewak, capture Kavieng and Manus Island for use as naval bases, and, having completed this, then move westwards across New Guinea to the Vogelkop Peninsula.

From there, he should prepare to invade Mindanao in the Philippines. And so MacArthur was granted his path to the Philippines! The rationale behind this was sound. The Allies, by maintaining potentially aggressive profiles in the Aleutians, the Central Pacific Area and the South West Pacific Area (SWPA), obliged the Japanese High Command to spread its defensive forces over an enormous perimeter and to maintain what proved to be an unachievable level of flexibility of movement across that perimeter. Secondly, the quantity of war resources, both human and material, now located at bases within Australia, New Zealand and elsewhere in the SWPA, would have posed a major logistics problem had an advance towards Japan only through the Central Pacific been decided—the sheer task of moving most of this northwards would have used up all transport resources for months!

It was agreed that the Central Pacific offensive towards Japan would start with a seaborne invasion of Tarawa in the Gilbert Islands, scheduled for November. So the 2nd Marine Division was transferred from Admiral Halsey's command to Admiral King's, thus signalling a change of plans in the Solomons. However, any increase in the intensity of bombing raids on Rabaul, and air cover for Allied forces progressing through New Guinea towards New Britain, required additional fighter bases closer to Rabaul. Following a number of revisions of proposals to land on Bougainville, Halsey and MacArthur decided to bypass the heavily defended Japanese positions around the southern part of the island and to surprise the Japanese by a landing, scheduled for 1 November 1943, on the lightly defended Cape Torokina in Empress Augusta Bay on the west coast of the island. This had been chosen as it would be a suitable site for airfield development. It was expected that there would be heavy Japanese counter-attacks on this site, using the strong air, land and sea forces available. So, for the first time in Operation Cartwheel, MacArthur and Halsey were offered a brief period of support from some of the newly created carrier task forces available to Nimitz. The carriers gave essential additional strength to the sustained attacks on local Japanese bases, and also made a decisive strike on Rabaul in early November, severely damaging three of the seven heavy cruisers that Admiral Koga had ordered down from Truk to attack the beachhead at Cape Torokina. In fact, Koga withdrew his cruisers back to Truk and never again sent capital ships of the IJN to Rabaul. Koga had also sent 170 carrier-based planes with aircrew down to Rabaul in order to attack Allied bases in the Solomons and New Guinea. These planes were now also directed at the US Cape Torokina landings. But Allied defence prevailed, and by 12 November, after heavy losses of planes and aircrew, Koga withdrew the remainder and did not again strengthen the airbase at Rabaul in this way. Thus, by mid-November, Rabaul, although not entirely reduced, had ceased to be capable of any further serious offensive operations.

23.4   The Western Prong of Cartwheel

MacArthur had neither carriers nor capital ships available to assist with the steady movement westwards along the northern coast of New Guinea from Buna. His plans required the very careful coordination of airborne, overland and amphibious operations aimed at steadily gaining Lae and the Markham and Ramu valleys in its hinterland. The first steps had been the successful defence of Wau and the Australian advances from there towards Salamaua. The next step was a successful landing on 30 June at Nassau Bay (some 300 km west of Buna) where a staging point was established for the marshalling of resources for the next moves west. This was in addition to previously constructed bases on Goodenough Island. General Kenney needed airfields in the Lae hinterland to support the westwards advance and to attack Japanese bases. A site was selected on the Watut River and by mid-August 3,000 troops and a fighter squadron were based there, without this being realised by the Japanese.

Alerted by Sigint, the Allies knew that a large number of aircraft had been flown into Wewak in early August, and Kenney now had the fighter cover necessary to launch major attacks on it. Raids on 17 and 18 August destroyed over 100 planes on the ground at Wewak, which continued to be bombed throughout August. This made its air power ineffective in opposing the landings around Salamaua and then Lae during August and September and the parachute drop on Nadzab on 5 September, which immediately yielded another airstrip. A radar-equipped US destroyer on patrol in the Vitiaz Strait supplied the necessary early warning of any Japanese aircraft approaching from Rabaul and thus minimised damage.

With orders to retreat through the hinterland towards Madang, 9,000 Japanese withdrew from Lae in mid-September. Defending Salamaua and Lae and the resulting difficult retreat cost the IJA some 2,600 dead.

MacArthur's forces pressed on near the Huon Peninsula. New landings in late September led to the capture of Finschhafen on 2 October, well before Japanese 20th Division troop reinforcements, sent overland from Bogadjim, some 300 km away towards Madang, could arrive. These troops were then ordered to retreat towards Sio, losing numbers due to the jungle conditions and from attack. In fact, Japanese strategy now focused on the defence of Wewak and all forces east of that stronghold were progressively ordered to withdraw to it. So, the Allies were able, in order, to occupy Sio, then Saidor, and then Madang, without facing the anticipated levels of resistance. Thus, by 24 April 1944, all the western prong objectives of Operation Cartwheel had been achieved, since landings across the Vitiaz Strait onto Cape Gloucester and Arawa on the western tip of New Britain during December resulted in their capture and conversion into Allied bases by the end of February.

23.5 The Coup at Sio, January 1944

When it failed to reach Finschhafen in time to prevent its fall, the 20th Division of the IJA was also ordered to retreat back towards Madang. Passing through Sio, a full set of Japanese Army high-level (mainline) code books and ancillary documentation was dumped into a water-filled hole. The officer responsible had assured Madang that all this material had been destroyed and hence that present codes were secure! The box was detected soon after by an Australian patrol checking for mines and booby traps.

The material was flown back to the Allied Translator and Interpreter Service in Brisbane. Its nature was quickly appreciated and it was passed on to Central Bureau. Here, after carefully drying out the pages, linguists and cryptanalysts confirmed the significance of this find. Two linguists from Frumel—Thomas Mackie and Tex Biard, each expert in reading telegraphic Japanese—were lent to CBB for this important task.[6] Photographic copies of the major code books were sent immediately to Washington. The decrypted messages included a report giving a detailed assessment of Japanese military strength along the New Guinea coast, of immense value in helping MacArthur plan to bypass Wewak in favour of Hollandia in his westwards advance and then to bypass Halmahera in favour of Morotai.

Changes to the IJA coding systems prevented the Allies from being able to read all subsequent intercepts of messages sent in the main operational codes of the IJA. However, the value of the Sio find was extended because of a subsequent find at Aitape of a charred copy of a later additive table. A build-up of personnel in the Bletchley Park and Arlington Hall sections devoted to Army codes resulted in a steadily increasing amount of Comint being derived from them.

As more and more Japanese enclaves in the Pacific became successively isolated an important intelligence by-product was derived. Not all of these could be supplied with new cryptographic materials immediately when changes were made to operational codes, This forced the use of old codes and so often provided cribs into the replacement codes. Here stereotyped messages were particularly useful to the cryptanalysts.

---

[6]This account of the recovery of cryptographic materials at Sio inevitably repeats part of Sect. 17.5. The situation in January 1944 was that Frumel had much more experience than Central Bureau in reading telegraphic Japanese. Some of its staff had been handling JN-25 decrypts since 1941.

23.6   Towards the Philippines

Finally, after high-level exchanges over several months between Halsey, other naval commanders in the Pacific and MacArthur's SWPA, it was agreed first that the Admiralty Islands should be invaded, in order to provide airfields and a seabase in Seeadler Harbour (Manus Island), and later that Kavieng in New Ireland could safely be bypassed in favour of a painless landing on Emirau, an island just north of it and suitable for airfield development. Thus, by May 1944, two additional important airbases and one major naval base were available for Allied forces in their moves westwards towards the Philippines.

The Central Pacific assault had, by now, obliterated Truk as an IJN base, captured Tarawa and Kwajalein, and was preparing to invade Saipan in the Marianas. Admiral Koga had been forced to move his Combined Fleet HQ westwards to Palau, Also, carriers from one of the USN's Pacific Task Forces had in April supported MacArthur's great leap westwards, leapfrogging Hansa Bay and Wewak to land at Aitape and Hollandia, respectively some 150 and 300 km west of Wewak, with airstrips at each and relatively lightly defended by comparison with the Wewak-Hansa Bay area.

General Kenney had produced yet another surprise to facilitate these landings. Hollandia and Aitape were well beyond the normal fighter range available to Kenney to support his bombers. Consequently, Japanese fighters and bombers had been accumulating around Hollandia, believed to be a safe base. But Kenney had helped to pioneer the use of additional long range drop tanks on fighters, doubling their range. When he had received enough of these, he was ready to attack. Starting on 30 March, his aircraft caught the Japanese by surprise and within days had destroyed some 400 planes. They then continued to wipe out all surface military installations, leaving the areas 'smoking wrecks' and eliminating all air opposition to MacArthur's landings there.

It had been hoped that the airfields captured at Hollandia would prove suitable for development as bomber bases to support both further advances along the coast and the coming USN onslaught on the Marianas planned by Admiral Nimitz for June. But the soil structure along this part of the New Guinea coast, combined with the heavy rainfall pattern, was found to make the projected airstrip upgrades impossible. So, although Hollandia itself proved suitable as the major seabase from which the eventual invasion of Leyte in the Philippines would start (a truly enormous logistics operation), MacArthur accelerated further advances westwards via Wakde and Sarmi on to the important island of Biak. Because Wakde was too small to operate as a major airbase great importance was attached to the rapid conquest of Biak and the build-up and operation of its satisfactory airstrips. However, poor intelligence estimates of the strength of its defenders and lack of knowledge of its terrain resulted in the Biak operation becoming costly in terms of casualties and so drawn out that Nimitz's Marianas offensive took place without MacArthur's promised support. This was the famous battle of the Philippine Sea during which USN carrier-based fighters destroyed so many Japanese planes that the encounter became known as 'the great Marianas turkey shoot'. (It was at this time that the Japanese officially sanctioned the use of *kamikaze* attacks on enemy ships. Kamikaze attacks became more common as the strength of Japanese air power diminished.)

In fact the Nimitz assault had a significant impact on the Biak operation. The IJN, fully aware of the value of air support from its Biak base in the coming engagement in the Central Pacific, and cognisant of the threat from Biak if the Allies had control of that base, determined to reinforce the troops there in order to delay or repulse the landings. Already, Sigint information had helped the Allies inflict heavy damage on a convoy from China carrying seasoned troops intended to reinforce Palau and western New Guinea. Some 4,000 Japanese troops were killed and the remainder lost most of their equipment. The *Take* (Bamboo) convoy episode, as it is known, forced the IJN to review all aspects of convoy security but once again found its mainline codes not seriously compromised. Three successive attempts were made by the Japanese to ferry troops to Biak, in Operation *Kon*. The final attempt had such strong naval support that it may well have overcome the defending naval force under Admiral Crutchley. But while in progress, the IJN decided to recall its warships north to help 'conquer' the US fleet in its 'decisive' battle of the Philippine Sea, and Biak eventually succumbed.

Next, MacArthur, again avoiding heavily defended Japanese positions on the Vogelkop peninsula of Dutch New Guinea, struck at Sansapor on its north coast, again a site suitable for development as an airbase. From there, another leap to the northwest avoided the heavily defended Halmahera to land on what Comint had confirmed as being an undefended area at the northern tip of Morotai,[7] again suitable for airstrip development. From there, the pathway to the Philippines lay open.

Throughout the 1943–1944 campaigns, and until the end of the Pacific War, the Allies benefited not only from a steadily growing volume of strategic Sigint, but also from growing numbers of mobile wireless interception units. Located close to the military front lines or on the flagships of USN Task Forces, they were able to intercept vital messages relevant to immediate tactical and operational use. As well as many US units, especially shipboard USN units, Australian Army interception units and Australian Air Force Wireless Units played important roles in the planning and conduct of Allied operations. For example, Air Force units at Nadzab, Biak and accompanying the invading troops on Leyte worked tirelessly to alert Allied forces of impending Japanese action. This was particularly important at Leyte, where plans to build airstrips at the landing site were completely frustrated by the soil and rainfall conditions and MacArthur had to rely on carrier-based air support for much longer than expected. The Japanese attacks were incessant, as were their efforts to reinforce

---

[7]For example, Canberra NAA item A9695 525 is a report on Morotai up to 4 October 1944. 'Plans were therefore made in July 1944 to bypass the mainland of Halmahera and to secure objectives on weakly-garrisoned Morotai Island, situated off the north tip of Halmahera. This non-familiar strategy had been employed with complete success in the operations culminating in the occupation and resultant domination of the northern coast of New Guinea and in the isolation of strong hostile forces occupying this island.'

Leyte through the port of Ormoc on its western side. Non-stop radio monitoring helped to prepare for air attacks and also alerted available aircraft and submarines to the positions of support convoys approaching Ormoc. Such warnings were in addition to Sigint information pertinent to the IJN's warship movements leading up to the Battle of Leyte Gulf, which, like the early days of the invasion, was a very close-run operation.

23.7 Developments Elsewhere in 1943 and Later

The Casablanca Conference had allocated priority to the bombing of Germany. This was the principal reason why the supply of heavy and medium bombers to other theatres, including the Pacific region, was constrained during 1943. Another confirmed target was the recapture of Burma by the British, but no firm timetable was proposed for that, as it had low priority. The monsoon season from May through to October remained a critical factor in war planning by both sides, and it was not until December 1943 that the British and their Allies were able to plan forward operations with a reasonable expectation of success.

The IJA in Burma launched an attack on India towards Imphal and Kohima early in 1944. This was repulsed after heavy fighting, and resulted in significant Japanese casualties, mainly from illness and starvation. The Japanese were steadily forced to retreat and were eventually evicted from Burma by August 1945, having lost some 200,000 of the 300,000 troops committed to the campaign.

The North African campaign had ended, Sicily had been invaded and conquered and the Allied forces in Italy were moving towards Rome. The Italian Government surrendered on 8 September 1943, leaving the German forces to defend against the advance northwards. On the Eastern Front, after their victories at Stalingrad and the Battle of Kursk, the Soviet Armies continued westwards reaching Kiev in November 1943, ending the siege of Leningrad in January 1944, and clearing the Crimea in May. Soon after the D-Day landings in France on 6 June 1944, the Soviets launched an immense assault (Operation Bagration) on the German forces west of Moscow which resulted in over 600,000 German casualties (including 160,000 captured) and some 170,000 on the Soviet side. From then until the end of the European War in May 1945 the German forces were generally in retreat.

Recommended Reading

1. Ballard, Geoffrey (1991) On Ultra Active Service: The Story of Australia's Signals Intelligence Operations During WW2. Spectrum Publications.
   This describes what the author saw, which was quite a lot.
2. Barker, Wayne (1979) Cryptanalysis of an Enciphered Code Problem—where an 'Additive' Method of Encipherment has been used. Aegean.
   Recommended for those who think breaking additive ciphers is easy!
3. Biard, Forest (2006) The Breaking of Japanese Naval Codes: Pre-Pearl Harbor to Midway in Cryptologia 30(2) 151–158.
   The text of a speech given by Tex Biard at National Cryptologic Museum in 2002.
4. Bleakley, Jack (1992) The Eavesdroppers. AGPS, Canberra.
   The definitive study of the Australian Air Force WW2 Wireless Units.
5. Boyd, Carl (1993) Hitler's Japanese Confidant: General Oshima and Magic Intelligence 1941–1945. University of Kansas Press.
   The 'Magic' intelligence was of great utility in the European Theatre of Operations.
6. Brion, Irene (1995) Lady GI. Presidio Press, Novato, California.
   Written in 1946, this is a fresh account of her career as a junior WW2 cryptanalyst.
7. Budiansky, Stephen (2000) Battle of Wits: The Complete Story of Codebreaking in World War Two. Free Press.
   Mostly European Theatre but still very useful.
8. Budiansky, Stephen (2000) Closing the Book on Pearl Harbor. Cryptologia 24(2) 119–130.
   Damages certain Pearl Harbor conspiracy theories.
9. Davies, Norman (2006) Europe at War 1939–1945. Macmillan.
   Reminds the reader that WW2 was mostly fought in Eastern Europe.
10. DeBrosse, Jim and Burke, Colin (2004) The Secret in Building 26. Random House.
    The story of the NCR plant at Dayton, Ohio, in WW2.
11. Drea, Edward (1992) MacArthur's Ultra. University of Kansas Press.
    Very detailed indeed on operations but no technical material.
12. Erskine, Ralph and Freeman, Peter (2003) Brig. John Tiltman: One of Britain's Finest Cryptologists. Cryptologia 27(4) 289–318.
13. Erskine, Ralph and Smith, Michael (2011) The Bletchley Park Codebreakers. Biteback, London.
    Tiltman's role in WW2 communications intelligence was extremely important.
14. Feldt, Eric (1946) The Coast Watchers. OUP, Melbourne.
    The definitive account of the Australian Coastwatching organisation.

© Springer International Publishing Switzerland 2014
P. Donovan, J. Mack, *Code Breaking in the Pacific*,
DOI 10.1007/978-3-319-08278-3

15. Hastings, Max (2007) Nemesis: The Battle for Japan 1944–45. Harper.
    'The treatment of subject peoples and prisoners described in this book is wholly unaccepted by most modern Japanese, even where supported by overwhelming evidence.'
16. Hinsley, F. Harry et al (1979–1990) History of British Intelligence in the Second World War, 6 volumes. HMSO, London.
    European Theatre only, but extremely well researched. Hinsley was at Bletchley Park.
17. Holmes, W. Jasper (1978) Double-Edged Secrets: US Naval Intelligence Operations in the Pacific in WW2. Naval Institute Press, Annapolis MD.
    Very useful in describing the key work of Joe Rochefort.
18. Kahn, David (1996) The Codebreakers (rev. edition). Simon & Schuster.
    Much background information is to be found here.
19. Layton, Edwin, et al (1985) "And I Was There". Morrow, NY.
    Layton was there as a key USN intelligence officer in Hawaii.
20. Long, Gavin (1973) The Six Year War. AWM and AGPS, Canberra.
    Excellent introduction to WW2 from an Australian perspective.
21. Mullins , Wayman (ed) (1994) 1942: 'Issue in Doubt'. Eakin Press.
    Yes, the issue was in doubt in 1942. The breaking of JN-25B8 was very important.
22. Overy, Richard (1995) Why the Allies Won. Cape.
    Concludes that communications intelligence was a significant but not decisive factor.
23. Parker, Frederick (1993) A Priceless Advantage: US Naval Communications Intelligence and the Battles of Coral Sea, Midway and the Aleutians. Center for Cryptologic History, National Security Agency.
    No technical material on JN-25 but traces other aspects carefully.
24. Potter, Elmer and Nimitz, Chester (eds) (1960) The Great Sea War. Prentice-Hall.
    There was only one great sea war!
25. Prados, John (1995) Combined Fleet Decoded. Random House, NY.
    No technicalities about ciphers but very good on operational matters.
26. Smith, Michael (2000) The Emperor's Codes. Bantam Press, London.
    Complements rather than competes with this book.
27. Straczek, Jozef (2001) The Empire is Listening: Naval Signals Intelligence in the Far East to 1942. AWM Electronic Journal 35.
    A key account of the preparation for WW2.
28. Webster, Donovan (2003) The Burma Road. Farrar, New York.
    A good modern account of the China-Burma-India front in WW2.
29. Winter, Barbara (1995) The Intrigue Master: Commander Long and Naval Intelligence in Australia. Boolarong Press, Brisbane.
    The Australian Naval Intelligence Dept. was of the greatest importance in 1941–1942.

Index

© Springer International Publishing Switzerland 2014
P. Donovan, J. Mack, *Code Breaking in the Pacific*,
DOI 10.1007/978-3-319-08278-3

Printed in the United States
By Bookmasters